COMPUTATIONAL METHODS FOR THE SOLUTION OF ENGINEERING PROBLEMS

C. A. Brebbia, *Reader, Department of Civil Engineering,*
University of Southampton
A. J. Ferrante, *Professor of Civil Engineering, Coppe Universidade*
Federal do Rio de Janeiro

Third revised edition

PENTECH PRESS
London

First published 1978
by Pentech Press Limited
Graham Lodge, Graham Road,
London NW4 3DG

2nd revised edition, 1979
3rd revised edition, 1986

British Library Cataloguing in Publication Data
Brebbia, C. A.
    Computational methods for the solution
    of engineering problems – 3rd ed.
    1. Engineering – Data processing
    I. Title   II. Ferrante, A. J.
    620'.0028'5   TA345

    ISBN 0-7273-0315-5

Printed and bound in Great Britain

PREFACE TO THIRD REVISED EDITION

The success of the previous editions of this book has motivated us to prepare a new revised edition. This new edition incorporates a series of revisions and corrections to the original printing and includes further computer codes, in particular those relevant to the solution of six node triangular finite elements.

The authors hope that this edition will be useful and continue to be well received by the readers.

C. A. Brebbia and A. J. Ferrante

PREFACE TO FIRST EDITION

The advent of modern digital computers has had a profound effect on all approaches to engineering, teaching and research, entrepreneurial and consultative. With the new computer-based methods, problems once considered intractable by traditional conventional methods can now be quickly solved. Indeed such computer methods are the new conventions and this book is concerned to promote their assimilation in advanced form at an earlier stage.

Nowadays engineering students learn to use computers at the beginning of their studies and then are exposed to the theoretical bases and applications of useful computer oriented methods. Normally these are studied in depth only in postgraduate courses, as for example is the case with the finite element method. However these methods are now so well developed and widely used that they should be taught to undergraduate engineering students with an emphasis on practical applications.

This book attempts to provide a simple and practical text on the bases, applications, and computer implementation of numerical methods for engineering problem solving. It begins with an introductory chapter on computers in engineering followed by a chapter on matrix algebra by computer, which includes several computer programs for basic mathematical procedures. Chapter 3 discusses the matrix analysis of simple structural systems. In Chapter 4 a review of the basic principles of solid mechanics is presented, while Chapter 5 contains an introduction to the Rayleigh-Ritz and Galerkin approximate methods. The presentation of the finite element method is done in Chapter 6, with applications to the solution of the Laplace equation and two-dimensional elasticity problems. Finally, Chapter 7 presents numerical formulations for fluid mechanics problems.

The main feature of this book is the integrated treatment given to the formal and computing aspects of the numerical methods discussed. Many useful complete computer programs are given, including programs for the solution of linear simultaneous equations, eigenvalue calculations, static and dynamic analysis of frames, and finite element programs for the solution of the Laplace equation and two-dimensional elasticity problems.

C. A. Brebbia and A. J. Ferrante

CONTENTS

LIST OF PROGRAMS

Chapter 2

Chapter 3.

Programs for static analysis of plane trusses

Chapter 4.

Programs for static analysis of plane frame systems

Programs for dynamic analysis of plane frame systems

Chapter 6

**Programs for the solution of the generalized Laplace's equation
using first order triangular elements**

Programs for the solution of bidimensional elasticity problems

**Programs for the solution of Laplace's equation using second
order triangular elements**

Chapter 1

Computers in engineering

1.1 ENGINEERING, COMPUTERS AND NUMERICAL METHODS

A definition of engineering is the 'science by which the properties of matter and the sources of energy in nature are made useful to man'. Thus, an engineer will have to study the properties and behaviour of physical systems, applying his knowledge to attain purposes useful to society. Such activity can be considered as consisting of the study and solution of *real physical problems*. The complexities involved in dealing with all the parameters relative to the properties of a real problem will, on many occasions, force him to study an equivalent *engineering problem* which can be mathematically defined. He will try to solve such a problem according to his knowledge and experience, and using the computational tools at hand.

Advances are greatly stimulated by the new computational tools made available as a result of technological progress itself. An extremely adequate example of this can be found by examining the effects of the introduction of the modern digital electronic computer.

The building of a computing machine was an ancient idea of man, as illustrated by the efforts of Pascal, Leibnitz, Babbagge, and others, but the era of the modern computer only began in the 1940s. The development of the MARK I machine by Aiken in 1944, at Harvard University, is often taken as its starting point. However, that was an electromechanical device. A year later, Eickert and Mauchly developed the ENIAC machine, at the University of Pennsylvania, which was a truly electronic device. Soon after, many researchers were engaged in similar development activities, producing a variety of different machines. Initially born as a part of the war effort, they required a large amount of space, were very cumbersome to use, and often gave unreliable results. Larger and larger machines were built, following the wrong conception that a few, gigantic computers would be able to satisfy all the computational needs of the world.

Industry soon recognized the immense possibilities behind such electronic devices, leading to their commercial introduction at the beginning of the 1950s. These computers were mostly oriented to data processing. They comprise

1

what is known as the first generation of computers, the designation given, from the standpoint of their technological characteristics, to the computers built using electronic valves. As their use became more common, their applications were extended to many different fields including, with the more advanced computers of this generation, engineering applications.

The second generation of computers was characterized by being completely transistorized. Their influence was felt in many of the activities of man, the volume of their engineering applications becoming considerable. They were designed for two different functions, i.e. commercial and scientific. The first type was conceived to handle large amounts of data, and little computation, while the second type handled little data, but was able to perform large quantities of complex computations.

The utilization of integrated electronic micro-circuits was the determinant characteristic of the third generation of computers, introduced by the mid 1960s. By then it was realized that it was often required to perform large amounts of complex computations in commercial applications, and to deal with voluminous data in scientific applications. This realization led to the concept of the general-purpose computer, typical of this generation, and of present computers.

The engineering applications of computers nowadays cover many disciplines, requiring both numeric and non-numeric information processing of different types. In many cases, the behaviour of the engineering problem at hand can be defined in terms of a system of differential equations, combined with given boundary conditions. Such representation can be considered as the *mathematical model* for the engineering problem. Being a continuous model it will have an infinite number of degrees of freedom. Excluding some simple cases, it is normally very difficult, if not impossible, to find an analytical or exact solution for a problem so defined. Then, the engineer will try to obtain an approximate solution. For this he will use some technique which allows the continuous mathematical model to be replaced by an approximate *discrete or numerical model,* having a finite number of degrees of freedom, and a simpler solution.

In the past, by hand computation or with the help of desk calculators, engineers were only able to analyse very small discrete models which either corresponded to very simple problems, or were rough approximations of more complex problems. The limitations of their computational capabilities had to be overcome by their experience, engineering intuition and the adoption of very conservative safety factors. In most cases, however, several hypotheses to simplify the problem had to be taken into account, to define a discrete model which could be solved numerically. Sometimes, an inadequate simplification could result in a discrete model which could not reasonably represent the real problem.

When we think of today's engineering problems, such as the construction of nuclear reactors, space satellites, oil rigs in the sea, etc. it is clear that they must be analysed using very refined discrete models, so as to closely approximate their real behaviour. Fortunately, with the help of the modern computer, it is possible today to treat discrete models with up to several thousands of degrees of freedom. Thus, the need for formulating larger and more complex discrete models, and the availability of computers which make feasible their analysis, are the causes of the considerable interest produced in relation to numerical methods and computer analysis techniques, some of which are the subject of this book.

A method of reducing a system with infinite degrees of freedom to one with a finite number of unknowns is the finite difference technique. Its origins may be traced to some problems solved by Euler and to the pioneers of differential calculus, but it was not until this century that the method was fully developed as a new numerical technique. Its main attraction is that it is simple to use when analysing problems with regular geometry. However the method may be difficult to generalize for problems with complex geometry and arbitrary boundary conditions. Currently the method has been superseded by finite elements for many engineering applications.

Finite elements are based on variational concepts presented in a discretized way. Variational techniques for solving engineering problems were introduced by Rayleigh, who proposed substituting an unknown function by a series of known ones with undetermined coefficients. The coefficients can then be obtained by minimizing a functional such as the total potential energy. Rayleigh used this technique for the solution of vibration problems and the method was developed further by Ritz, who applied it to a larger range of problems. The main shortcoming of the approach is the need to possess a known functional. In many practical problems the governing equations are known but it is impossible or extremely difficult to formulate a functional. For these cases we can apply an error minimization method, such as Galerkin's, which needs only the governing equations and boundary conditions. This technique presents a more general basis for finite elements and although it is basically different from Rayleigh—Ritz one can find, under certain circumstances, that both methods are equivalent (i.e. Rayleigh—Ritz can be seen as a special case of Galerkin's method).

Finite element techniques have now become standard tools of engineering analysis. The method started at the end of the 1950s with a paper by Turner et al.[1] In a few years the technique became a favourite topic of research and was rapidly developed in the sixties. At the same time work on several general-purpose finite element programs was started, the first of them the STRUDL system of MIT. This and some other systems are still in use and many of them are widely applied in engineering. By the beginning of the 1970s the finite element method was well established and several comprehensive books started to appear.

Figures 1.1 to 1.3 give some idea of the complexity of the problems currently analysed by the engineers using finite elements. In Figure 1.1 a

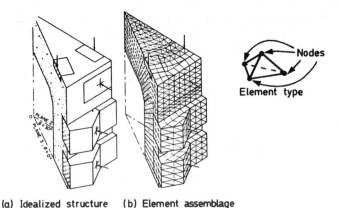

(a) Idealized structure   (b) Element assemblage

Figure 1.1   Finite element idealization for 3—dimensional analysis of a pressure vessel.

nuclear pressure vessel model is shown together with its finite element discretization[2]. The structure is subjected to internal pressure and prestressing loads. Due to symmetry it was possible to work with only a 30° sector. This sector was divided into 4178 tetrahedron elements with 2678 nodes. As there are three unknowns per node, the total number of degrees of freedom was 8034, before applying any boundary conditions. The solution of this system of equations required a special 'iterative' technique. The user controls the accuracy of the solution by specifying a maximum number of iterations and the allowed out-of-balance internal forces versus the external ones.

Finite elements have now become accepted in many fluid mechanics applications. Figure 1.2(a) shows the discretization of the channel flow around an obstruction to study the influence of inertia, drag and lift forces[3]. A grid with approximately 500 nodes was used and at each of them the stream function and vorticity were alternatively taken as unknowns. The fluid was instantaneously accelerated by applying a uniform velocity up-stream of the channel. For a Reynolds number of 100, instabilities inherent in the numerical solution of the momentum equations, and propagated by computer round-off-error, eventually increased until vortex shedding occurred. Figure 1.2(b) shows this phenomenon. In order to show the vortices more clearly the free-flow streamlines values were subtracted from the computer streamlines values (this gives the so-called perturbation streamlines).

Interaction problems, involving fluids and structures can also be studied using finite elements[4]. Figure 1.3(a) shows a gravity type oil rig, which has been discretized as a first approximation using a few elements, Fig 1.3(b). The rig was analysed with random vibrations theory, and once the designer had decided the final shape using the simplified model, the rig was fully analysed using a large number of elements. Random vibrations in this case consists of working with the wave spectrum obtained using probabilistic theory, to be able to predict the maximum results for stresses, displacements, etc. of the system, within a known confidence level.

Structures of complex geometry such as the poliedric folded-plate theatres shown in Figure 1.4 can be studied using triangular finite elements[5]. The larger structure (theatre A) was built on an approximately elliptical plant, with 3000 m^2 covered surface. The finite element mesh consisted of about 750 triangular elements and 500 nodes. (There were 6 unknowns per node, which gives a total number of degrees of freedom of around 3000). Special beam elements were introduced in the model to simulate support frames and other beam types. This structure was analysed by one of the authors using a generalized version of STRUDL-II finite element program.

1.2   AN ENGINEER'S VIEW OF A COMPUTER

From the user point of view, we can define the computer as an 'electronic device for high speed automatic, *information processing*'. This definition is in accordance with the 'black box' concept, considering the computer as a

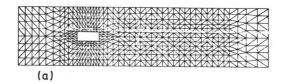

(a)

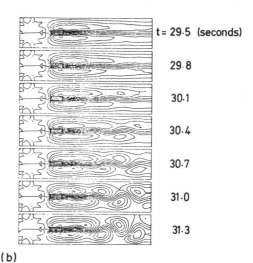

t = 29.5 (seconds)

29.8

30.1

30.4

30.7

31.0

31.3

Figure 1.2   (a) Finite element mesh. (b) Vortex (street) developments for Re = 100 (perturbation streamlines)

(b)

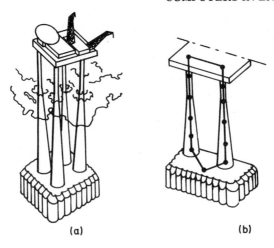

Figure 1.3   *Simple discretization of an offshore platform*

device which receives the information provided by the user, operates upon it, and produces new information, some of which is delivered back to the user. For a better understanding of this process of acquisition, processing, and delivering of information, we should identify the basic components of a computer, and analyse their functions. These components are shown by the simplified scheme, Figure 1.5, where the 'black box' is indicated by the dashed lines.

The information supplied by the user is received by the *input unit* or *reader*. The user does not give that information in a completely arbitrary manner, but will have to follow some specific rules, to allow for its interpretation. In addition, the information will have to be materialized in some physical form, which can be assimilated by the input unit. The physical media used to register information is known as *information support*. Examples of common information support are punched cards, punched paper tapes, magnetic tapes, etc. For each type of information support there is a type of input unit which can assimilate it. On the other hand, the reading being a 'slow' process, it is common for a computer system to have more than one input unit, of the same or different kind, to avoid creating a bottleneck in the flow of information.

The user supplies the computer with information, which can be of a different nature. Generally, that information is composed of *data*, relative to a problem to be solved, and *instructions* concerning how the computer should operate upon that data to obtain the desired solution. The information produced by the computer, defining that solution, is known as the *results* of the

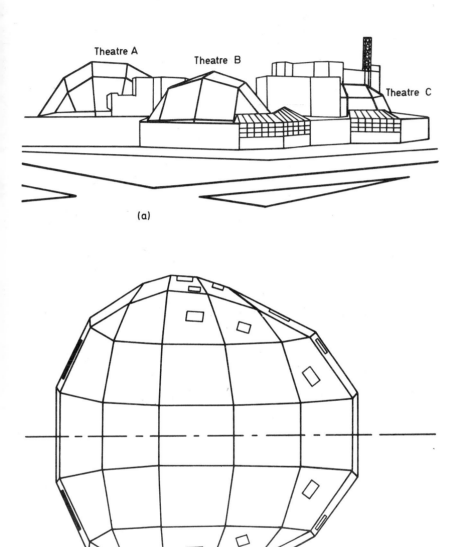

Theatre A

Theatre B

Theatre C

(a)

(b)

Figure 1.4   *(a) Folded-plate structures. (b) Theatre A plan view*

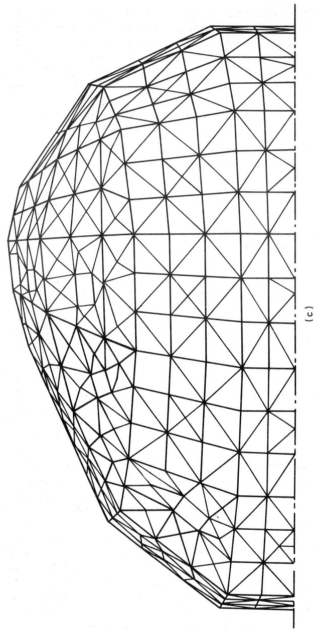

(c)

Figure 1.4. (c) Theatre A finite element mesh

process. The set of instructions given to the computer, defining the operations required to achieve the solution is known as a *program* or *computer program*. A program is written according to certain rules, which are part of a so called *computer language*. The user will write his program using a language, which he selects from the several languages which the computer he is using might accept. The data for the program will be written in the corresponding information support according to the rules set by the user, when writing the program. Program and data are not, in general, read simultaneously or intermixed. The program is normally read first. Then the computer operates according to the instructions included in the program which, among other things, indicates to the computer when and how to read the data. This process is normally called *execution* or *processing* of the program. In many cases the program is already stored in the computer, and the user limits himself to simply providing the data. It is a common case that a program will be written by one person, or group of persons, and be used by many others.

The information which enters the computer is immediately transferred to the *computer memory*. A computer can have several memories, including a *primary memory,* which will always exist, and one or more *secondary memories,* which may or may not exist. The function of a memory is to store information, for short or long periods of time, *without transforming* it. The primary memory can be considered as the 'heart' of the computer, since all the information that it receives, or generates, must pass through the primary memory. The efficiency of the primary memory is a determinant factor of the overall or global efficiency of the computer. The capacity of the primary memory for storing information is another fundamental parameter.

The primary memory can be conceived as a large collection of small 'cells', each able to store a unit of information. Normally these cells are called *memory positions*. Each memory position is identified by a number, called *memory address*. The information read by the computer, through an input unit, or that internally-generated, is subdivided into information units, called *words*. A word is further subdivided into *characters*. The number of characters in a word defines the *word length*. The content of a memory position is a word. Every time a word needs to be put into, or taken from, a memory position this is accessed through its address. Most primary memories are magnetic core devices. Because of this the primary memory is often called *core memory*.

The extremely high utilization of the primary memory and its cost, does not permit large amounts of information to be stored in it for long periods of time, especially if the information is not required for current processing. For this, much cheaper secondary memories are used with a capacity many times greater than that of the primary memory. When the amount of information to be handled is too large for the primary memory, it is stored in the secondary memory devices and the parts of that information required for the processing, are transferred to the primary memory, as necessary. The informa-

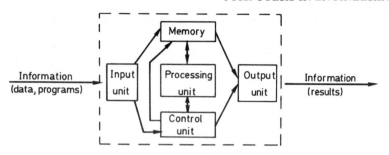

Figure 1.5   Basic components of a computer

tion is transferred back to the secondary memories when no longer required, so that the primary memory can accept the fresh information, which is to be operated upon. This mechanism of moving information between the secondary and primary memories slows down the computer processing, but on the other hand it allows large amounts of information to be processed, which could not be handled otherwise. In computers up to the third generation, information is transferred from secondary to primary memory, and vice versa, according to instructions included in the program provided by the user, making programming a very complex task. In the computers of the fourth generation, it is the computer itself, to a great extent, which performs that task. Such capability is normally given the name of *virtual memory* capability.

Secondary memory devices can be fixed or removable magnetic disks, magnetic tapes, magnetic drums, etc., the first two types being the more common ones. Magnetic tapes and removable magnetic disks are also used as information support, to store large amounts of information for long periods of time, in a very compact way. It is interesting to note that in a small volume, corresponding to one magnetic tape or disk, several millions of characters can be stored.

The processing of the information stored in the primary memory is carried out by the *processing unit.* This unit has the capacity to perform arithmetic operations, and to take logical decisions. It receives information from the primary memory, subdivided in words, operates upon that information, and places in the primary memory the new information generated.

The computer delivers to the user the results of a solution to a problem through the output unit. Again an information support will be used to register that information. The information supports for output can be continuous sheets of paper obtained from the printer, magnetic tapes, punched cards, microfilm, paper with drawings obtained from a plotter, etc. When the results are to be man-readable, in most cases they are printed on paper. It is also possible to have an output unit consisting of a screen where results can be displayed, for instance in the form of diagrams. Sometimes, these devices produce a 'hard copy' of that information, for instance on paper. In general, a computer will have more than one output unit, one or more of which will be printers.

All the operations which are carried out inside a computer are controlled by its *control unit*. This unit is the brain of the computer, in the sense that it initiates, supervises, and directs the activities of the remaining units. The control unit is the one 'giving the orders'.

A computer can be used in different ways, depending on its capabilities, and on how the computation centre is organized. Most computation centres are organized according to the 'closed shop' operation concept. The user will prepare his program and data, for instance on cards, with the required control cards placed where needed. The organized collection of cards is called a *job*. He then *submits* his job for processing. Later when processed, he will receive the computer results. This is called *batch processing*. The time spent from the moment of submitting the program until the results become available for the user, is called '*turn-around*' time, and it is of great concern to most users. It can take from a   few minutes to several days, depending on a number of factors.

To use a computer one need not be physically close to it. Some installations allow for remote processing through *terminals* connected to the central computer, in the same way as telephones are connected to a telephone exchange. These may be *remote job entry* terminals, in which case the user submits his programs through the terminal, taking this as a normal input unit, or they can be terminals with the possibility of *time-sharing processing*. In this case the user will be competing with other time-sharing and batch processing users for the utilization of the computer resources. The turn-around time is practically nil, and it may be possible to operate in a *conversational mode,* allowing a very fruitful man—machine iteration.

In summary, a computer can perform several simple tasks. It can receive and store information, perform elementary arithmetic operations, take simple logical decisions, and deliver information. However, the *speed* at which a computer can perform those operations, and the fact that, due to the *stored program* concept, it can do it in an *automatic* way, transforms it into a device of extraordinary importance.

1.3   PROGRAMMING LANGUAGES

A computer can perform elementary operations using stored information. The user can order the computer to perform an operation, giving the proper instruction, as a part of his program. The set of all the instructions which a computer accepts constitutes the *machine language,* or *absolute language* of that computer. The machine language instructions are always expressed by numeric codes. Each instruction normally consists of one operation code, and one or more parameters or arguments. For instance, the sequence

60099098100
65100097101

can represent two instructions written according to a given machine language.

Let us assume that the operation code 60 (first two digits of the first instruction) indicates addition. Then the first instruction could be a request to add the content of the memory position of address 099 (first argument), with the content of the memory position of address 098 (second argument), placing the result in the memory position of address 100 (third argument). That would be a three-arguments instruction. The second instruction has an operation code 65. Let us assume that it is used to request a multiplication. In that case, the second instruction might mean that the contents of the memory positions of addresses 100 and 97 are to be multiplied, placing the result in the memory position of address 101.

This hypothetical example attempts to illustrate how a machine language program is written. The difficulties implicit are evident. The numeric operation codes require memorizing, which may be hard for an extensive instruction set. Each instruction refers to one or more memory positions, designated by their numeric addresses, which requires careful planning of the memory utilization, before writing the program. In this type of programming, errors are frequently made. Their detection and correction is difficult especially when operating with *absolute addressing,* that is, referring to each memory position by its numeric address. On the other hand, working at the level of elementary operations, it would obviously be very hard to program the solution of an engineering problem of practical interest. Also, machine languages are not the same for different computers.

To solve some of these problems, *symbolic* languages were introduced. In these languages symbolic names are used for the operation codes. The memory positions are also referred to by symbolic names, which is called *symbolic addressing.* For instance, in a hypothetical symbolic language the two previous machine language instructions could be written as

    ADD   A, B, C
    MUL   C, D, E

where the first instruction would indicate addition (code ADD) of the contents of the memory positions A and B, placing the results in the memory position C. Similarly, the second instruction would indicate multiplication (code MUL) of the contents of the memory positions C and D, placing the results in the memory position E.

Clearly, this type of programming is much simpler than machine language programming. The symbolic operation codes can be remembered easily, and the symbolic addressing simplifies considerably the coding. However, a computer cannot directly accept this type of language, since it only 'understands' its machine language. It is then necessary to translate the symbolic language program to a machine language program before it can be processed. This translation can be done directly by the computer, in an automatic way, using a special translation program, which is normally called an *assembler*. The assembler, residing in the computer memory, is a program which receives a symbolic language program as data, and produces as a result an equivalent machine language program which is processed by the computer.

The basic task of an assembler consists of replacing the symbolic codes by their equivalent numeric codes, and assigning memory addresses to the symbolic names appearing in the arguments.

In general, a one-to-one correspondence exists between the symbolic and machine language instructions, excluding the case of some special *macro-instructions* available in most symbolic languages, which translate into several machine language instructions. The symbolic languages are also sometimes called *assembler languages*.

The programming into symbolic language of a practical engineering problem, comprising some complex operations, will be still very hard. The same is true for other areas. Thus, the need to count with more powerful languages is evident. As far as engineering is concerned, this need is satisfied with the introduction of the *algorithmic languages*.

Algorithmic languages enable the solution of a problem to be defined as a sequence of algebraic formulae, including some logical conditions and instructions for miscellaneous operations. For instance, in a hypothetical algorithmic language the previous instructions will be written as

$$E = D * (A + B)$$

which is called a statement.

For an engineer, this type of programming is highly efficient since it allows a convenient notation to be used, similar to that of algebra. The rules relative to algorithmic language programming can be easily learned, and it is no longer necessary to work with only elementary operations.

It is obvious that a computer will not accept such languages directly, but require translation to machine language. This translation, far more complicated than in the case of symbolic languages, is also done automatically, using a special program called a *compiler*. In this case, each statement can be translated into many machine instructions, by a translation process called *compilation*. These type of languages also referred to as *compiler languages*.

It is important to note that while a computer has its own machine language, and eventually its own symbolic language, a given algorithmic language can be used in a variety of different computers providing the required compiler program is available.

Of the algorithmic languages, the best known is FORTRAN, which is

almost universally available. Most computers will have at least some version
of FORTRAN. Other relatively common algorithmic languages are ALGOL,
BASIC, and *PL/1*.

Even when there are differences between the various algorithmic languages,
they all include a basic set of capabilities. From the point of view of programming an engineering solution they must permit data input and output, the
evaluation of algebraic formulae, and decision making on the basis of a numeric or logical value. They should also allow repeated loops (such as the
FORTRAN DO loop) to be performed. The programmer should be able to
subdivide his computer solution into a set of relatively independent programs
or *subprograms* (also called *routines*), such that they can be linked together
using the proper calling sequences, and that data can be transferred among
them. The language should also allow working with *integer, real,* and *logical*
constants or variables, including the case of *single variables* and *subscripted
variables* or *arrays*. Statements for reading-from, and writing-into, secondary
memory devices should also exist.

Some languages of this type are not completely translated or compiled
before they are executed by the computer. Rather, each statement is translated and immediately executed in the sequence given in the program. This
type of processing can be extremely useful when operating in a time-sharing
environment, because it allows for the conversational mode of processing.
The translation program is, in this case, called *interpreter.*

Finally, at the other extreme of the spectrum of the programming languages, we find the *problem oriented languages.* These languages, designed
and implemented mostly by engineers, are oriented to provide computer
solutions for problems in a given engineering area. Their fundamental
characteristic is that in using them the man—machine communication can be
established at a high level, and that they employ the same terms as the engineer uses in his technical vocabulary. They are suitable for a wide range of
applications, and are very easy to use and to learn. Some well known problem
oriented languages for engineering applications are COGO, STRESS, and
those included in the ICES system, such as STRUDL, ROADS, PROJECT,
etc., all implemented at the Massachusetts Institute of Technology.

When an engineer decides to use a computer to solve a given problem, and
he does not have an adequate program already available, or cannot use a problem oriented language, normally he will select an algorithmic language to program the problem solution. In most cases he will use FORTRAN, which is
the language adopted for the applications presented in this book. Some of the
basic details about the implementation and successful use of a program, are
given in the following paragraphs.

1.4   PROGRAMMING A PROBLEM SOLUTION

The first step in programming the solution of a problem, is to make an *analysis* of the problem. The problem is studied in all its detail, and an appropriate solution scheme is selected. Then the computer solution to be implemented is schematized graphically, by means of a diagram called a *flow-chart*. These diagrams can be detailed to a greater or lesser extent, according to circumstances, but they are fundamental to the analysis and understanding of all aspects of the computer solution. The person in charge of this task is normally called an *analyst*.

Once the *flow-chart* is completed, the computer solution is coded, writing the computer program using the language selected. This task is carried out by a *programmer*.

With the program registered in some information support, and the data and proper control commands added, the program is tested by submitting it for processing. At the beginning the program will include some errors. These can be *sintax* errors and *logic* errors. Sintax errors appear because some of the rules of the language are violated. Normally these can be easily detected and corrected. Logic errors correspond to the case that a sintax error-free program is submitted for execution, and it does not produce the desired results. The following situations can be encountered.

(1)   Execution of the program stops before producing results.
(2)   Execution of the program does not stop and results are not produced. The program is finally discontinued after exceeding the allowed time.
(3)   Results are produced, but they are incorrect.
(4)   The program works successfully.

In the first processing of a program it is very rare to have the situation (4). In the other cases, the programmer will have to make a careful examination to detect and correct the existing errors, referred to as *bugs*. This task is usually called *debugging* the program. Its length depends on several factors, it being important to note that a good programmer is not the one who transforms the definition of the problem into a program the quickest, but the one who is fastest in having the program working properly.

Debugging can be simplified by a careful selection of the test data. These should not be chosen at random. It is better to use test data for which results are available, starting with simple cases, but finally testing all the possible alternatives. Also it is very helpful to include in the program statements for the intermediate output. These will indicate where the computation is going wrong. When the program is correct these statements can be removed.

Once the program is working it can be used to solve practical problems, thus entering into its *production phase*. Even then the user should not accept the results blindly. Consistency checks should be performed on the results to evaluate their correctness. Many times they will be found to be incorrect because the data was specified erroneously. It is also possible that the programmer declared the program to be correct after a few tests, without checking all the alternatives. Sometimes the user is working with a large general programming system. The implementation of these general systems can be so complicated, and the tests required so extensive, that they are rarely 'bug-free'. In all cases the user should never accept the correctness of computer results, until after analysing them and being reasonably sure that they are not erroneous.

1.5   PROGRAMS PRESENTED IN THE BOOK

A large number of computer programs for different applications are included in this book, all written using the FORTRAN language.

Several versions of FORTRAN exist, some more powerful than others. The programs presented, however, were written using only basic FORTRAN statements, since these can be used in most machines. The only feature used in these programs which might not be available in some machines is variable dimensioning of arrays. Where that is the case, the dimension specifications can be easily changed to fixed numeric values. Enough details are included, to allow for this.

An attempt was made for the different programs to be compatible. The programs of Chapter 2, which perform matrix operations, including solution of systems of equations, matrix inversion, and eigenvalue and eigenvector calculation, were designed to be useful in structural analysis and finite element applications developed in the remaining chapters. In that sense, they are used extensively.

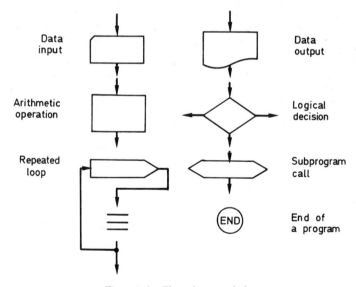

Figure 1.6   Flow-chart symbols

The structural analysis and finite element programs share several general routines. A modular program for static analysis of trusses is developed in Chapter 3. Then it is modified for static and dynamic analysis of frame systems, in Chapter 4. In Chapter 5, programs for the solution of the Laplace equation and finite element analysis of plane stress/strain problems are also presented, in each showing the modifications to be made in the original programs.

The programs have been kept as simple as possible, bearing in mind efficiency and usefulness. Thus, most of the programs could be somehow generalized, for instance to include the consideration of temperature effects in truss systems, lateral loads in plane stress/strain problems, etc. It is felt, however, that the solution scheme adopted for all programs allows for easy modification, and that the information included both in the theoretical derivation and program implementation will minimize the effort required to do so.

When advisable flow charts are also presented, using the symbols given in Figure 1.6 for the basic operations.

Complete listings are included for most programs, and in most cases illustrative examples, showing computer results, are also included.

References

1. Turner, M. S., Clough, R. W., Martin, H. C. and Topp, L. J., "Stiffness and deflection analysis of complex structures". *J. Aero. Soc.*, **23**, No.9, pp 805–823 (1956)
2. Corum, J. M. and Krishnamurthy, N., "A three-dimensional finite element analysis of a prestressed concrete reactor vessel model", in Proceedings Symposium on Application of Finite Element Methods in Civil Engineering, Nashville, Tennessee, (1970)
3. Smith, S. L. and Brebbia, C. A., "Finite-element solution of Navier–Stokes equations for transient two-dimensional incompressible flow", *J. Computational Physics*, **17**, No.3, (1975)
4. Brebbia, C. A., Ferrante, A. J. and de Lima, F e P., "Structural analysis of fixed offshore platforms", in Proceedings 18th South-American Conference on Structural Engineering, Salvador, Brazil, (1976)
5. Ferrante, A. J., Halbritter, A. L. and Groehs, A. G., "Analysis of laminar structures by the finite element method", in Proceedings 18th South-American Conference on Structural Engineering, Caracas, Venezuela, (1975)

Chapter 2

Matrix algebra by computers

2.1 INTRODUCTION

This chapter presents the basic mathematical tools needed for the solution of a wide variety of engineering problems, some of which are studied in later chapters.

Many engineering problems are more conveniently described using matrix notation. A summary of definitions and rules concerning matrix operations is given in Section 2.2, sufficient for understanding the matrix notation used in subsequent chapters.

The solution of linear systems of simultaneous equations is treated in Section 2.3. Even though a large number of different methods of solving linear systems of simultaneous equations exist, no attempt has been made to present an extensive study of this subject. Rather, the emphasis is placed on the few methods most used in practice for engineering applications, giving a less detailed treatment to other methods of lesser interest, included for consistency.

Section 2.4 is an introduction to the study of eigenvalue and eigenvector problems and their methods of solution. The methods presented are selected because of their simplicity, recognizing the existence of other more efficient but considerably more complex methods.

Sections 2.5 and 2.6 treat the subjects of quadratic forms and matrix representation of extremum problems, a knowledge of which is required for a better understanding of the engineering formulations introduced in later chapters.

This chapter includes 25 computer routines for matrix operations, solution of linear systems of simultaneous equations, and eigenvalue and eigenvector computation. They were programmed using mostly only the basic features of the FORTRAN language, to allow for their implementation in the largest possible number of different computers. The only feature which may not be universally available is variable dimensioning, used in some programs, although sufficient explanation is provided, to allow for their proper modification when required. In some cases the computational procedures discussed are first presented in flow chart form, before showing their corresponding FORTRAN code. It should be noticed that the flow charts are not an exact graphical reproductions of the FORTRAN codes but, on the contrary, they are an aid to analyzing a problem before proceeding to its programming. Therefore, all the symbols and operations included in a flow chart do not necessarily follow exactly the FORTRAN language rules. Also, to avoid confusion with the letter 'o', the number zero is represented by ϕ.

Finally, it is important to mention that even when efficiency was a major consideration in preparing the computer codes included in this chapter, in some cases it was decided to make the programming as simple as possible, to facilitate a clearer understanding of the processes involved. Some of the programs presented could therefore be optimized by the more advanced reader.

The sequence of programs presented include some which are not used in later chapters in order to present a consistent sequence of programs, progressively becoming more involved, and because some of these programs could be useful in other applications which the reader may wish to develop by himself.

2.2 ELEMENTARY MATRIX OPERATIONS

Matrix algebra is the set of rules which defines operations performed with matrices. A matrix is a bi-dimensional array of numbers organized according to n rows and m columns, which we refer to as a matrix of order nxm. In what follows, a matrix will be indicated by bold type. Thus, $\mathbf{R}$ will be a matrix, while T is a scalar number. The elements or coefficients of a matrix will have two sub-indices, indicating row and column position, respectively. For instance

$$\mathbf{A} = \begin{bmatrix} a_{11} & a_{12} & a_{13} \\ a_{21} & a_{22} & a_{23} \end{bmatrix} \tag{2.1}$$

is a rectangular matrix, of order 2×3, that is having 2 rows and 3 columns. The sub-indices of the coefficients indicate their position within the matrix. Thus, the coefficient a_{12} is placed in the first row and the second column.

A matrix having the same number of rows and columns is called a *square matrix,* such as

$$\mathbf{B} = \begin{bmatrix} b_{11} & b_{12} & b_{13} \\ b_{21} & b_{22} & b_{23} \\ b_{31} & b_{32} & b_{33} \end{bmatrix} \tag{2.2}$$

which is a square matrix of 3 rows and 3 columns, or of order 3×3. In cases such as this, when speaking specifically of square matrices, we can simply say that **B** is of order 3. The coefficient b_{11}, b_{22}, and b_{33}, or in general b_{ij} for $i = j$, are the coefficients of the *principal* or *main diagonal* of the matrix.

Several types of square matrices can be considered. When the coefficients of a square matrix **B** satisfy the condition

$$b_{ij} = b_{ji}, \text{ for } i = 1, 2, \ldots, n, \text{ and } j = 1, 2, \ldots, n \qquad (2.3)$$

B is said to be a *symmetric matrix*. In particular, if **B** is a symmetric matrix of order 3, it will have the following structure:

$$\mathbf{B} = \begin{bmatrix} b_{11} & b_{12} & b_{13} \\ b_{12} & b_{22} & b_{23} \\ b_{13} & b_{23} & b_{33} \end{bmatrix} \qquad (2.4)$$

A skew-symmetric matrix, on the other hand, is such that

$$b_{ij} = -b_{ji}, \text{ for } i \neq j$$

$$b_{ii} = 0, \text{ for } i = 1, 2, \ldots, n \qquad (2.5)$$

as, for example,

$$\mathbf{B} = \begin{bmatrix} 0 & b_{12} & b_{13} \\ -b_{12} & 0 & b_{23} \\ -b_{13} & -b_{23} & 0 \end{bmatrix} \qquad (2.6)$$

When all the coefficients of a square matrix **B** are null, except for the main diagonal, such as,

$$\mathbf{B} = \begin{bmatrix} b_{11} & 0 & 0 \\ 0 & b_{22} & 0 \\ 0 & 0 & b_{33} \end{bmatrix} \qquad (2.7)$$

B is called a *diagonal matrix*. If, in particular, all the coefficients of the main diagonal are equal to 1, the matrix is called a *unit matrix,* and is symbolized by **I**. For instance, the following is the unit matrix of order 3.

$$I = \begin{bmatrix} 1 & 0 & 0 \\ 0 & 1 & 0 \\ 0 & 0 & 1 \end{bmatrix} \qquad (2.8)$$

A matrix having only one column is called a *column vector*. Normally, the coefficients of a column vector will be given only one subscript, corresponding to the row position, and will be enclosed by curled brackets instead of square brackets, placed either vertically or horizontally (for convenience), as in

$$C = \begin{Bmatrix} c_1 \\ c_2 \\ c_3 \end{Bmatrix} = \{c_1 \quad c_2 \quad c_3\} \qquad (2.9)$$

where C is a vector of order 3.

When a matrix has only one row it will be called a *row vector* and, again, each of its coefficients will be given only one subscript.

For instance,

$$D = [d_1 \, d_2 \, d_3 \, d_4] \qquad (2.10)$$

is a row vector of order 4.

In what follows the use of the term 'vector' alone will always be taken to mean 'column vector'.

The *transpose* of a matrix is obtained by interchanging rows and columns, of the matrix, and is indicated by an upper-index T. For instance, the transpose of A, given by (2.1) is.

$$A^T = \begin{bmatrix} a_{11} & a_{21} \\ a_{12} & a_{22} \\ a_{13} & a_{23} \end{bmatrix} \qquad (2.11)$$

while that of matrix B, given by (2.2), is

$$B^T = \begin{bmatrix} b_{11} & b_{21} & b_{31} \\ b_{12} & b_{22} & b_{32} \\ b_{13} & b_{23} & b_{33} \end{bmatrix} \qquad (2.12)$$

It can be noticed that the transpose of a column vector is a row vector, and viceversa. Thus, the transpose of (2.9) is the row matrix

$$\mathbf{C}^T = [c_1 \, c_2 \, c_3] \tag{2.13}$$

while the transpose of row matrix (2.10) is the vector

$$\mathbf{D}^T = \begin{Bmatrix} d_1 \\ d_2 \\ d_3 \\ d_4 \end{Bmatrix} \tag{2.14}$$

The transpose of the transpose gives the original matrix, that is

$$(\mathbf{A}^T)^T = \mathbf{A} \tag{2.15}$$

Condition (2.3), defining the particular property of a symmetric matrix, can be equivalently written as

$$\mathbf{B} = \mathbf{B}^T \tag{2.16}$$

Another special matrix which can be defined is the *null matrix*, represented by **0**, whose coefficients are all equal to zero. For instance,

$$\mathbf{0} = \begin{bmatrix} 0 & 0 & 0 \\ 0 & 0 & 0 \end{bmatrix} \tag{2.17}$$

is a null matrix of order 2 × 3.

Program 1   Transpose of a rectangular matrix

The following routine is a very simple example of a computer program operating with matrices which can be used to obtain the transpose of a rectangular

```
      SUBROUTINE TRANR(A,B,N,M)
C
C PROGRAM 1
C
C THIS PROGRAM COMPUTES THE TRANSPOSE
C OF A MATRIX "A" STORING IT IN THE
C ARRAY "B"
C
```

```
C  N :  ACTUAL NUMBER OF ROWS OF A
C  M :  ACTUAL NUMBER OF COLUMS OF A
C
        DIMENSION A(10,10),B(10,10)
        DO 10 I=1,N
        DO 10 J=1,M
    10  B(J,I)=A(I,J)
        RETURN
        END
```

matrix. The array **A** contains the matrix to be transposed, the integer n contains its number of rows, and m the number of columns. The transposed matrix is returned in the array **B**.

Program 2   Transpose of a square matrix

In the case of a square matrix it is possible to transpose the matrix in itself, without creating a new one. The following routine receives the original matrix

```
        SUBROUTINE TRANS(A,N)
C
C PROGRAM 2
C
C THIS PROGRAM COMPUTES THE TRANSPOSE
C OF A SQUARE MATRIX "A" STORING IT
C IN ITSELF
C
C N :  ACTUAL NUMBER OF ROWS AND COLUMS OF "A"
C
        DIMENSION A(10,10)
        N1=N-1
        DO 10 I=1,N1
        I1=I+1
        DO 10 J=I1,N
        S=A(I,J)
        A(I,J)=A(J,I)
    10  A(J,I)=S
        RETURN
        END
```

in the array **A**, of order N, and after operating returns the transposed matrix in the same array **A**. Note that an auxiliary variable S is used to allow for the transposition in place.

Two different matrices can be added, or subtracted, provided they are of the same order. Let us consider the following two matrices

$$\mathbf{A} = \begin{bmatrix} a_{11} & a_{12} & a_{13} \\ a_{21} & a_{22} & a_{23} \end{bmatrix}, \qquad \mathbf{B} = \begin{bmatrix} b_{11} & b_{12} & b_{13} \\ b_{21} & b_{22} & b_{23} \end{bmatrix} \qquad (2.18)$$

By definition the sum of **A** and **B** will give another matrix **C**, also of the same order, ie

$$\mathbf{C} = \begin{bmatrix} c_{11} & c_{12} & c_{13} \\ c_{21} & c_{22} & c_{23} \end{bmatrix} \tag{2.19}$$

whose coefficients are

$$c_{11} = a_{11} + b_{11}$$

$$c_{12} = a_{12} + b_{12}$$

$$c_{13} = a_{13} + b_{13}$$

$$c_{21} = a_{21} + b_{21}$$

$$c_{22} = a_{22} + b_{22}$$

$$c_{23} = a_{23} + b_{23} \tag{2.20}$$

In general

$$c_{ij} = a_{ij} + b_{ij}, \text{ for } i = 1, 2, \ldots, n; j = 1, 2, \ldots, m \tag{2.21}$$

Matrix subtraction is defined in a similar manner. Thus, if we have the matrix equation

$$\mathbf{C} = \mathbf{A} - \mathbf{B} \tag{2.22}$$

the coefficients of $\mathbf{C}$ will be computed by

$$c_{ij} = a_{ij} - b_{ij}, \text{ for } i = 1, 2, \ldots, n; j = 1, 2, \ldots, m \tag{2.23}$$

Program 3   Addition

A routine for the addition and subtraction of matrices is very easily programmed. For instance, the following routine can be used for addition or subtraction of two matrices. The arrays $\mathbf{A}$ and $\mathbf{B}$ contain the matrices to be added or subtracted, which have N rows and M columns. The array $\mathbf{C}$ will contain the result matrix. When the integer $L = 1$ is specified the routine will perform the addition operation, but when $L = 2$ the routine will perform the subtraction operation.

```
      SUBROUTINE ADSUB(C,A,B,N,M,L)
C
C PROGRAM 3
C
C THIS PROGRAM COMPUTES THE MATRIX OPERATION
C C = A + B , WHEN L = 1 AND
C C = A - B , WHEN L = 2
```

```
C
C N : ACTUAL NUMBER OF ROWS OF A, B, AND C
C M : ACTUAL NUMBER OF COLUMNS OF A, B, AND C
C
      DIMENSION A(10,10),B(10,10),C(10,10)
      DO 10 I=1,N
      DO 10 J=1,M
      GO TO(2,4),L
    2 C(I,J)=A(I,J)+B(I,J)
      GO TO 10
    4 C(I,J)=A(I,J)-B(I,J)
   10 CONTINUE
      RETURN
      END
```

Multiplication between matrices can also be performed. Let us consider a matrix A of order 2×3, and a matrix B of order 3×2, such as

$$A = \begin{bmatrix} a_{11} & a_{12} & a_{13} \\ a_{21} & a_{22} & a_{23} \end{bmatrix}, \qquad B = \begin{bmatrix} b_{11} & b_{12} \\ b_{21} & b_{22} \\ b_{31} & b_{32} \end{bmatrix} \qquad (2.24)$$

The matrix multiplication equation

$$C = AB \qquad (2.25)$$

gives as a result a new matrix C, with coefficients

$$c_{11} = a_{11} b_{11} + a_{12} b_{21} + a_{13} b_{31}$$

$$c_{12} = a_{11} b_{12} + a_{12} b_{22} + a_{13} b_{32}$$

$$c_{21} = a_{21} b_{11} + a_{22} b_{21} + a_{23} b_{31}$$

$$c_{22} = a_{21} b_{12} + a_{22} b_{22} + a_{23} b_{32} \qquad (2.26)$$

or, in general

$$c_{ij} = \sum_{k=1}^{3} a_{ik} b_{kj} \qquad (2.27)$$

It is easily seen that two matrices can be multiplied only if the number of columns of the first matrix is equal to the number of rows of the second matrix. The result matrix will have the same number of rows as the first

matrix, and the same number of columns as the second matrix. Notice that
according to its definition, matrix multiplication is not commutative

$$AB \neq BA \qquad\qquad (2.28)$$

and that, in particular, if the left-hand side is defined, the right-hand side will
not be defined unless the number of rows of **A** is also equal to the number of
columns of **B**. Even in that case, however, the results of the left-hand side,
and the right-hand side will be different, excluding very particular cases.

Program 4   Multiplication of two matrices

The multiplication of two matrices can be readily programmed. The following
routine receives the first matrix of order $N \times M$, in the array **A**, and the second
matrix, of order $M \times L$, in the array **B**. The result matrix, of order $N \times L$, will
be stored in the array **C**. Note that since the implementation of Equation
(2.26) requires accumulating the partial multiplications in the array **C**, we
have been careful in previously setting $c_{ij} = 0$.

```
      SUBROUTINE ATIMB(C,A,B,N,M,L)
C
C PROGRAM 4
C
C THIS PROGRAM COMPUTES THE MATRIX
C OPERATION C = A * B
C
C N : ACTUAL NUMBER OF ROWS OF A AND C
C M : ACTUAL NUMBER OF COLUMNS OF A AND ROWS OF B
C L : ACTUAL NUMBER OF COLUMNS OF B AND C
C
      DIMENSION A(10,10),B(10,10),C(10,10)
      DO 10 I=1,N
      DO 10 J=1,L
      C(I,J)=0.
      DO 10 K=1,M
   10 C(I,J)=C(I,J)+A(I,K)*B(K,J)
      RETURN
      END
```

Program 5   Multiplication of two matrices, storing the results in the first matrix

In some cases, to save storage space, it is interesting to implement the matrix
multiplication in such a way that the result matrix is stored in the array originally containing the first matrix. Since according to the parameters passed
the results might be stored in the array originally containing the coefficients
of **A**, the rows of the result matrix can be stored temporarily in an auxiliary
vector V, so as not to destroy those coefficients. After each row is computed,
it is transferred to array **C**, which may coincide with array **A**. A routine performing these operations is the following:

```
      SUBROUTINE MATMA(C,A,B,N,M,L)
C
C PROGRAM 5
C
C THIS PROGRAM COMPUTES THE MATRIX
C OPERATION C = A * B ALLOWING FOR
C C BEING STORED IN THE ARRAY
C ORIGINALLY CONTAINING A
C N : ACTUAL NUMBER OF ROWS OF A AND C
C M : ACTUAL NUMBER OF COLUMNS OF A AND ROWS OF B
C L : ACTUAL NUMBER OF COLUMNS OF B AND C
C
      DIMENSION C(10,10),A(10,10),B(10,10),V(10)
      DO 10 I=1,N
      DO 5 J=1,L
      V(J)=0
      DO 5 K=1,M
    5 V(J)=V(J)+A(I,K)*B(K,J)
      DO 10 J=1,L
   10 C(I,J)=V(J)
      RETURN
      END
```

For the results to be stored in array **A**, the calling sequence should be of the type CALL MATMA (A, A, B, N, M, L).

Program 6 Multiplication of two matrices, storing the results in the second matrix

We also present a routine for the multiplication of two matrices, such that the result matrix is stored in the array originally containing the second matrix. We will restrict this routine in the case of square matrices, and will use variable dimensioning. The parameters passed are the arrays **A** and **B**, containing the first and second matrices in the multiplication, the auxiliary vector array **V**, the order of the matrices N, and the row and column dimension NX. If the FORTRAN compiler used does not allow for variable dimensioning, **V** and NX can be eliminated as parameters, and numeric dimensions must be used for arrays **A**, **B**, and **V**.

Notice that in this case we compute the result matrix by columns rather than by rows, not to destroy coefficients of **B** still needed. This is done by simply interchanging the loops on I and J.

```
      SUBROUTINE MATMB(A,B,V,N,NX)
C
C PROGRAM 6
C
C THIS PROGRAM PERFORMS THE MATRIX OPERATION
C B = A * B
C
C N : ACTUAL NUMBER OF ROWS AND COLUMS OF A AND V
C NX: ROW AND COLUMN DIMENSION OF A AND B
C
C V : AUXILIARY ARRAY
```

```
C
      DIMENSION A(NX,NX),B(NX,NX),V(NX)
      DO 20 J=1,N
      DO 16 I=1,N
      V(I)=0.
      DO 16 K=1,N
   16 V(I)=V(I)+A(I,K)*B(K,J)
      DO 20 I=1,N
   20 B(I,J)=V(I)
      RETURN
      END
```

Program 7   Evaluation of $C = B^T AB$ using previous routines

In many engineering applications it is required to evaluate a matrix expression such as

$$C = B^T AB \tag{2.29}$$

where A must be a square matrix of order $N \times N$, and B a rectangular matrix of order $N \times M$. The result matrix C will be of order $M \times M$. Using some of the routines previously described, we can write the following routine to evaluate (2.29)

```
      SUBROUTINE BTAB1(C,A,B,N,M)
C
C PROGRAM 7
C
C THIS PROGRAM COMPUTES THE MATRIX
C OPERATION C = TRANSPOSE(B) * A * B
C WHERE A IS A SQUARE MATRIX
C
C N : ACTUAL NUMBER OF ROWS AND COLUMNS
C     OF A AND OF ROWS OF B
C M : ACTUAL NUMBER OF COLUMNS OF B
C
C BT: AUXILIARY ARRAY
C
C
      DIMENSION A(10,10),B(10,10),C(10,10),BT(10,10)
      CALL TRANR(B,BT,N,M)
      CALL MATMA(BT,BT,A,M,N,N)
      CALL MATMA(C,BT,B,M,N,M)
      RETURN
      END
```

Notice that the sequence of operations carried out are

(1) $B^T \rightarrow BT$

(2) $B^T A \rightarrow BT$

(3) $(B^T A)B \rightarrow C$

which are materialized by three subroutine calls.

Program 8   Evaluation of $C = B^T AB$ without using previous routines

If we wish to evaluate the matrix Equation (2.29) more efficiently, avoiding the subroutine calls, we can program a new routine including explicitly the operations required, Figure 2.1.

```
      SUBROUTINE BTAB2(C,A,B,N,M)
C
C PROGRAM 8
C
C
C THIS PROGRAM COMPUTES THE MATRIX
C OPERATION C = TRANSPOSE(B) * A * B
C WHERE A IS A SQUARE MATRIX, WITHOUT
C CALLING OTHER SUBPROGRAMS
C
C N : ACTUAL NUMBER OF ROWS AND COLUMNS
C     OF A AND OF ROWS OF B
C M : ACTUAL NUMBER OF COLUMNS OF B
C BB: AUXILIARY ARRAY
C
      DIMENSION C(10,10),A(10,10),B(10,10),BB(10,10)
C
C COMPUTE A * B AND STORE IN BB
C
      DO 10 I=1,N
      DO 10 J=1,M
      BB(I,J)=0.
      DO 10 K=1,N
   10 BB(I,J)=BB(I,J)+A(I,K)*B(K,J)
C
C COMPUTE TRASPOSE(B) * BB AND STORE IN C
C
      DO 20 I=1,M
      DO 20 J=1,M
      C(I,J)=0.
      DO 20 K=1,N
   20 C(I,J)=C(I,J)+B(K,I)*BB(K,J)
      RETURN
      END
```

In this case we use the following sequence of operations

(1) $AB \rightarrow BB$

(2) $B^T(AB) \rightarrow C$

Notice that we are evaluating (1) and placing the result matrix in the array *BB*. Then, to premultiply the matrix in **BB** by the transpose of **B** we simply alter the subscripts of the array **B** when evaluating (2).

Program 9   Evaluation of $C = B^T AB$, storing the result in A

If we wish to save storage space, we might program the matrix operation (2.29) such that the result matrix is stored in the array originally containing the

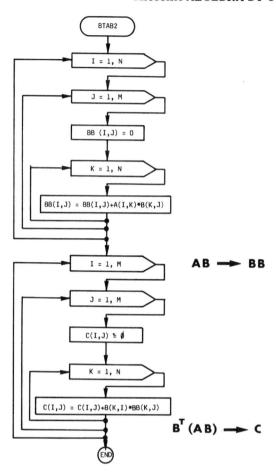

Figure 2.1   Flow-chart for BTAB2 program

matrix **A**. Let us consider only the case of square matrices. Again we use an auxiliary vector array **V** to store temporarily the columns of the result matrix, such that coefficients of **A** still needed are preserved. The order of the matrices involved is N, while NX is used for variable dimensioning.

```
      SUBROUTINE BTAB3(A,B,V,N,NX)
C
C PROGRAM 9
C
C THIS PROGRAM COMPUTES THE MATRIX
C OPERATION A = TRANSPOSE(B) * A * B,
C WHERE A AND B ARE SQUARE MATRICES
C N : ACTUAL ORDER OF A AND B
```

```
C NX: ROW AND COLUMN DIMENSION OF A AND B
C
C V : AUXILIARY VECTOR
C
      DIMENSION A(NX,NX),B(NX,NX),V(NX)
C
C COMPUTE A * B AND STORE IN A
C
      DO 10 I=1,N
      DO  5 J=1,N
      V(J)=0.
      DO 5 K=1,N
    5 V(J)=V(J)+A(I,K)*B(K,J)
      DO 10 J=1,N
   10 A(I,J)=V(J)
C
C COMPUTE TRANSPOSE(B) * A AND STORE IN A
C
      DO 20 J=1,N
      DO 15 I=1,N
      V(I)=0.
      DO 15 K=1,N
   15 V(I)=V(I)+B(K,I)*A(K,J)
      DO 20 I=1,N
   20 A(I,J)=V(I)
      RETURN
      END
```

A special case of matrix multiplication is encountered when the second matrix is a column matrix or vector. Let us consider a matrix A, of order $n \times m$, and a vector B of order m. The matrix multiplication $C = AB$ gives as a result a vector C of order n, whose coefficients are computed by

$$c_i = \sum_{j=1}^{m} a_{ij}b_j \text{ for } i = 1, 2, \ldots, n \qquad (2.30)$$

Program 10   Multiplication of a matrix by a vector

We can now program a routine for the multiplication of a matrix of order $N \times M$, which is stored in an array A, by vector of order M which is stored in an array B, producing a result vector of order N, which will be stored in an array C.

```
      SUBROUTINE MULTV(C,A,B,N,M)
C
C PROGRAM 10
C
C THIS PROGRAM COMPUTES TH MATRIX
C OPERATION C = A * B, WHERE B IS
C A VECTOR
C
C N : ACTUAL NUMBER OF ROWS OF A
C M : ACTUAL NUMBER OF COLUMNS OF A
```

```
C     AND OF ROWS OF B
C
      DIMENSION C(10),A(10,10),B(10)
      DO 10 I=1,N
      C(I)=0
      DO 10 J=1,M
   10 C(I)= C(I)+A(I,J)*B(J)
      RETURN
      END
```

Another special type of multiplication is the product of a scalar by a matrix. If we have a scalar number d, and a matrix A of order $n \times m$, the product

$$B = dA \tag{2.31}$$

is another matrix B, whose coefficients are computed by

$$b_{ij} = da_{ij} \text{ for } i = 1, 2, \ldots, n; j = 1, 2, \ldots, m \tag{2.32}$$

that is, multiplying all the coefficients of A by the scalar d.

Another definition of great importance is that of the *inverse* of a square matrix. A square matrix B, of order n, will be the inverse of another square matrix A, also of order n, when pre- or post-multiplying A by B, the result is a unit matrix, of order n,

$$BA = AB = I \tag{2.33}$$

In such a case B is written as A^{-1}, and is called the inverse of A. A square matrix A will have an inverse A^{-1}. When the inverse can be obtained, then it is said that A is a *regular matrix*. When it cannot be obtained, then it is said that A is a *singular matrix*. It can be verified that in the last case the determinant of A is null.

So far we have considered the matrices expressed in terms of their scalar coefficients. Sometimes, however, it is convenient to write a matrix in partitioned form, that is subdivided into submatrices. For instance, the matrix

$$A = \begin{bmatrix} 4 & 1 & 3 & 5 & 0 \\ 6 & 2 & 7 & 1 & 1 \\ 4 & 8 & 3 & 9 & 3 \end{bmatrix}$$

can be partitioned into 4 submatrices, as follows:

$$A = \begin{bmatrix} A_{11} & A_{12} \\ -- & -- \\ A_{21} & A_{22} \end{bmatrix} = \begin{bmatrix} 4 & 1 & 3 & 5 & 0 \\ 6 & 2 & 7 & 1 & 1 \\ ------ & ----- \\ 4 & 8 & 3 & 9 & 3 \end{bmatrix}$$

where

$$A_{11} = \begin{bmatrix} 4 & 1 & 3 \\ 6 & 2 & 7 \end{bmatrix} \qquad A_{12} = \begin{bmatrix} 5 & 0 \\ 1 & 1 \end{bmatrix}$$

$$A_{21} = \begin{bmatrix} 4 & 8 & 3 \end{bmatrix} \qquad A_{22} = \begin{bmatrix} 9 & 3 \end{bmatrix}$$

As for a matrix element, the submatrices of a partitioned matrix are identified by row and column indices. A matrix can be partitioned as desired. It should be obvious, however, that partitioning does not alter a matrix, since it is only written in a different way without changing its nature.

Matrix operations can be performed between matrices expressed in partitioned form, in the same way as when they are written in terms of their scalar coefficients.

For instance, considering the previous matrix A, and

$$B = \begin{bmatrix} 1 & 2 \\ 0 & 1 \\ 2 & 3 \\ ---- \\ 1 & 4 \\ 3 & -1 \end{bmatrix} = \begin{bmatrix} B_1 \\ --- \\ B_2 \end{bmatrix}$$

where

$$B_1 = \begin{bmatrix} 1 & 2 \\ 0 & 1 \\ 2 & 3 \end{bmatrix} \qquad B_2 = \begin{bmatrix} 1 & 4 \\ 3 & -1 \end{bmatrix}$$

we can perform the matrix product **AB** in partitioned form as follows

$$\mathbf{AB} = \begin{bmatrix} \mathbf{A}_{11} & \mathbf{A}_{12} \\ -- & -- \\ \mathbf{A}_{21} & \mathbf{A}_{22} \end{bmatrix} \begin{bmatrix} \mathbf{B}_1 \\ -- \\ \mathbf{B}_2 \end{bmatrix} = \begin{bmatrix} \mathbf{A}_{11}\,\mathbf{B}_1 + \mathbf{A}_{12}\,\mathbf{B}_2 \\ ---------- \\ \mathbf{A}_{21}\,\mathbf{B}_1 + \mathbf{A}_{22}\,\mathbf{B}_2 \end{bmatrix}$$

using the same rules and of course obtaining the same results as matrix muliplication for matrices written in terms of their scalar coefficients. It is required, however, that when performing an operation with partitioned matrices, their partitions should be compatible relative to the operation being carried out. Thus, in the previous case the number of columns of $\mathbf{A}_{11}$ should be equal to the number of rows of $\mathbf{B}_1$, the same applying to $\mathbf{A}_{12}$ and $\mathbf{B}_2$, $\mathbf{A}_{21}$ and $\mathbf{B}_1$, and $\mathbf{A}_{22}$ and $\mathbf{B}_2$, or otherwise their multiplication would not be defined.

Other matrix operations, such as addition and subtraction, can be similarly performed with partitioned matrices, provided the matrix partitions are compatible.

2.3 SOLUTION OF SYSTEMS OF SIMULTANEOUS LINEAR EQUATIONS

The solution of many engineering problems can be reduced to finding the solution of a system of simultaneous linear algebraic equations. Let us remember that such systems can be written as

$$a_{11} x_1 + a_{12} x_2 + \ldots a_{1n} x_n = b_1$$

$$a_{21} x_1 + a_{22} x_2 + \ldots a_{2n} x_n = b_2$$

$$\ldots$$

$$a_{n1} x_1 + a_{n2} x_2 + \ldots a_{nn} x_n = b_n \tag{2.34}$$

which can also be written as,

$$\sum_{j=1}^{n} a_{1j} x_j = b_1$$

$$\sum_{j=1}^{n} a_{2j} x_j = b_2$$

$$\sum_{j=1}^{n} a_{nj} x_j = b_n \tag{2.35}$$

or, simply as

$$\sum_{j=1}^{n} a_{ij} x_j = b_i, i = 1, 2, \ldots, n \tag{2.36}$$

in more compact notation. Comparing (2.36) with (2.30) we conclude that (2.34) can be equivalently written in matrix form as

$$\mathbf{AX} = \mathbf{B} \tag{2.37}$$

where $\mathbf{A}$ is a matrix of order n having coefficients a_{ij}, and $\mathbf{X}$ and $\mathbf{B}$ are vectors of order n, having coefficients x_i and b_i respectively. The problem consists of finding the unknown coefficients of $\mathbf{X}$, for known coefficients of $\mathbf{A}$ and $\mathbf{B}$. We will show how to solve linear systems of equations by the Gauss elimination, Cholesky, and iterative methods.

2.3.1 Solution of linear systems of equations by the Gauss elimination method

In order to better explain the solution of linear systems of equations by the Gauss elimination method, let us examine the following example.

Example 2.1
Consider the following system of 3 equations with 3 unknowns.

$$6x_1 + 3x_2 + 6x_3 = 30 \tag{a.1}$$

$$2x_1 + 3x_2 + 3x_3 = 17 \tag{a.2}$$

$$x_1 + 2x_2 + 2x_3 = 11 \tag{a.3}$$

which can also be written

$$\begin{bmatrix} 6 & 3 & 6 \\ 2 & 3 & 3 \\ 1 & 2 & 2 \end{bmatrix} \begin{Bmatrix} x_1 \\ x_2 \\ x_3 \end{Bmatrix} = \begin{Bmatrix} 30 \\ 17 \\ 11 \end{Bmatrix} \qquad (b)$$

Dividing Equation (a.1) by 6, we obtain

$$x_1 + \tfrac{1}{2}x_2 + x_3 = 5 \qquad\qquad\qquad\qquad (c.1)$$

$$2x_1 + 3x_2 + 3x_3 = 17 \qquad\qquad\qquad\qquad (c.2)$$

$$x_1 + 2x_2 + 2x_3 = 11 \qquad\qquad\qquad\qquad (c.3)$$

We can now substitute the value of x_1, as given by the first equation, into Equations (c.2) and (c.3) to obtain

$$x_1 + \tfrac{1}{2}x_2 + x_3 = 5 \qquad\qquad\qquad\qquad (d.1)$$

$$2x_2 + x_3 = 7 \qquad\qquad\qquad\qquad (d.2)$$

$$\tfrac{3}{2}x_2 + x_3 = 6 \qquad\qquad\qquad\qquad (d.3)$$

Divide Equation (d.2) by 2.

$$x_1 + \tfrac{1}{2}x_2 + x_3 = 5 \qquad\qquad\qquad\qquad (e.1)$$

$$x_2 + \tfrac{1}{2}x_3 = \tfrac{7}{2} \qquad\qquad\qquad\qquad (e.2)$$

$$\tfrac{3}{2}x_2 + x_3 = 6 \qquad\qquad\qquad\qquad (e.3)$$

We can next substitute x_2 as given by (e.2) into (e.3) and obtain

$$x_1 + \tfrac{1}{2}x_2 + x_3 = 5 \qquad\qquad\qquad\qquad (f.1)$$

$$x_2 + \tfrac{1}{2}x_3 = \tfrac{7}{2} \qquad\qquad\qquad\qquad (f.2)$$

$$\tfrac{1}{4}x_3 = \tfrac{3}{4} \qquad\qquad\qquad\qquad (f.3)$$

Now we find the value of x_3 from Equation (f.3), the value of x_2 from (f.2) and x_1 from (f.1). This procedure is called 'back-substitution', and gives

$$x_3 = 3 \qquad \qquad \text{(g.1)}$$

$$x_2 = \tfrac{7}{2} - \tfrac{3}{2} = 2 \qquad \text{(g.2)}$$

$$x_1 = 5 - 1 - 3 = 1 \qquad \text{(g.3)}$$

Note that in the Gauss elimination process, we have divided each equation by its leading coefficient. This operation was done 3 times, where 3 is the number of equations. Thus, the final determinant, which is 1, should be multiplied by the divisors 6, 2, ¼. The value of the determinant is,

$$6 \times 2 \times \tfrac{1}{4} = 3 \qquad \text{(h)}$$

which can be indicated as

$$|\mathbf{A}| = |a_{ij}| = \prod_{i=1}^{n} d_i \qquad \text{(i)}$$

where d_i are the divisors and Π indicates product. The value of the determinant may be of interest in some applications. For instance to indicate when the determinant of the matrix $\mathbf{A}$ changes sign.

In this example we can also verify that each elimination step requires dividing the coefficients of a row by the diagonal coefficient. If that diagonal coefficient happens to be zero, the division cannot be performed, and the system may or may not be solved. In a great number of engineering problems, the governing systems of equation are such that a null diagonal coefficient will never be encountered during the elimination procedure. Such is the case for *positive definite matrices* (see Section 2.5 for a definition of these matrices).

Example 2.2
Let us now consider the systems

$$6x_1 + 3x_2 + 6x_3 = 30 \qquad \text{(a.1)}$$

$$2x_1 + x_2 + 3x_3 = 17 \qquad \text{(a.2)}$$

$$x_1 + 2x_2 + 2x_3 = 11 \qquad \text{(a.3)}$$

Dividing Equation (a.1) by 6 we obtain

$$x_1 + \tfrac{1}{2}x_2 + x_3 = 5 \qquad \text{(b.1)}$$

$$2x_1 + x_2 + 3x_3 = 17 \tag{b.2}$$

$$x_1 + 2x_2 + 2x_3 = 11 \tag{b.3}$$

Eliminating x_1, as in the previous example, we obtain

$$x_1 + \tfrac{1}{2}x_2 + x_3 = 5 \tag{c.1}$$

$$x_3 = 7 \tag{c.2}$$

$$\tfrac{3}{2}x_2 + x_3 = 6 \tag{c.3}$$

The elimination process cannot now continue, because the diagonal coefficient of the second row is null. However, if we interchange the second and third row, we obtain the equivalent system

$$x_1 + \tfrac{1}{2}x_2 + x_3 = 5 \tag{d.1}$$

$$\tfrac{3}{2}x_2 + x_3 = 6 \tag{d.2}$$

$$x_3 = 7 \tag{d.3}$$

The elimination process can now proceed as normally, leading to the solution

$$\mathbf{x} = \begin{Bmatrix} -5/3 \\ -2/3 \\ 7 \end{Bmatrix} \tag{e}$$

It is possible to compute the determinant of the system matrix which is negative, i.e.

$$|\mathbf{A}| = -9 \tag{f}$$

This example shows that even in the case in which a null diagonal coefficient is found, the Gauss elimination process can still be applied, provided we adequately interchange a given number of rows. When this is possible, the system matrix is not positive definite but it is still a regular matrix.

Example 2.3
We will now analyse the following system

$$6x_1 + 3x_2 + 6x_3 = 30 \tag{a.1}$$

$$2x_1 + x_2 + 3x_3 = 17 \tag{a.2}$$

$$x_1 + \tfrac{1}{2}x_2 + 2x_3 = 11 \tag{a.3}$$

If we divide the first row by its diagonal coefficient, and then eliminate the unknown x_1, we obtain the following

$$x_1 + \tfrac{1}{2}x_2 + x_3 = 5 \tag{b.1}$$

$$x_3 = 7 \tag{b.2}$$

$$x_3 = 6 \tag{b.3}$$

We see that the diagonal coefficient of the second row is again null. However, in this case the interchange of the second and third rows will not help. The Gauss elimination procedure cannot proceed, implying that there is no solution for this system of equations. It can be verified that the determinant of the system matrix is null

$$|\mathbf{A}| = 0 \tag{c}$$

which means that the matrix $\mathbf{A}$ is singular. Thus, we conclude that when the system matrix is singular, the system of equations does not have a solution.

After these specific examples we can now generalize the Gauss elimination procedure. For this, we will consider the general system

$$a_{11}^0 x_1 + a_{12}^0 x_2 + a_{13}^0 x_3 + \ldots + a_{1n}^0 x_n = b_1^0$$

$$a_{21}^0 x_1 + a_{22}^0 x_2 + a_{23}^0 x_3 + \ldots + a_{2n}^0 x_n = b_2^0$$

$$\ldots$$

$$a_{n1}^0 x_1 + a_{n2}^0 x_2 + a_{n3}^0 x_3 + \ldots + a_{nn}^0 x_n = b_n^0 \tag{2.38}$$

where the superscript '0' indicates they are the original values.

First we divide by a^0_{11} obtaining

$$x_1 + a_{12}^1 x_2 + a_{13}^1 x_3 + \ldots + a_{1n}^1 x_n = b_1^1$$

where

$$a_{1j}^1 = a_{1j}^0 / a_{11}^0 \quad j = 2, n$$

$$b_1^1 = b_1^0 / a_{11}^0 \tag{2.39}$$

We can now eliminate x_1 from the second, third, etc., of Equations (2.38) and obtain a modified system of $(n-1)$ equations

$$a_{22}^1 x_2 + a_{23}^1 x_3 + \ldots + a_{2n}^1 x_n = b_2^1$$

$$a_{32}^1 x_2 + a_{33}^1 x_3 + \ldots + a_{3n}^1 x_n = b_3^1$$

. . .

$$a_{n2}^1 x_2 + a_{n3}^1 x_3 + \ldots + a_{nn}^1 x_n = b_n^1 \tag{2.40}$$

where

$$\left.\begin{array}{l} a_{ij}^1 = a_{ij}^0 - a_{i1}^0 a_{1j}^1 \\[2mm] b_i^1 = b_i^0 - a_{i1}^0 b_1^1 \end{array}\right\} \quad ; \; i,j = 2, \ldots, n \tag{2.41}$$

A similar procedure is then used to eliminate x_2 from Equations (2.40), and so on. A general algorithm for the elimination of x_k may now be written as

$$\left.\begin{array}{l} a_{kj}^k = a_{kj}^{k-1}/a_{kk}^{k-1} \\[2mm] b_k^k = b_k^{k-1}/a_{kk}^{k-1} \end{array}\right\} \quad ; \; j = k+1, \ldots, n \tag{2.42}$$

and

$$\left.\begin{array}{l} a_{ij}^k = a_{ij}^{k-1} - a_{ik}^{k-1} a_{kj}^k \\[2mm] b_i^k = b_i^{k-1} - a_{ik}^{k-1} b_k^k \end{array}\right\} \quad i,j = k+1, \ldots, n \tag{2.43}$$

After the above procedure has been applied $(n-1)$ times, the original set of equations is reduced to the following single equation,

$$a_{nn}^{n-1} x_n = b_n^{n-1} \tag{2.44}$$

which is solved directly for x_n

$$x_n = \frac{b_n^{n-1}}{a_{nn}^{n-1}} \tag{2.45}$$

After the elimination process is completed, the original system of equations has been transformed into an upper triangular system, with unit diagonal coefficients. The general form for row k is now

$$x_k + a_{k,k+1}^k x_{k+1} + \ldots + a_{k,n}^k x_n = b_k^k$$

Thus, once x_n is computed from Equation (2.45), we can compute the remaining unknowns successively, applying in reversed order, the formula

$$x_k = b_k^k - \sum_{j=k+1}^{n} a_{kj}^k x_j \qquad (2.46)$$

Program 11   Solution of simultaneous linear systems of equations, by the Gauss elimination method, without allowing for row interchange (SLPDS)

The Gauss elimination method can be easily implemented, as shown by the flow-chart in Figure 2.2. The elimination process is applied within the outer loop on k. First, the possibility of encountering a null diagonal coefficient is tested; for a null diagonal coefficient the processing is discontinued; otherwise the coefficients of row k and the kth term of **B** are divided by the diagonal coefficient of that row. Then unknown x_k is eliminated from rows $k + 1$ to n. This process is repeated until completing the outer do loop, thus triangularising the system.

After checking that the diagonal coefficient of the last row is different from zero, unknown x_n is computed and stored in the last position of vector array **B**. Then the backsubstitution process is applied, to compute the remaining unknowns, in reversed order, storing them also in array **B**.

Finally the value of the determinant of the system matrix is computed. This is done multiplying the diagonal of the transformed matrix **A**, which contain the divisors of each row.

It is important to note that the original system matrix is destroyed, after the application of the Gauss elimination procedure. If needed, it should be saved in an auxiliary array, before starting. The same happens with the vector of independent coefficients. In particular, we see that the solution vector returns in the array originally containing the independent coefficients. It should also be remarked that this scheme is only useful when the system matrix is positive definite, since no interchange of rows is considered when a zero diagonal coefficient is detected.

We can now transform the flow-chart in Figure 2.2 into the following FORTRAN code, for the solution of systems of linear equations, considering the full positive definite system matrix.

```
      SUBROUTINE SLPDS(A,B,D,N,NX)
C
C PROGRAM 11
C
C SOLUTION OF LINEAR SYSTEMS OF EQUATIONS
C BY THE GAUSS ELIMINATION METHOD WITHOUT
C PROVIDING FOR ROW INTERCHANGES
```

```
C
C A : SYSTEM MATRIX
C B : ORIGINALLY IT CONTAINS THE INDEPENDENT
C       COEFFICIENTS. AFTER SOLUTION IT CONTAINS
C       THE VALUES OF THE SYSTEM UNKNOWNS
C
      DIMENSION A(NX,NX),B(NX)
      N1=N-1
      DO 100 K=1,N1
      C=A(K,K)
      K1=K+1
      IF(ABS(C)-0.000001)1,1,3
    1 WRITE(6,2) K
    2 FORMAT(' **** SINGULARITY IN ROW',I5)
      D=0.
      GO TO 300
C
C DIVIDE ROW BY DIAGONAL COEFFICIENT
C
    3 DO 4 J=K1,N
    4 A(K,J)=A(K,J)/C
      B(K)=B(K)/C
C
C ELIMINATE UNKNOWN X(K) FROM ROW I
C
      DO 10 I=K1,N
      C=A(I,K)
      DO 5 J=K1,N
    5 A(I,J)=A(I,J)-C*A(K,J)
   10 B(I)=B(I)-C*B(K)
  100 CONTINUE
      IF(ABS(A(N,N))-0.000001)1,1,101
C
C COMPUTE LAST UNKNOWN
C
  101 B(N)=B(N)/A(N,N)
C
C APPLY BACKSUBSTITUTION PROCESS TO COMPUTE REMAINING UNKNOWNS
C
      DO 200 L=1,N1
      K=N-L
      K1=K+1
      DO 200 J=K1,N
  200 B(K)=B(K)-A(K,J)*B(J)
C
C COMPUTE VALUE OF THE DETERMINANT
C
      D=1.
      DO 250 I=1,N
  250 D=D*A(I,I)
  300 RETURN
      END
```

It should be noticed that the diagonal coefficients are not tested to see if they are 'exactly zero', but a very small value. It happens that they might be a 'computed zero', different from an 'exact zero' only because of round-off errors. Thus, if we test for 'exact zeros' we may not detect a matrix singularity, in which case the Gauss elimination scheme would be applied only to obtain absurd results.

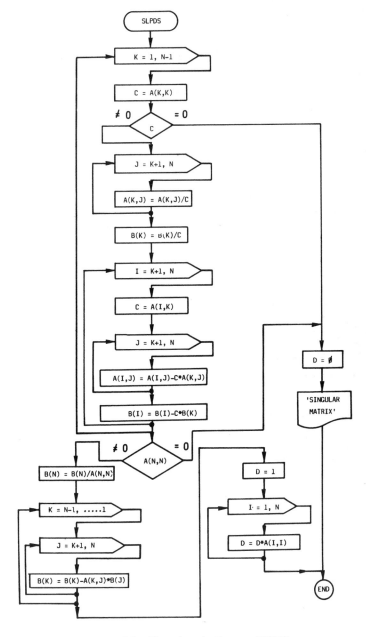

Figure 2.2   Flow-chart for Program SLPDS

Program 12   Solution of simultaneous systems of equations, by the Gauss elimination method, allowing for row interchange (SLNPD)

The previous program, with very few modifications, can be extended for the consideration of non positive-definite system matrices. For this, we need to perform row interchanges when a zero diagonal coefficient is detected, deciding that the system matrix is singular only when no row interchange will produce a non-zero diagonal coefficient, for a given row position.

The modifications needed in the flow-chart of Figure 2.2 are indicated in Figure 2.3, where the statements not included are identical to those of routine SLPDS.

The complete FORTRAN code for this case is given in Program 12.

```
      SUBROUTINE SLNPD(A,B,D,N,NX)
C
C PROGRAM 12
C
C SOLUTION OF LINEAR SYSTEMS OF EQUATIONS
C BY THE GAUSS ELIMINATION METHOD PROVIDING
C FOR INTERCHANGING ROWS WHEN ENCOUNTERING A
C ZERO DIAGONAL COEFICIENT
C
C A : SYSTEM MATRIX
C B : ORIGINALLY IT CONTAINS THE INDEPENDENT
C     COEFFICIENTS. AFTER SOLUTION IT CONTAINS
C     THE VALUES OF THE SYSTEM UNKNOWNS.
C
C N : ACTUAL NUMBER OF UNKNOWNS
C NX: ROW AND COLUMN DIMENSION OF A
C
      DIMENSION A(NX,NX),B(NX)
      N1=N-1
      DO 100 K=1,N1
      K1=K+1
      C=A(K,K)
      IF(ABS(C)-C.000001)1,1,3
    1 DO 7 J=K1,N
C
C TRY TO INTERCHANGE ROWS TO GET NON ZERO DIAGONAL COEFFICIENT
C
      IF(ABS(A(J,K))-0.000001)7,7,5
    5 DO 6 L=K,N
      C=A(K,L)
      A(K,L)=A(J,L)
    6 A(J,L)=C
      C=B(K)
      B(K)=B(J)
      B(J)=C
      C=A(K,K)
      GO TO 3
    7 CONTINUE
    8 WRITE(6,2) K
    2 FORMAT(' **** SINGULARITY IN ROW',I5)
      D=0.
      GO TO 300
C
C DIVIDE ROW BY DIAGONAL COEFFICIENT
C
    3 C=A(K,K)
```

```
       DO 4 J=K1,N
     4 A(K,J)=A(K,J)/C
       B(K)=B(K)/C
C
C ELIMINATE UNKNOWN X(K) FROM ROW I
C
       DO 10 I=K1,N
       C=A(I,K)
       DO 9 J=K1,N
     9 A(I,J)=A(I,J)-C*A(K,J)
    10 B(I)=B(I)-C*B(K)
   100 CONTINUE
C
C COMPUTE LAST UNKNOWN
C
       IF(ABS(A(N,N))-0.000001)8,8,101
   101 B(N)=B(N)/A(N,N)
C
C APPLY BACKSUBSTITUTION PROCESS TO COMPUTE REMAINING UNKNOWNS
C
       DO 200 L=1,N1
       K=N-L
       K1=K+1
       DO 200 J=K1,N
   200 B(K)=B(K)-A(K,J)*B(J)
C
C COMPUTE VALUE OF DETERMINANT
C
       D=1.
       DO 250 I=1,N
   250 D=D*A(I,I)
   300 RETURN
       END
```

Program 13   Solution of simultaneous systems of equations, by the Gauss elimination method, for positive definite symmetric matrices (SLSIM)

In many engineering problems the system matrix is a symmetric matrix. This characteristic can be used to solve the system of equations with a greater efficiency. Therefore, we can implement a computer program for symmetric systems of equations, where the array A will originally contain only the upper triangular portion of the matrix, including the diagonal coefficients. The coefficients located in the lower triangular portion, when needed, can be obtained from those in the upper triangular portion. On the other hand, knowing that the Gauss elimination scheme is a process which conserves symmetry, even for the transformed coefficients, we can perform the row elimination starting from the diagonal coefficient of the corresponding row.

A computer program implemented taking into account those considerations is given below.

```
       SUBROUTINE SLSIM(A,B,D,N,NX)
C
C PROGRAM 13
C
C SOLUTION OF LINEAR SYSTEMS OF EQUATIONS
C BY THE GAUSS ELIMINATION METHOD, FOR
```

```
C SYMMETRIC SYSTEMS
C
C A : ARRAY CONTAINING THE UPPER
C       TRIANGULAR PART OF THE SYSTEM
C       MATRIX
C B : ORIGINALLY IT CONTAINS THE INDEPENDENT
C       COEFFICIENTS. AFTER SOLUTION IT CONTAINS
C       THE VALUES OF THE SYSTEM UNKNOWNS.
C
C N : ACTUAL NUMBER OF UNKNOWNS
C NX: ROW AND COLUMN DIMENSION OF A
C
      DIMENSION A(NX,NX),B(NX)
      N1=N-1
      DO 100 K=1,N1
      C=A(K,K)
      K1=K+1
      DO 11 J=K1,N . . . . . . . . . . . . . . . . . (I)
   11 A(J,K)=A(K,J). . . . . . . . . . . . . . . . . (II)
      IF(ABS(C)-0.000001)1,1,3
    1 WRITE(6,2) K
    2 FORMAT(' **** SINGULARITY IN ROW',I5)
      D=0.
      GO TO 300
C
C DIVIDE ROW BY DIAGONAL COEFFICIENT
C
    3 DO 4 J=K1,N
    4 A(K,J)=A(K,J)/C
      B(K)=B(K)/C
C
C ELIMINATE UNKNOWN X(K) FROM ROW I
C
      DO 10 I=K1,N
      C=A(I,K)
      DO 5 J=I,N. . . . . . . . . . . . . . . . . . (IIL)
    5 A(I,J)=A(I,J)-C*A(K,J)
   10 B(I)=B(I)-C*B(K)
  100 CONTINUE
C
C COMPUTE LAST UNKNOWN
C
      IF(ABS(A(N,N))-0.000001)1,1,101
  101 B(N)=B(N)/A(N,N)
C
C APPLY BACKSUBSTITUTION PROCESS TO COMPUTE REMAINING UNKNOWNS
C
      DO 200 L=1,N1
      K=N-L
      K1=K+1
      DO 200 J=K1,N
  200 B(K)=B(K)-A(K,J)*B(J)
C
C COMPUTE VALUE OF DETERMINANT
C
      D=1.
      DO 250 I=1,N
  250 D=D*A(I,I)
  300 RETURN
      END
```

Notice that the only changes with relation to the program SLPDS are the statements marked I, II, and III. The first two are new statements required to obtain coefficients of the lower triangular portion from coefficients of the upper triangular portion of the system matrix. Statement III is a modification of an existing statement, to start the elimination from the diagonal coefficient of the corresponding row.

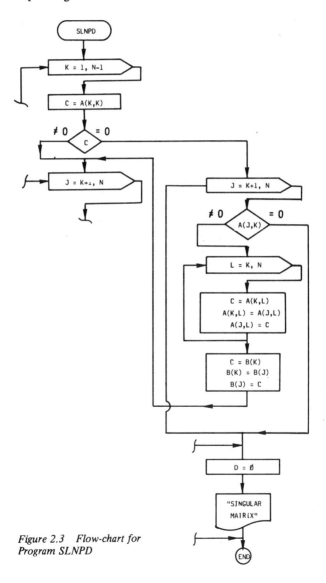

Figure 2.3   Flow-chart for Program SLNPD

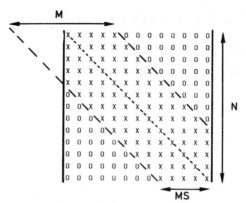

Figure 2.4   Banded system of equations

2.3.2 Symmetric and banded system of equations

The systems of equations obtained in most practical problems are not only
symmetric but also banded. This property means that the matrix has the form
of the one shown in Figure 2.4. We could store only the terms on the diagonal
plus the terms on the right hand side up to M. The number M is called the
'bandwidth' of the matrix. MS is called 'semi-bandwidth' being $M = 2*MS - 1$.
Notice that all the non zero coefficients are within the band. Outside of it
there are only null coefficients.

A computer program can be implemented, taking into account the special
characteristics of this type of system, to obtain their solution with greater
efficiency. To better explain the solution scheme to be used, we first examine
a simple example.

Example 2.4
Consider the case of the continuous beam shown in Figure 2.5.

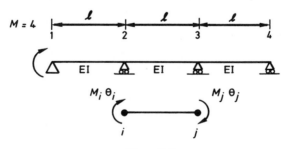

Figure 2.5

For each individual member the stiffness matrix is,

$$\frac{IE}{l} \begin{bmatrix} 4 & 2 \\ 2 & 4 \end{bmatrix} \begin{Bmatrix} \theta_i \\ \theta_j \end{Bmatrix} = \begin{Bmatrix} M_i \\ M_j \end{Bmatrix} \tag{a}$$

where E is the modulus of elasticity, I the moment of inertia, θ_i rotations and M_i moments applied at the ends of the element.

By superposition we can obtain the following system of equations

$$\begin{bmatrix} 4 & 2 & 0 & 0 \\ 2 & 8 & 2 & 0 \\ 0 & 2 & 8 & 2 \\ 0 & 0 & 2 & 4 \end{bmatrix} \begin{Bmatrix} x_1 \\ x_2 \\ x_3 \\ x_4 \end{Bmatrix} = \begin{Bmatrix} 4.0 \\ 0.0 \\ 0.0 \\ 0.0 \end{Bmatrix} \tag{b}$$

where

$$x_i = \theta_i \frac{EI}{l}$$

The elimination gives

$$\begin{bmatrix} 1.0 & 0.500 & 0 & 0 \\ 0 & 1.0 & 0.285 & 0 \\ 0 & 0 & 1.0 & 0.269 \\ 0 & 0 & 0 & 1.0 \end{bmatrix} \begin{Bmatrix} x_1 \\ x_2 \\ x_3 \\ x_4 \end{Bmatrix} = \begin{Bmatrix} 1.0 \\ -0.285 \\ 0.077 \\ -0.044 \end{Bmatrix} \tag{c}$$

After back-substitution we obtain,

$$\begin{Bmatrix} x_1 \\ x_2 \\ x_3 \\ x_4 \end{Bmatrix} = \begin{Bmatrix} 1.155 \\ -0.311 \\ 0.089 \\ -0.044 \end{Bmatrix} \tag{d}$$

In the above example we have noticed that it is not necessary to work with the zero terms outside the band width M. Evidently, if we use a program such as SLPDS or SLSIM we are wasting computer time. Therefore, it is convenient to implement a computer program taking into account the special characteristics of this type of system, both to save time and memory.

To save memory, we can store the non-zero coefficients in a rectangular array, as suggested in Figure 2.6.

Since in general M will be considerable smaller than N, the storage saving may be important, particularly for large matrices. Notice that in the banded organization the original rows remain rows, but the original columns become diagonals. On the other hand, the original main or principal diagonal, and those parallel to it, become columns. The important point here is that a coefficient originally located in the position (I, J) will occupy the position $(I, MS + J - I)$ in the new organization.

If, in addition, the banded system matrix is also symmetric, the new storage scheme need only to take into account the upper triangular part of the matrix, including the diagonal coefficients. Figure 2.7 suggests how the coefficients

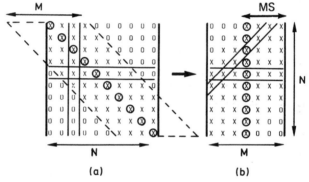

Figure 2.6   *(a) Conventional organisation*   *(b) Banded organisation*

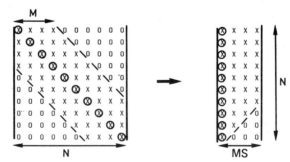

Figure 2.7   *(a) Conventional organisation*   *(b) Symmetric banded organisation*

of the banded symmetric matrix can be organized, for additional savings in storage space. In this case the storage savings are even greater. On the other hand we can easily see that a coefficient originally located in the position (I, J), will be located in the position $(I, J - I + 1)$ in the new organization.

Program 14 Solution of simultaneous systems of equations, by the Gauss elimination method, for banded matrices (SLBNS)

The program SLPDS requires very few modifications to be transformed into another program which takes into account the banded characteristics of the system matrix, included below.

```
      SUBROUTINE SLBNS(A,B,N,M,NX,MX)
C
C PROGRAM 14
C
C SOLUTION OF LINEAR SYSTEMS OF EQUATIONS
C BY THE GAUSS ELIMINATION METHOD, FOR
C NON SYMMETRIC BANDED SYSTEMS
C
C A : CONTAINS THE SYSTEM MATRIX, STORED
C     ACCORDING TO THE NON SYMMETRIC
C     BANDED SCHEME
C ARRAY B CONTAINS THE INDEPENDENT COEFFICIENTS.
C         AFTER SOLUTION IT CONTAINS THE UNKNOW
C         VALUES.
C
C N IS THE NUMBER OF UNKNOWS
C M IS THE BANDWIDTH
C NX IS THE ROW DIMENSION OF ARRAYS A AND B
C MX IS THE COLUMN DIMENSION OF ARRAY A
C
      DIMENSION A(NX,MX),B(NX)
      N1=N-1
      MS=(M+1)/2
      DO 100 K=1,N1
      C=A(K,MS)
      K1=K+1
      IF(ABS(C)-0.000001)1,1,3
    1 WRITE(6,2) K
    2 FORMAT(' **** SINGULARITY IN ROW',I5)
      GO TO 300
C
C DIVIDE ROW BY DIAGONAL COEFFICIENT
C
    3 NI=K1+MS-2
      L=MIN(NI,N)
      DO 4 J=K1,L
      K2=MS+J-K
    4 A(K,K2)=A(K,K2)/C
      B(K)=B(K)/C
C
C ELIMINATE UNKNOWN X(K) FROM ROW I
C
      DO 10 I=K1,L
      K2=MS+K-I
      C=A(I,K2)
      DO 5 J=K1,L
```

```
          K2=MS+J-I
          K3=MS+J-K
        5 A(I,K2)=A(I,K2)-C*A(K,K3)
       10 B(I)=B(I)-C*B(K)
      100 CONTINUE
C
C COMPUTE LAST UNKNOWN
C
          IF(ABS(A(N,MS))-0.000001)1,1,101
      101 B(N)=B(N)/A(N,MS)
C
C APPLY BACKSUBSTITUTION PROCESS TO COMPUTE REMAINING UNKNOWNS
C
          DO 200 I=1,N1
          K=N-I
          K1=K+1
          NI=K1+MS-2
          L=MIN(NI,N)
          DO 200 J=K1,L
          K2=MS+J-K
      200 B(K)=B(K)-A(K,K2)*B(J)
      300 RETURN
          END
```

This program makes use of a library function MIN, to obtain the minimum of the values passed in the argument list.

By comparison of this program with SLPDS we can see that most of the modifications are introduced to properly compute the subscripts of the matrix coefficients, for the banded storage scheme adopted. Also, the limits of some loops are altered, to operate only within the band. The value of the determinant is not computed.

Program 15   Solution of simultaneous systems of equations, by the Gauss elimination method, for symmetric banded matrices (SLBSI)

The following program, for the solution of linear systems of equations, takes into account the symmetric banded storage scheme of Figure 2.7.

```
          SUBROUTINE SLBSI(A,B,D,N,MS,NX,MX)
C
C PROGRAM 15
C
C SOLUTION OF LINEAR SYSTEM OF EQUATIONS
C BY THE GAUSS ELIMINATION METHOD,FOR
C SYMMETRIC BANDED SYSTEMS
C
C A : ARRAY CONTAINING THE UPPER TRIANGULAR
C     PART OF THE SYSTEM MATRIX, STORED
C     ACCORDING TO THE SYMMETRIC BANDED
C     SCHEME
C B : ORIGINALLY IT CONTAINS THE INDEPENDENT
C     COEFFICIENTS. AFTER SOLUTION IT CONTAINS
C     THE VALUES OF THE SYSTEM UNKNOWNS.
C
C N : ACTUAL NUMBER OF UNKNOWNS
C N : ACTUAL NUMBER OF UNKNOWNS
```

```
C MS: ACTUAL HALF BANDWIDTH
C NX: ROW DIMENSION OF A AND B
C MX: COLUMN DIMENSION OF A
C
C D : AUXILIARY VECTOR
C
C
      DIMENSION A(NX,MX),B(NX),D(MX)
      N1=N-1
      DO 100 K=1,N1
      C=A(K,1)
      K1=K+1
      IF(ABS(C)-0.000001)1,1,3
    1 WRITE(6,2) K
    2 FORMAT(' **** SINGULARITY IN ROW',I5)
      GO TO 300
C
C DIVIDE ROW BY DIAGONAL COEFFICIENT
C
    3 NI=K1+MS-2
      L=MIN(NI,N)
      DO 11 J=2,MS
   11 D(J)=A(K,J)
      DO 4 J=K1,L
      K2=J-K+1
    4 A(K,K2)=A(K,K2)/C
      B(K)=B(K)/C
C
C ELIMINATE UNKNOWN X(K) FROM ROW I
C
      DO 10 I=K1,L
      K2=I-K1+2
      C=D(K2)
      DO 5 J=I,L
      K2=J-I+1
      K3=J-K+1
    5 A(I,K2)=A(I,K2)-C*A(K,K3)
   10 B(I)=B(I)-C*B(K)
  100 CONTINUE
C
C COMPUTE LAST UNKNOWN
C
      IF(ABS(A(N,1))-0.000001)1,1,101
  101 B(N)=B(N)/A(N,1)
C
C APPLY BACKSUBSTITUTION PROCESS TO COMPUTE REMAINING UNKNOWNS
C
      DO 200 I=1,N1
      K=N-I
      K1=K+1
      NI=K1+MS-2
      L=MIN(NI,N)
      DO 200 J=K1,L
      K2=J-K+1
  200 B(K)=B(K)-A(K,K2)*B(J)
  300 RETURN
      END
```

This program can be easily understood, by comparison with SLBNS and SLSIM.

2.3.3 Solution of systems of equations using the inverse matrix

A system of equations can be written in matrix form as

$$AX = B \tag{2.47}$$

Pre-multiplying both sides by A^{-1}, we obtain

$$A^{-1}AX = A^{-1}B \tag{2.48}$$

or

$$IX = A^{-1}B \tag{2.49}$$

which is equivalent to

$$X = A^{-1}B \tag{2.50}$$

Thus, we can obtain the desired solution by premultiplying the vector **B** by the inverse of **A**. If we choose to do so, it will be necessary to compute the inverse matrix A^{-1}.

The simplest approach to find the inverse of a $n \times n$ matrix consists of solving n systems of equations, with the same left-hand side (the matrix to be inverted), but whose right-hand sides are,

$$\mathbf{b}^1 = \begin{Bmatrix} 1 \\ 0 \\ 0 \\ . \\ . \\ . \\ 0 \end{Bmatrix}, \quad \mathbf{b}^2 = \begin{Bmatrix} 0 \\ 1 \\ 0 \\ . \\ . \\ . \\ 0 \end{Bmatrix} \dots \mathbf{b}^n = \begin{Bmatrix} 0 \\ 0 \\ . \\ . \\ . \\ 0 \\ 1 \end{Bmatrix} \tag{2.51}$$

Thus we have

$$Ax^1 = b^1, Ax^2 = b^2 \dots Ax^n = b^n \tag{2.52}$$

Which can be written

$$A(x^1 \quad x^2 \ldots x^n) = (b^1 \quad b^2 \ldots b^n) = I \tag{2.53}$$

If we multiply Equation (2.53) by A^{-1} we obtain,

$$(x^1 \quad x^2 \ldots x^n) = A^{-1} \tag{2.54}$$

so that the vectors $x^1, x^2, \ldots, x^n$, which can be obtained applying one of the previous procedures, are the columns of A^{-1}. A more convenient approach is to consider the following augmented matrix

$$(AI) = \begin{bmatrix} a_{11} & a_{12} \ldots a_{1n} & 1 & 0 \ldots 0 \\ a_{21} & a_{22} \ldots a_{2n} & 0 & 1 \ldots 0 \\ . & . & . & . & . & . \\ . & . & . & . & . & . \\ . & . & . & . & . & . \\ a_{n1} & a_{n2} \ldots a_{nn} & 0 & 0 \ldots 1 \end{bmatrix} \tag{2.55}$$

We apply to it the elimination process, which is the same as before, but for a rectangular matrix of n rows and $2n$ columns. Thus the division of a row by its diagonal coefficient is carried out until column $2n$, and the elimination of the unknowns is also carried out up to column $2n$. After the matrix A, in the first $n \times n$ positions of the augmented matrix is triangularized, we apply the backsubstitution process to the matrix in rows 1 to n and columns $n + 1$ to $2n$. For this we simply consider each column of that matrix as a vector of independent coefficients. The final result is that where the unit matrix was originally we obtain A^{-1}.

Program 16   Inverse of a matrix by the Gauss elimination method (INVER)

To obtain the inverse matrix according to the scheme suggested above, we need to introduce a few modifications in a program for solution of systems of equations, using the Gauss elimination method.

The first parameter in the argument list is the array A, which should be dimensioned to allow the storing in it of as many rows and twice as many columns, as those of the matrix to be inverted. The matrix to be inverted is passed in the first $N \times N$ positions of array A. The first operation performed by the program is to place a unit matrix in rows 1 to N, and columns $N + 1$ to $2 \times N$ of array A. The inversion procedure is then applied as described above. The following FORTRAN code can be easily understood, especially when compared with program SLPDS.

```
      SUBROUTINE INVER(A,D,N,NX,MX)
C
C PROGRAM 16
C
C THIS PROGRAM COMPUTES THE INVERSE OF A MATRIX USING THE
C GAUSS ELIMINATION METHOD
C
C A : RECTANGULAR ARRAY OF DIMENSIONS N X 2*N.
C     IN THE FIRST N X N POSITIONS IT ORIGINALLY CONTAINS THE
C     MATRIX TO BE INVERTED. IN ROWS FROM 1 TO N AND COLUMNS
C     FROM N+1 TO 2*N IT STARTS WITH A UNIT MATRIX.AFTER
C     COMPLETION THE INVERSE MATRIX IS PLACED IN THE FIRST N X N
C     POSITIONS OF ARRAY A.THE ORIGINAL MATRIX IS DESTROYED
C D : VALUE OF THE DETERMINANT OF THE MATRIX
C N : ACTUAL ORDER OF A
C NX: ROW AND COLUMN DIMENSION OF A,AND
C     ROW DIMENSION OF B.
C MX: COLUMN DIMENSION OF B, WHICH SHOULD
C     BE AT LEAST 2*N
C
      DIMENSION A(NX,MX)
      N1=N-1
      N2=2*N
C
C     PUT A UNIT MATRIX IN THE AUGMENTED PART OF A
C
      DO 2 I=1,N
      DO 1 J=1,N
      J1=J+N
    1 A(I,J1)=0
      J1=I+N
    2 A(I,J1)=1.
C
C     APPLY THE ELIMINATION PROCESS
C
      DO 10 K=1,N1
      C=A(K,K)
      IF(ABS(C)-0.000001)3,3,5
    3 WRITE(6,4) K
    4 FORMAT(' **** SINGULARITY IN ROW',I5)
      D=0.
      GO TO 300
    5 K1=K+1
      DO 6 J=K1,N2
    6 A(K,J)=A(K,J)/C
      DO 10 I=K1,N
      C=A(I,K)
      DO 10 J=K1,N2
   10 A(I,J)=A(I,J)-C*A(K,J)
      NP1=N+1
      IF(ABS(A(N,N))-0.000001)3,3,19
   19 DO 20 J=NP1,N2
   20 A(N,J)=A(N,J)/A(N,N)
C
C     APPLY THE BACKSUBSTITUTION PROCESS
C
      DO 200 L=1,N1
      K=N-L
      K1=K+1
      DO 200 I=NP1,N2
      DO 200 J=K1,N
  200 A(K,I)=A(K,I)-A(K,J)*A(J,I)
C
```

```
C     PUT THE INVERSE IN THE FIRST N X N POSITIONS OF ARRAY A
C
      DO 250 I=1,N
      DO 250 J=1,N
      J1=J+N
  250 A(I,J)=A(I,J1)
C
C     COMPUTE THE VALUE OF THE DETERMINANT
C
      D=1.
      DO 220 I=1,N
  220 D=D*A(I,I)
  300 RETURN
      END
```

2.3.4 Solution of symmetric systems by Choleski's method

In the case that the system matrix is symmetric a very efficient solution scheme is provided by Choleski's method, sometimes also called the Banachiewicz method. This method uses the fact that a symmetric matrix can be expressed as the product of two triangular matrices, as

$$A = S^T S \tag{2.56}$$

or

$$
\begin{bmatrix}
a_{11} & a_{12} & a_{13} & \cdots & a_{1n} \\
a_{21} & a_{22} & a_{23} & \cdots & a_{2n} \\
\cdot & \cdot & \cdot & & \cdot \\
\cdot & \cdot & \cdot & & \cdot \\
\cdot & \cdot & \cdot & & \cdot \\
a_{n1} & a_{n2} & a_{n3} & \cdots & a_{nn}
\end{bmatrix}
=
$$

$$
\begin{bmatrix}
s_{11} & 0 & 0 & \cdots & 0 \\
s_{12} & s_{22} & 0 & \cdots & 0 \\
s_{13} & s_{23} & s_{33} & \cdots & 0 \\
\cdot & \cdot & \cdot & & \cdot \\
\cdot & \cdot & \cdot & & \cdot \\
\cdot & \cdot & \cdot & & \cdot \\
s_{1n} & s_{2n} & s_{3n} & \cdots & s_{nn}
\end{bmatrix}
\begin{bmatrix}
s_{11} & s_{12} & s_{13} & \cdots & s_{1n} \\
0 & s_{22} & s_{23} & \cdots & s_{2n} \\
0 & 0 & s_{33} & \cdots & s_{3n} \\
\cdot & \cdot & \cdot & & \cdot \\
\cdot & \cdot & \cdot & & \cdot \\
0 & \cdots\cdots\cdots\cdots\cdots & s_{nn}
\end{bmatrix}
\tag{2.57}
$$

From the previous expression, and considering the rules for matrix multiplication, we see that

$$a_{ij} = s_{1i} s_{1j} + s_{2i} s_{2j} + \ldots + s_{ii} s_{ij}, \quad i < j \tag{2.58}$$

$$a_{ii} = s_{1i}^2 + s_{2i}^2 + \ldots + s_{ii}^2, \quad i = j \tag{2.59}$$

Therefore, we can determine the coefficients of the first row of **S** by

$$s_{11} = \sqrt{a_{11}}; \quad s_{1j} = \frac{a_{1j}}{\sqrt{a_{11}}} \tag{2.60}$$

and in general

$$s_{ii} = \left(a_{ii} - \sum_{l=1}^{i-1} s_{li}^2 \right)^{1/2}; \quad s_{ij} = \frac{a_{ij} - \sum_{l=1}^{j-1} s_{li} s_{lj}}{s_{ii}}, \quad j > i \tag{2.61}$$

Furthermore, we can verify that the solution of the system

$$AX = B \tag{2.62}$$

reduces to finding the solution of the two equivalent systems

$$S^T C = B \tag{2.63}$$

and

$$SX = C \tag{2.64}$$

The elements of **C** are determined by the formulae

$$c_1 = \frac{b_1}{s_{11}} \tag{2.65}$$

and

$$c_i = \frac{b_i - \sum_{l=1}^{i-1} s_{li} c_l}{s_{ii}}, \quad i > 1 \tag{2.66}$$

Once C is known we can find x from Equation (2.64) using the same scheme as in the back-substitution employed in the Gauss elimination method, that is

$$x_n = \frac{c_n}{s_{nn}} \tag{2.67}$$

and

$$x_i = \frac{c_i - \sum\limits_{l=i+1}^{n} s_{il} x_l}{s_{ii}}, \quad i < n \tag{2.68}$$

Program 17   Decomposition of a symmetric matrix using Choleski's method (DECOG)

We can imagine the Choleski's method as divided in two parts: (a) decomposition, to find **S**, and (b) back substitution, to compute **X**. The following is a routine to perform the decomposition of a symmetric matrix (Figure 2.8).

```
      SUBROUTINE DECOG(A,N,NX)
C
C PROGRAM 17
C
C THIS PROGRAM PERFORMS THE DECOMPOSITION
C OF A SYMMETRIC MATRIX, INTO AN UPPER
C TRIANGULAR MATRIX,FOR POSITIVE DEFINITE
C MATRICES.
C
C A : ARRAY ORIGINALLY CONTAINING THE
C       MATRIX TO BE DECOMPOSED. AT THE
C       END IT CONTAINS THE UPPER
C       TRIANGULAR MATRIX
C
C N : ORDER OF A
C NX: ROW AND COLUMN DIMENSION OF A
C
      DIMENSION A(NX,NX)
      IF(A(1,1))1,1,3
    1 WRITE(6,2)
    2 FORMAT(' ZERO OR NEGATIVE RADICAND')
      GO TO 200
    3 A(1,1)=SQRT(A(1,1))
      DO 10 J=2,N
   10 A(1,J)=A(1,J)/A(1,1)
      DO 40 I=2,N
      I1=I-1
      D=A(I,I)
      DO 20 L=1,I1
   20 D=D-A(L,I)*A(L,I)
      IF(A(I,I))1,1,21
```

```
   21 A(I,I)=SQRT(D)
      IF(N-I) 45,45,47
   47 I2=I+1
      DO 40 J=I2,N
      D=A(I,J)
      DO 30 L=1,I1
   30 D=D-A(L,I)*A(L,J)
   40 A(I,J)=D/A(I,I)
   45 DO 50 I=2,N
      I1=I-1
      DO 50 J=1,I1
   50 A(I,J)=0.
  200 RETURN
      END
```

Program 18   Decomposition of a symmetric banded matrix using Choleski's method (DECOB)

From the previous routine, with few modifications, we obtain the following routine for decomposition of a symmetric banded matrix, stored as in Figure 2.7.

```
      SUBROUTINE DECOB(A,N,MS,NX,MX)
C
C PROGRAM 18
C
C THIS PROGRAM PERFORMS THE DECOMPOSITION
C OF A SYMMETRIC MATRIX, STORED ACCORDING
C TO THE BANDED SYMMETRIC SCHEME, INTO AN
C UPPER TRIANGULAR MATRIX, FOR POSITIVE
C DEFINITE MATRICES
C
C MS: ACTUAL HALF BANDWIDTH
C MX: COLUMN DIMENSION OF A
C
      DIMENSION A(NX,MX)
      IF(A(1,1))1,1,3
    1 WRITE(6,2)
    2 FORMAT(' ZERO OR NEGATIVE RADICAND')
      GO TO 200
    3 A(1,1)=SQRT(A(1,1))
      DO 10 J=2,MS
   10 A(1,J)=A(1,J)/A(1,1)
      DO 40 I=2,N
      I1=I-1
      D=A(I,1)
      DO 20 L=1,I1
      I3=I-L+1
   20 D=D-A(L,I3)*A(L,I3)
      IF(A(I,1))1,1,21
   21 A(I,1)=SQRT(D)
      I2=I+1
      DO 40 J=I2,N
      I3=J-I+1
      D=A(I,I3)
      DO 30 L=1,I1
      I4=I-L+1
      I5=J-L+1
   30 D=D-A(L,I4)*A(L,I5)
   40 A(I,I3)=D/A(I,1)
  200 RETURN
      END
```

Program 19   Solution of equations using Choleski's method (CHOLE)

Using the first decomposition routine we can program the following FORTRAN
code to apply Choleski's method (Figure 2.9).

```
      SUBROUTINE CHOLE(A,B,N,NX)
C
C PROGRAM 19
C
C SOLUTION OF LINEAR SYSTEMS OF EQUATIONS
C BY THE CHOLESKI METHOD FOR SYMMETRIC
C POSITIVE DEFINITIVE MATRICES.
C
C A : ARRAY CONTAINING THE SYSTEM MATRIX
C B : ARRAY CONTAINING THE VECTOR OF INDEPENDENT COEFFICIENTS
C C : AUXILIARY VECTOR
C N : ORDER OF A
C NX: ROW AND COLUMN DIMENSION OF A
C
      DIMENSION A(NX,NX),B(NX)
C
C COMPUTE UPPER TRIANGULAR MATRIX FROM A AND
C STORE ALSO IN A
C
      CALL DECOG(A,N,NX)
C
C COMPUTE THE C VECTOR AND STORE IN ARRAY B
C
      B(1)=B(1)/A(1,1)
      DO 10 I=2,N
      D=B(I)
      I1=I-1
      DO 5 L=1,I1
    5 D=D-A(L,I)*B(L)
   10 B(I)=D/A(I,I)
      B(N)=B(N)/A(N,N)
C
C COMPUTE THE SYSTEM UNKNOWNS AND STORE IN ARRAY B.
C
      N1=N-1
      DO 30 L=1,N1
      K=N-L
      K1=K+1
      DO 20 J=K1,N
   20 B(K)=B(K)-A(K,J)*B(J)
   30 B(K)=B(K)/A(K,K)
      RETURN
      END
```

Program 20   Inverse of an upper triangular matrix (INVCH)

In some cases we may need to invert an upper triangular matrix, such as S of
expression (2.56). Obviously we could use a program such as INVER, but not
taking advantage of the special structure of the upper triangular matrix we
would be wasting computer time. It is, thus, interesting to find a special algorithm to compute the inverse of an upper triangular matrix. To do so, let us
consider the matrix equation

$$
\begin{bmatrix} s_{11} & s_{12} & s_{13} & s_{14} \\ 0 & s_{22} & s_{23} & s_{24} \\ 0 & 0 & s_{33} & s_{34} \\ 0 & 0 & 0 & s_{44} \end{bmatrix} \begin{bmatrix} a_{11} & a_{12} & a_{13} & a_{14} \\ 0 & a_{22} & a_{23} & a_{24} \\ 0 & 0 & a_{33} & a_{34} \\ 0 & 0 & 0 & a_{44} \end{bmatrix} =
$$

$$
\begin{bmatrix} 1 & 0 & 0 & 0 \\ 0 & 1 & 0 & 0 \\ 0 & 0 & 1 & 0 \\ 0 & 0 & 0 & 1 \end{bmatrix} \tag{a}
$$

where the matrix A is the inverse of S. It can be easily verified that A is also an upper triangular matrix.

By multiplying the ith row of S by the ith column of A we obtain

$$
s_{ii} a_{ii} = 1 \tag{b}
$$

so that the formula

$$
a_{ii} = \frac{1}{s_{ii}} \tag{c}
$$

permits us to compute the terms of the main diagonal of A.

By multiplying the ith row of S by the jth column of A, with $j = i + 1$, we see that

$$
s_{ii} a_{ij} + s_{ij} a_{jj} = 0 \tag{d}
$$

so that for computing the terms of the diagonal of A following the main diagonal we can use the formula

$$
a_{ij} = -\frac{1}{s_{ii}} s_{ij} a_{jj} \tag{e}
$$

If we continue this way we will find out that to compute the terms of the diagonal of A displaced k positions from the main diagonal, we can use the formula

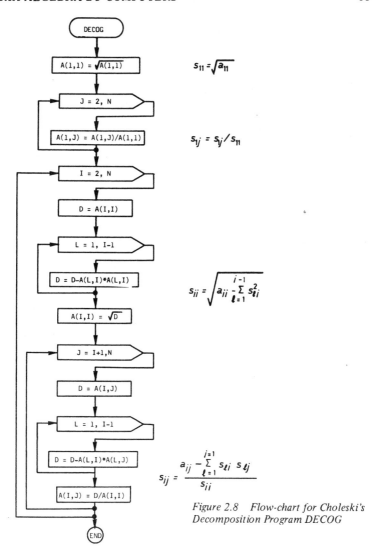

$$s_{11} = \sqrt{a_{11}}$$

$$s_{1j} = s_{1j} / s_{11}$$

$$s_{ii} = \sqrt{a_{ii} - \sum_{\ell=1}^{i-1} s_{\ell i}^2}$$

$$s_{ij} = \frac{a_{ij} - \sum_{\ell=1}^{j-1} s_{\ell i} \, s_{\ell j}}{s_{ii}}$$

Figure 2.8   Flow-chart for Choleski's Decomposition Program DECOG

$$a_{ij} = -\frac{1}{s_{ii}} \sum_{l=i+1}^{i+k} s_{il} \, a_{lj}, \quad j = i + k \qquad (f)$$

According to this recurrence formula, we can program routine 20, which receives the original upper triangular matrix in array S and computes its inverse which is returned in the array A.

```
      SUBROUTINE INVCH(S,A,N,NX)
C
C PROGRAM 20
C
C THIS PROGRAM COMPUTES THE INVERSE OF AN
C UPPER TRIANGULAR MATRIX, STORED IN "S",
C PLACING THE RESULTS IN "A".
C
C N : ACTUAL ORDER OF A AND S
C NX: ROW AND COLUMN DIMENSION OF A AND S
C
      DIMENSION A(NX,NX),S(NX,NX)
C
C COMPUTE DIAGONAL TERMS OF A
C
      DO 10 I=1,N
   10 A(I,I)=1./S(I,I)
C
C COMPUTE THE TERMS OF KTH DIAGONAL OF A
C
      N1=N-1
      DO 100 K=1,N1
      NK=N-K
      DO 100 I=1,NK
      J=I+K
      D=0.
      I1=I+1
      IK=I+K
      DO 20 L=I1,IK
   20 D=D+S(I,L)*A(L,J)
  100 A(I,J)=-D/S(I,I)
      RETURN
      END
```

2.3.5 Solution of simultaneous linear equations by iteration

The Gauss elimination methods are part of the so called direct methods.
Systems of linear equations can also be solved using iterative methods under
certain conditions. To illustrate the characteristics of one of these methods,
we have selected the Gauss–Seidel method.

Consider the system

$$a_{11} x_1 + a_{12} x_2 + \ldots a_{1k} x_k + \ldots a_{1n} x_n = b_1$$

$$a_{21} x_1 + a_{22} x_2 + \ldots a_{2k} x_k + \ldots a_{2n} x_n = b_2$$

$$\cdot \qquad \cdot \qquad \cdot \qquad \cdot \qquad \cdot$$

$$a_{k1} x_1 + a_{k2} x_2 + \ldots a_{kk} x_k + \ldots a_{kn} x_n = b_k$$

$$\cdot \qquad \cdot \qquad \cdot \qquad \cdot \qquad \cdot$$

$$a_{n1} x_1 + a_{n2} x_2 + \ldots a_{nk} x_k + \ldots a_{nn} x_n = b_n \qquad (2.69)$$

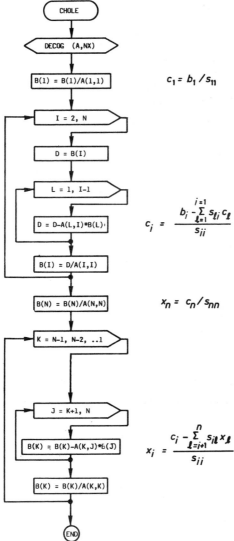

The flow-chart contains the following elements:

CHOLE

DECOG (A,NX)

B(1) = B(1)/A(1,1) $c_1 = b_1 / s_{11}$

I = 2, N

D = B(I)

L = 1, I-1

D = D-A(L,I)*B(L) $c_i = \dfrac{b_i - \sum\limits_{\ell=1}^{i=1} s_{\ell i} c_\ell}{s_{ii}}$

B(I) = D/A(I,I)

B(N) = B(N)/A(N,N) $x_n = c_n / s_{nn}$

K = N-1, N-2, ..1

J = K+1, N

B(K) = B(K)-A(K,J)*B(J) $x_i = \dfrac{c_i - \sum\limits_{\ell=i+1}^{n} s_{i\ell} x_\ell}{s_{ii}}$

B(K) = B(K)/A(K,K)

END

Figure 2.9   Flow-chart for Program CHOLE

Solving the kth equation for x_k we have

$$x_k = \frac{1}{a_{kk}} \left[b_k - \sum_{i=1}^{k-1} a_{ki} x_i - \sum_{i=k+1}^{n} a_{ki} x_i \right] \tag{2.70}$$

where $k = 1$ to n.

The basic concept of the iterative method is to select a trial solution and using formula (2.70) for each unknown, to compute a new solution. This solution, if not satisfactory, is taken as a new trial solution, and formula (2.70) is again used for each unknown to compute another solution. This procedure is repeated again and again, until it is observed that the differences between the trial and the computed solution are sufficiently small.

Thus we have the general equation for the r-iteration cycle.

$$x_k^{(r)} = \frac{1}{a_{kk}} \left[b_k - \sum_{i=1}^{k-1} a_{ki} x_i^{(r)} - \sum_{i=k+1}^{n} a_{ki} x_i^{(r-1)} \right] \tag{2.71}$$

Example 2.5
Consider the system of equations of Example 2.4

$$\begin{bmatrix} 4 & 2 & 0 & 0 \\ 2 & 8 & 2 & 0 \\ 0 & 2 & 8 & 2 \\ 0 & 0 & 2 & 4 \end{bmatrix} \begin{Bmatrix} x_1 \\ x_2 \\ x_3 \\ x_4 \end{Bmatrix} = \begin{Bmatrix} 4.0 \\ 0. \\ 0. \\ 0. \end{Bmatrix} \tag{a}$$

We can now solve the last equation of x_1, the second for x_2, etc. using formula (2.71).

$$x_1 = \tfrac{1}{4}(4.0 - 2x_2)$$

$$x_2 = \tfrac{1}{8}(-2x_1 - 2x_3)$$

$$x_3 = \tfrac{1}{8}(-2x_2 - 2x_4)$$

$$x_4 = \tfrac{1}{4}(-2x_3) \tag{b}$$

The first cycle gives (starting with $x_i = 0$)

$$x_1 = \tfrac{1}{4}(4.0) = 1$$

$$x_2 = \tfrac{1}{8}(-2.) = -\tfrac{1}{4}$$

$$x_3 = \tfrac{1}{8}(2\tfrac{1}{4}) = \tfrac{1}{16}$$

$$x_4 = \tfrac{1}{4}(-2\tfrac{1}{16}) = -\tfrac{1}{32} \tag{c}$$

For the second cycle,

$$x_1 = \tfrac{1}{4}(4 + \tfrac{2}{4}) = 1.12$$

$$x_2 = \tfrac{1}{8}(-2.24 + \tfrac{2}{16}) = -0.26$$

$$x_3 = \tfrac{1}{8}(0.52 + \tfrac{1}{16}) = 0.144$$

$$x_4 = \tfrac{1}{4}(-2. + 0.144) = -0.072 \tag{d}$$

etc. The process will converge to the correct values

$$x_1 = 1.155, \quad x_2 = -0.311, \quad x_3 = 0.089, \quad x_4 = -0.044$$

after several more iterations.

Program 21   Solution of equations using Gauss–Seidel iterative method (SLGSG)

The following program, for iterative solution of linear systems of equations, operates according to the Gauss–Seidel method described above (Figure 2.10).

```
      SUBROUTINE SLGSG(A,B,X,N,NIM,T)
C
C PROGRAM 21
C
C SOLUTION OF LINEAR SYSTEMS OF EQUATIONS
C BY THE GAUSS-SEIDEL ITERATIVE METHOD
C
C A : ARRAY CONTAINING THE SYSTEM MATRIX
C B : ARRAY CONTAINING THE INDEPENDENT COEFFICIENTS
C X : ARRAY WHICH, AFTER SOLUTION, WILL
C     CONTAIN THE VALUES OF THE SYSTEM
C     UNKNOWNS
C N : ORDER OF THE SYSTEM
C NIM:MAXIMUM NUMBER OF ITERATIONS ALLOWED
C T : TOLERANCE
C
      DIMENSION A(50,50),B(50),X(50),Y(50)
      DO 1 I=1,N
    1 X(I)=0
      ITER=0
    2 DO 3 I=1,N
    3 Y(I)=X(I)
      ITER=ITER+1
      DO 10 K=1,N
```

```
         D=0.
         DO 5   I=1,N
         IF(K-I)4,5,4
   4     D=D+A(K,I)*X(I)
   5     CONTINUE
  10     X(K)=(B(K)-D)/A(K,K)
         DO 20  I=1,N
         IF(ABS(X(I)-Y(I))-T)20,20,22
  20     CONTINUE
  21     RETURN
  22     IF(ITER-NIM)2,23,23
  23     WRITE(6,24)
  24     FORMAT('MAXIMUM NUMBER OF ITERATIONS  REACHED')
         GO TO 21
         END
```

In these solutions it is necessary to adopt a criterion for convergence which indicates when the iteration should be finished, and in our case we used a simple one. In general it is usual to take the norm of the difference between $X^{(r-1)}$, placed in array X, and $X^{(r)}$, in array Y, as the criterion, (for instance $<1\%$) and at the same time to indicate the maximum number of iterations to be allowed, in case the solution does not converge.

For some systems the convergence may be relatively slow. In such cases, by using an over-relaxation factor we can increase the rate of convergence of the Gauss–Seidel iteration. In order to do so we first calculate the change of the unknowns between cycles

$$\Delta x^{(r)} = x_k^{(r)} - x_k^{(r-1)} \tag{2.72}$$

Written in terms of the iterative equation,

$$\Delta x_k^{(r)} = \frac{1}{a_{kk}} \left[b_k - \sum_{i=1}^{k-1} a_{ki} x_i^{(r)} - \sum_{i=k}^{n} a_{ki} x_i^{(r)} \right] \tag{2.73}$$

The new value of the unknown can then be approximated as

$$x_k^{(r)} = x_k^{(r-1)} + \gamma \Delta_k^{(1)} \tag{2.74}$$

where γ is the over-relaxation factor, usually between 1.4 to 1.9.

The Gauss–Seidel method, although simple to apply, is unfortunately very sensitive to small variations in the matrix coefficients. The method will converge very slowly for ill-conditioned systems, thus, even for non-singular systems convergence can be slow or the system fail to converge. In practice, the criterion for convergence is that,

$$|a_{ii}| > \sum_{\substack{j=1 \\ j \neq i}}^{n} |a_{ij}| \tag{2.75}$$

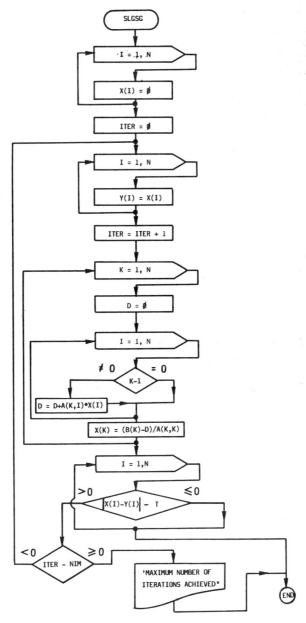

Figure 2.10   Flow-chart for Program SLGSG

That is, the system should be strongly diagonal.

Example 2.6

Let us consider, as an example, the case of the following singular system of equations,

$$2x_1 + x_2 = 6$$

$$4x_1 + 2x_2 = 10 \qquad\qquad\qquad (a)$$

We have the following recurrence relationships

$$x_1 = \frac{6 - x_2}{2}$$

$$x_2 = \frac{10 - 4x_1}{2} \qquad\qquad\qquad (b)$$

Then for the different cycles we obtain Table 2.1, which shows that the results diverge. If the system was nearly singular the convergences would be too slow for our purposes.

Table 2.1

Cycle	x_1	x_2
1	3	−1
2	3.5	−2
3	4	−3
4	4.5	−4
5	5	−5
6	5.5	−6
.	.	.
.	.	.
.	.	.

Program 22   Solution of symmetric banded matrices using Gauss–Seidel (SLESS)

Let us consider the case of a symmetric banded matrix stored as shown in Figure 2.6. We have N number of equations, M bandwidth, NIM maximum number of iterations to be allowed, T tolerance required. We use a **Y** vector to store the results of the last iteration in order to calculate the error from one iteration to the next.

```
        SUBROUTINE SLESS(A,B,X,N,MS,NIM,T)
C
C PROGRAM 22
C
C SOLUTION OF LINEAR SYSTEMS OF EQUATIONS
C BY THE GAUSS-SEIDEL ITERATIVE METHOD,
C FOR SYMMETRIC SYSTEMS.
C
C A : SYSTEM MATRIX, STORED ACCORDING TO
C      THE SYMMETRIC BANDED SCHEME
C B : ARRAY CONTAINING THE INDEPENDENT COEFFICIENTS
C X : ARRAY WHICH, AFTER SOLUTION, WILL
C      CONTAIN THE VALUES OF THE SYSTEM
C      UNKNOWNS
C Y : AUXILIARY VECTOR
C N : ORDER OF THE SYSTEM
C MB: HALF BAND WIDTH OF MATRIX A
C MX: COLUMN DIMENSION OF A
C NIM:MAXIMUM NUMBER OF ITERATIONS ALLOWED
C T : TOLERANCE
C
C
        DIMENSION A(50,20),B(50),X(50),Y(50)
        DO 1 I=1,N
    1   X(I)=0
        ITER=0
    2   DO 3 I=1,N
    3   Y(I)=X(I)
        ITER=ITER+1
        DO 100 I=1,N
        S=0
        DO 7 J=2,MS
        K=I+J-1
        IF(K-N)4,4,5
    4   S=S+A(I,J)*X(K)
    5   K=I-J+1
        IF(K-1)7,6,6
    6   S=S+A(K,J)*X(K)
    7   CONTINUE
  100   X(I)=(B(I)-S)/A(I,1)
        DO 110 I=1,N
        IF(ABS(X(I)-Y(I))-T)110,110,120
  110   CONTINUE
  115   RETURN
  120   IF(ITER-NIM)2,125,125
  125   WRITE(6,130)
  130   FORMAT(' MAXIMUM NUMBER OF ITERATIONS REACHED')
        GO TO 115
        END
```

2.3.6 A comparison of several programs for solution of linear systems of equations

To compare their efficiency, some of the programs presented earlier in the chapter were run in the same computer, to solve systems of 10, 20, 50 and 100 unknowns. In the case of special programs for banded systems, different bandwidths were used. The times required in each case, in seconds, are shown in Table 2.2. The advantage of considering the special characteristics of the systems to be solved is evident.

Table 2.2

N	SLBSI		SLBNS		SLSIM (s)	SLPDS (s)	CHOLE (s)	INVER (s)
	MS	(s)	M	(s)				
10	2	0.007	3	0.007	0.022	0.024	0.025	0.098
	3	0.010	5	0.010				
	6	0.022	11	0.023				
20	3	0.020	5	0.020	0.123	0.173	0.115	0.698
	5	0.037	9	0.043				
	11	0.111	21	0.148				
50	6	0.125	11	0.155	1.471	2.391	1.354	10.242
	11	0.315	21	0.454				
	26	1.275	51	1.986				
100	11	0.666	21	0.962	10.653	18.776	11.071	80.997
	21	1.973	41	3.127				
	51	9.065	101	15.118				

2.4   EIGENVALUE AND EIGENVECTOR PROBLEMS

In many engineering problems, such as linearized buckling or vibrations, we have to solve a linear system of equations of the form

$$a_{11} x_1 + a_{12} x_2 + \ldots a_{1n} x_n = \lambda x_1$$

$$a_{21} x_1 + a_{22} x_2 + \ldots a_{2n} x_n = \lambda x_2$$

$$\begin{array}{cccc} \cdot & \cdot & \cdot & \cdot \\ \cdot & \cdot & \cdot & \cdot \\ \cdot & \cdot & \cdot & \cdot \end{array}$$

$$a_{n1} x_1 + a_{n2} x_2 + \ldots a_{nn} x_n = \lambda x_n \tag{2.76}$$

or, equivalently

$$(a_{11} - \lambda) x_1 + a_{12} x_2 + \ldots a_{1n} x_n = 0$$

$$a_{21} x_1 + (a_{22} - \lambda) x_2 + \ldots a_{2n} x_n = 0$$

$$\begin{array}{cccc} \cdot & \cdot & \cdot & \cdot \\ \cdot & \cdot & \cdot & \cdot \\ \cdot & \cdot & \cdot & \cdot \end{array}$$

$$a_{n1} x_1 + a_{n2} x_2 + \ldots (a_{nn} - \lambda) x_n = 0 \tag{2.77}$$

Since (2.77) is an homogeneous system, it will have a non-trivial solution only if the determinant

$$\Delta_\lambda = \begin{vmatrix} (a_{11} - \lambda) & a_{12} & \cdots & a_{1n} \\ a_{21} & (a_{22} - \lambda) & \cdots & a_{2n} \\ \cdot & & & \cdot \\ \cdot & & & \cdot \\ \cdot & & & \cdot \\ a_{n1} & a_{n2} & \cdots & (a_{nn} - \lambda) \end{vmatrix}$$

(2.78)

is null. Clearly, the expression

$$\Delta_\lambda = 0 \tag{2.79}$$

gives a nth order polynomial in λ, such that its roots will be the n values of λ which satisfy (2.79). Those values of λ are the *eigenvalues* of system (2.76). The set of values for the unknowns $x_1, x_2, \ldots, x_n$, associated to each eigenvalue and defining a solution for system (2.76) are called *eigenvector*.

Example 2.7
As a simple example, to illustrate the case of an eigenvalue and eigenvector problem, let us consider the vibration of the system of two masses m_1 and m_2, and three springs of stiffness k_1, k_2 and k_3, shown in Figure 2.11. The quantities u_1 and u_2 are the displacements of the two masses, measured from their unperturbed positions.

The equations of motion for the two masses are

$$m_1 \ddot{u}_1 + (k_1 + k_2)u_1 - k_2 u_2 = 0$$
$$m_2 \ddot{u}_2 - k_2 u_1 + (k_2 + k_3)u_2 = 0 \tag{a}$$

where $\ddot{u}_1$ and $\ddot{u}_2$ are the accelerations of m_1 and m_2 respectively. The solution of (a) has the following harmonic form

$$u_1 = x_1 \sin \omega t$$
$$u_2 = x_2 \sin \omega t \tag{b}$$

where x_1 and x_2 are amplitudes, ω is the frequency of the vibration, and t is time. Differentiating twice with regard to time we obtain

$$\ddot{u}_1 = -x_1\omega^2 \sin \omega t$$

$$\ddot{u}_2 = -x_2\omega^2 \sin \omega t \tag{c}$$

Considering Equations (b) and (c), Equation (a) is transformed into

$$(k_1 + k_2 - m_1\omega^2)x_1 - k_2 x_2 = 0$$

$$- k_2 x_1 + (k_2 + k_3 - m_2\omega^2)x_2 = 0 \tag{d}$$

or, dividing the first equation by m_1, and the second by m_2,

$$\left(\frac{k_1 + k_2}{m_1} - \omega^2\right)x_1 - \frac{k_2}{m_1}x_2 = 0$$

$$- \frac{k_2}{m_2}x_1 + \left(\frac{k_2 + k_3}{m_2} - \omega^2\right)x_2 = 0 \tag{e}$$

We see that system (e) is equivalent to system (2.77), with $\omega^2 = \lambda$. To find its solution we enforce the condition

$$\begin{vmatrix} \dfrac{k_1 + k_2}{m_1} - \omega^2 & -\dfrac{k_2}{m_1} \\[2ex] -\dfrac{k_2}{m_2} & \dfrac{k_2 + k_3}{m_2} - \omega^2 \end{vmatrix} = 0 \tag{f}$$

or

$$\left(\frac{k_1 + k_2}{m_1} - \omega^2\right)\left(\frac{k_2 + k_3}{m_2} - \omega^2\right) - \frac{k_2^2}{m_1 m_2} = 0 \tag{g}$$

so that

$$\omega^4 - \left(\frac{k_1 + k_2}{m_1} + \frac{k_2 + k_3}{m_2}\right)\omega^2 + \left(\frac{k_1 + k_2}{m_1} \cdot \frac{k_2 + k_3}{m_2} - \frac{k_2^2}{m_1 m_2}\right) = 0 \tag{h}$$

Taking $\omega^2 = \lambda$, Equation (h) becomes the following quadratic algebraic equation

$$\lambda^2 - \left(\frac{k_1 + k_2}{m_1} + \frac{k_2 + k_3}{m_2}\right)\lambda + \frac{k_1 k_2 + k_2 k_3 + k_1 k_3}{m_1 m_2} = 0 \tag{i}$$

whose two roots are the eigenvalues $\lambda_1 = \omega_1^2$ and $\lambda_2 = \omega_2^2$. Once the eigenvalues are known we can compute the eigenvectors from Equation (e). These will be known except for a multiplicative constant. For instance, from the first equation of (e) we obtain

$$\left(\frac{x_1}{x_2}\right)_i = \frac{k_2}{(k_1 + k_2 - m_1 \omega_i^2)}, \quad i = 1,2 \tag{j}$$

so that assigning a value of x_1 we obtain the corresponding x_2.

Let us now consider the following numerical values for the relevant magnitudes of the system under study

$$k_1 = 2; \quad k_2 = k_3 = m_1 = m_2 = 1$$

In this case Equation (i) becomes

$$\lambda^2 - 5\lambda + 5 = 0 \tag{k}$$

with roots:

$$\lambda_1 = \omega_1^2 = 1.382$$
$$\lambda_2 = \omega_2^2 = 3.618 \tag{l}$$

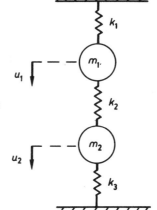

Figure 2.11   Two degrees of freedom system

Having the eigenvalues we obtain the eigenvectors from Equation (j)

$$X_1 = \{1 \qquad 1.618\}$$

$$X_2 = \{1 \qquad -0.618\} \tag{m}$$

We might, as is commonly done, normalize the eigenvectors by dividing by their length, so that

$$X_1 = \{0.526 \qquad 0.850\}$$

$$X_2 = \{0.850 \qquad -0.526\} \tag{n}$$

from which we can verify that the normalized eigenvectors are orthogonal vectors, i.e. they satisfy the condition

$$X_i^T X_i = 1$$

$$X_j^T X_i = 0$$

This property is generally valid, provided that the system matrix A is a symmetric matrix, as it will be shown later on.

While the analytical solution approach illustrated in the previous example can eventually be applied to very small eigenvalue problems, it is clear that it would not be convenient for practical cases, where the system of equations may involve tens or hundreds of unknowns. Various different methods exist for the eigenvalue and eigenvector computation, some of which involve rather complex techniques. In what follows, three simple but efficient methods are presented, including the corresponding computer programs.

2.4.1 The Vianello–Stodola method

Let us consider Equation (2.76) written in matrix form as

$$AX = \lambda X \tag{2.80}$$

where A is a real symmetric matrix of order n, X is the vector of unknown variables, and the scalar parameter λ is representative of the eigenvalue. Next we take a trial vector $X^{(1)}$, such that

$$AX^{(1)} = Y^{(1)} \tag{2.81}$$

If the vector $X^{(1)}$ is an eigenvector then

$$Y^{(1)} = \lambda X^{(1)} \tag{2.82}$$

for which a constant ratio should exist between the coefficients of $Y^{(1)}$ and $X^{(1)}$, such that

$$\frac{Y_i^{(1)}}{X_i^{(1)}} = \lambda, \quad i = 1, n \tag{2.83}$$

In general, however, Equation (2.82) will be only approximately satisfied. Then we can take

$$X^{(2)} = Y^{(1)} \tag{2.84}$$

and proceed to compute

$$Y^{(2)} = AX^{(2)} \tag{2.85}$$

and so on, so that

$$X^{(r)} = Y^{(r-1)}$$

$$Y^{(r)} = AX^{(r)} \tag{2.86}$$

and the iteration stops when all the corresponding coefficients of $Y^{(r)}$ and $X^{(r)}$ are in a constant ratio within the desired precision.

It can be shown that when A is a real symmetric matrix this iterative method converges to the dominant or highest eigenvalue, which is computed from

$$\frac{Y_i^{(r)}}{X_i^{(r)}} = \lambda, \quad \text{for any } i \tag{2.87}$$

The vector $X^{(r)}$ is the corresponding eigenvector.

If we wish to compute the smallest rather than the highest eigenvalue, Equation (2.80) can be written as

$$A^{-1}X = \frac{1}{\lambda} X \tag{2.88}$$

We then proceed to apply the iterative scheme outlined above. In this case it will converge to the largest $1/\lambda$ value, thus providing the inverse of the smallest eigenvalue λ.

In either case, when applying this method in practice, it is common to take $X^{(r)}$ not exactly equal to $Y^{(r-1)}$ as in Equation (2.86), but to normalize it such that its leading coefficient becomes 1, as illustrated in the next example.

Example 2.8

Let us consider again the system of Example 2.7.

The governing system of equations can be written as

$$\begin{bmatrix} 3 & -1 \\ -1 & 2 \end{bmatrix} \begin{Bmatrix} x_1 \\ x_2 \end{Bmatrix} = \omega^2 \begin{Bmatrix} x_1 \\ x_2 \end{Bmatrix} \tag{a}$$

To start the iterative procedure let us take

$$X^{(1)} = \{1 \qquad 1\} \tag{b}$$

Then

$$Y^{(1)} = AX^{(1)} = \{2 \qquad 1\} \tag{c}$$

Dividing by the leading coefficient we take

$$X^{(2)} = \tfrac{1}{2}Y^{(1)} = \{1 \qquad 0.5\} \tag{d}$$

and proceed to compute $Y^{(2)}$. After repeating this scheme several times the ratio between the coefficients of $X^{(r)}$ and $Y^{(r)}$ becomes constant, and it can be taken as the dominant eigenvalue.

The results obtained for this numerical example are summarized in Table 2.3 from which we conclude that the highest eigenvalue is

$$\lambda = 3.6179 \tag{e}$$

corresponding to the eigenvector

$$X = \{1 \qquad -0.6179\} \tag{f}$$

Notice that these results coincide with those of Example 2.7.

Program 23   First eigenvalue and eigenvector, by the Stodolla-Vianello method (EIGIS)

The computer implementation of this iterative scheme can be readily carried out, as suggested by the flow-chart of Figure 2.12, which corresponds to the following FORTRAN code.

```
      SUBROUTINE EIGIS(A,X,N,NIM,TOL,D)
C
C PROGRAM 23
C
C  THIS PROGRAM APPLIES THE ITERATIVE
```

```
C    STODOLLA-VIANELLO METHOD TO COMPUTE
C    THE HIGHEST EIGENVALUE
C    A  : SYSTEM MATRIX
C    D  : EIGENVECTOR
C    :  A * X
C    XX : AUXILIARY VECTOR
C    N  : ORDER OF THE SYSTEM MATRIX
C    NIM: MAXIUM NUMBER OF INTERATIONS ALLOWED
C    TOL: TOLERANCE
C
C    NOTICE THAT X SHOULD ORIGINALLY CONTAIN
C    THE FIRST TRIAL EIGENVECTOR. UPON THE
C    SOLUTION IS COMPLETED II WILL CONTAIN
C    THE ACTUAL EIGENVECTOR.
C
      DIMENSION A(50,50),X(50),XX(50),Y(50)
      ITER=0
      DO 10 I=1,N
 10   X(I)=1.
  1   DO 2 I=1,N
  2   XX(I)=X(I)
      ITER=ITER+1
      DO 3 I=1,N
      Y(I)=0.
      DO 3 J=1,N
  3   Y(I)=Y(I)+A(I,J)*X(J)
      D=Y(1)
      DO 4 I=1,N
  4   X(I)=Y(I)/D
      DO 5 I=1,N
      IF(ABS(XX(I)-X(I))-TOL)5,5,6
  5   CONTINUE
      GO TO 9
  6   IF(ITER-NIM)1,7,7
  7   WRITE(6,8)
  8   FORMAT(' MAXIMUM NUMBER OF ITERATIONS ACHIEVED')
  9   RETURN
      END
```

Table 2.3

Iteration	X_1	X_2	Y_1	Y_2	λ
1	1	1.0000	2.0000	1.0000	2.0000
2	1	0.5000	2.5000	0.0000	2.5000
3	1	0.0000	3.0000	−1.0000	3.0000
4	1	−0.3333	3.3333	−1.6666	3.3333
5	1	−0.5000	3.5000	−2.0000	3.5000
6	1	−0.5714	3.5714	−2.1428	3.5714
7	1	−0.6000	3.6000	−2.2000	3.6000
8	1	−0.6111	3.6111	−2.2222	3.6111
9	1	−0.6153	3.6153	−2.2307	3.6153
10	1	−0.6170	3.6170	−2.2340	3.6170
11	1	−0.6176	3.6176	−2.2352	3.6176
12	1	−0.6178	3.6178	−2.2357	3.6178
13	1	−0.6179	3.6179	−2.2359	3.6179

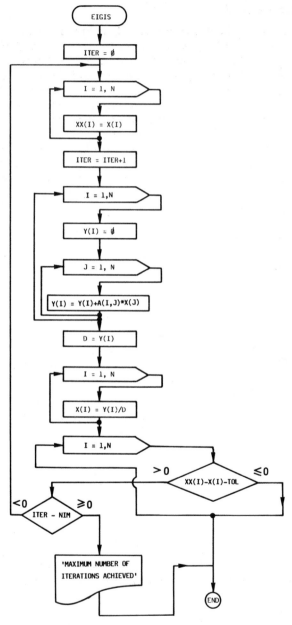

Figure 2.12   Flow-chart for Program EIGIS

2.4.2 Jacobi's method

In some engineering problems it is only of interest to know the largest eigenvalue and the associated eigenvector. In such cases it can be advantageous to
use the Stodola–Vianello method. If, on the other hand, it is necessary to
know the smallest eigenvalue, this method could still be used, but it requires
the inversion of the system matrix which is a time consuming process in a
computer. Many engineering problems, however, involve a more refined
analysis, such that all or several eigenvalues and eigenvectors need to be known.
In such cases a method other than the previously discussed should be used.

Jacobi's method provides a convenient scheme to compute all eigenvalues
and eigenvectors of a system such as Equation (2.80), when the system matrix
A is a real symmetric matrix. To better explain this method let us consider
the n matrix equations corresponding to each eigenvalue, in terms of the normalized eigenvectors

$$AX_1 = \lambda_1 X_1$$

$$AX_2 = \lambda_2 X_2$$

$$\begin{matrix} . & . \\ . & . \\ . & . \end{matrix}$$

$$AX_n = \lambda_n X_n \tag{2.89}$$

or, in compact form

$$A(X_1 \quad X_2 \dots X_n) = (\lambda_1 \quad X_1, \lambda_2 \quad X_2, \dots, \lambda_n \quad X_n) \tag{2.90}$$

Calling

$$Q = (X_1 \quad X_2 \dots X_n) \tag{2.91}$$

we can easily see that

$$(\lambda_1 X_1, \lambda_2 X_2, \dots, \lambda_n X_n) = Q\lambda \tag{2.92}$$

where λ is the diagonal matrix of eigenvalues

$$\lambda = \begin{bmatrix} \lambda_1 & 0 & \ldots & 0 \\ 0 & \lambda_2 & \ldots & 0 \\ \cdot & \cdot & & \cdot \\ \cdot & \cdot & & \cdot \\ \cdot & \cdot & & \cdot \\ 0 & 0 & \ldots & \lambda_n \end{bmatrix} \tag{2.93}$$

Equation (2.90) can then be written as

$$AQ = Q\lambda \tag{2.94}$$

Next we examine the matrix product

$$B = Q^T Q \tag{2.95}$$

The coefficient of B located on the ith row and jth column will be given by

$$b_{ij} = X_i^T X_j \tag{2.96}$$

corresponding to the scalar product of the ith and jth eigenvectors, associated with the two different eigenvalues λ_i, and λ_j by the expressions

$$AX_i = \lambda_i X_i \tag{2.97a}$$

$$AX_j = \lambda_j X_j \tag{2.97b}$$

Postmultiplying the transpose of (2.97a) by X_j, and premultiplying (2.97b) by X_i^T we obtain

$$X_i^T A^T X_j = \lambda_i X_i^T X_j \tag{2.98a}$$

$$X_i^T A X_j = \lambda_j X_i^T X_j \tag{2.98b}$$

Subtracting Equation (2.98b) from Equation (2.98a), and considering that $A = A^T$ for a symmetric matrix A, the result is

$$(\lambda_j - \lambda_i)X_i^T X_j = 0 \tag{2.99}$$

which shows that

$$X_i^T X_j = 0 \tag{2.100}$$

for two different eigenvalues λ_i and λ_j. In the case that $i = j$ we have

$$X_i^T X_i \neq 0 \tag{2.101}$$

and, in particular, if the eigenvectors are normalized

$$X_i^T X_i = 1 \tag{2.102}$$

From this discussion we conclude that

$$b_{ij} = \begin{cases} 0, \text{ for } i \neq j \\ 1, \text{ for } i = j \end{cases} \tag{2.103}$$

so that

$$B = I \tag{2.104}$$

and

$$Q^T = Q^{-1} \tag{2.105}$$

A matrix having the property (2.105) is called an orthogonal matrix.
Going back to Equation (2.94) and premultiplying by Q^T the result is

$$Q^T A Q = Q^T Q \lambda = \lambda \tag{2.106}$$

which shows that if we find an orthogonal matrix Q, such that applying to A the orthogonal transformation $Q^T ( ) Q$, produces a diagonal matrix, then the diagonal coefficients of that matrix are the eigenvalues of system (2.80), and the columns of matrix Q are the corresponding eigenvectors. Therefore, our problem reduces to try to diagonalize the matrix A.

Jacobi's method provides a scheme to eliminate, in turn, selected off-diagonal terms of matrix A by performing a sequence of elementary orthogonal transformations.

Consider the symmetric matrix of order 4.

$$A = \begin{bmatrix} a_{11} & a_{12} & a_{13} & a_{14} \\ a_{12} & a_{22} & a_{23} & a_{24} \\ a_{13} & a_{23} & a_{33} & a_{34} \\ a_{14} & a_{24} & a_{34} & a_{44} \end{bmatrix} \tag{2.107}$$

and assume that the term a_{24} is to be eliminated. Working with the orthogonal transformation matrix

$$
R_1 = \begin{bmatrix} 1 & 0 & 0 & 0 \\ 0 & C & 0 & -S \\ 0 & 0 & 1 & 0 \\ 0 & S & 0 & C \end{bmatrix}
\tag{2.108}
$$

where $C = \cos \theta$ and $S = \sin \theta$, with θ being a rotation angle to be determined, the result of a matrix operation of type (2.106) is:

$$
R_1^T A R_1 = \begin{bmatrix} a_{11} & ca_{12} + sa_{14} & a_{13} & -sa_{12} + ca_{44} \\ ca_{12} + sa_{14} & c^2 a_{22} + s^2 a_{44} + 2sca_{24} & ca_{23} + sa_{34} & -cs(a_{22} - a_{44}) + a_{24}(c^2 - s^2) \\ a_{13} & ca_{32} + sa_{34} & a_{33} & -sa_{32} + ca_{34} \\ -sa_{12} + ca_{14} & -cs(a_{22} - a_{44}) + a_{24}(c^2 - s^2) & -sa_{23} + ca_{34} & s^2 a_{22} + c^2 a_{44} - 2sca_{24} \end{bmatrix}
\tag{2.109}
$$

To eliminate the term in the second row and fourth column it must be

$$
-\cos \theta \sin \theta (a_{22} - a_{44}) + a_{24}(\cos^2 \theta - \sin^2 \theta) = 0
\tag{2.110}
$$

which can be transformed into

$$
a_{24} \tan^2 \theta + \tan \theta (a_{22} - a_{44}) - a_{24} = 0
$$

The roots of this second order equation are

$$
\tan \theta = \frac{-(a_{22} - a_{44}) \pm [(a_{22} - a_{44})^2 + 4a_{24}^2]^{1/2}}{2a_{24}}
\tag{2.111}
$$

Let us restrict ourselves to one of the roots, for instance

$$
\tan \theta = \frac{-(a_{22} - a_{44}) + [(a_{22} - a_{44})^2 + 4a_{24}^2]^{1/2}}{2a_{24}}
\tag{2.112}
$$

Notice that the other root will be $180°$ out of phase and would not affect the results. Working with root (2.112) is equivalent to considering only the $-\pi/2 < \theta < \pi/2$ interval.

Having $\tan \theta$ we can compute

$$\cos \theta = (1 + \tan^2 \theta)^{-1/2}$$

$$\sin \theta = \cos \theta \, \tan \theta \tag{2.113}$$

Jacobi's method consists in applying the above transformation to all the off-diagonal terms until all of them are, to a small error, equal to zero. Normally one starts with the off-diagonal term with the largest absolute value. Assuming that it occupies the location (I, J) the expression (2.112) becomes

$$\tan \theta = \frac{-(a_{ii} - a_{jj}) + [(a_{ii} - a_{jj})^2 + 4a_{ij}^2]^{1/2}}{2a_{ij}} \tag{2.114}$$

from which we can evaluate $\cos \theta$ and $\sin \theta$ using Equations (2.113). Next we build $\mathbf{R}_1$ taking a unit matrix and placing $\cos \theta$ in location (I, I) and (J, J), $-\sin \theta$ in location (I, J) and $\sin \theta$ in location (J, I). We can then perform the orthogonal transformation $\mathbf{R}_1^T \mathbf{A} \mathbf{R}_1$ which is equivalent to modify the ith and jth rows and columns of $\mathbf{A}$ according to the following scheme

Row i $a_{ii} = \cos^2 \theta a_{ii} + \sin^2 \theta a_{jj} + 2\sin \theta \, \cos \theta a_{ij}$

$a_{ij} = -\cos \theta \, \sin \theta (a_{ii} - a_{jj}) + a_{ij}(\cos^2 \theta - \sin^2 \theta) = 0$

$a_{ik} = \cos \theta a_{ik} + \sin \theta a_{jk}$

$$k = 1, n \text{ but } k \neq i, k \neq j \tag{2.115}$$

Row j $a_{jj} = \sin^2 \theta a_{ii} + \cos^2 \theta a_{jj} - 2\cos \theta \, \sin \theta a_{ij}$

$a_{ji} = -\cos \theta \, \sin \theta (a_{ii} - a_{jj}) + a_{ji}(\cos^2 \theta - \sin^2 \theta) = 0$

$a_{jk} = -\sin \theta a_{ik} + \cos \theta a_{jk}$

$$k = 1, n \text{ but } k \neq i, k \neq j \tag{2.116}$$

Column i $a_{ki} = \cos \theta a_{ki} + \sin \theta a_{kj}$

$$k = 1, n \text{ but } k \neq i, k \neq j \tag{2.117}$$

Column j $a_{kj} = -\sin \theta a_{ki} + \cos \theta a_{kj}$

$$k = 1, n \text{ but } k \neq i, k \neq j \tag{2.118}$$

Notice that the orthogonal transformation preserves symmetry which allows us to reduce the number of operations required by Equations (2.115) to (2.118).

We then again select the largest absolute value off-diagonal term, from those that remain different than zero and repeat the transformations outlined above. These transformations are repeatedly applied until no other than zero off-diagonal terms remain. Notice, however, that when applying the transformations (2.115) to (2.118) at a given stage, we might obtain a non-zero value for a term previously eliminated, which indicates the iterative nature of this method. Nevertheless it can be shown that this method is always convergent, and that is completely stable against round-off errors.

Assuming that we needed n iterations to diagonalize A, after all transformations have been applied, we obtain

$$R_n^T \ldots R_3^T R_2^T R_1^T A R_1 R_2 R_3 \ldots R_n = Q^T A Q = \lambda \qquad (2.119)$$

so that the eigenvector matrix Q is given by

$$Q = R_1 R_2 R_3 \ldots R_n \qquad (2.120)$$

Program 24   Computation of eigenvalues and eigenvectors, for a system of the type $AX = \lambda X$, by the Jacobi method (JACOB)

The following routine applies the Jacobi method to compute the eigenvalues and eigenvectors of a symmetric matrix stored in array A. After completion the diagonal terms of A are the problem eigenvalues. Its off-diagonal terms should be zero within the tolerance given by the variable ERR (normally 10^{-6} to 10^{-8}). The eigenvector are returned in array V. The variable ITM is the maximum number of iterations allowed.

```
      SUBROUTINE JACOB(A,V,ERR,N,NX)
C
C PROGRAM 24
C
C COMPUTATION OF EIGENVALUES AND EIGENVECTORS
C BY THE JACOBI'S METHOD
C
C A : SYSTEM MATRIX. AFTER THE COMPUTATIONS
C     ARE COMPLETED ITS DIAGONAL TERMS WILL
C     BE THE EIGENVALUES
C V : EACH COLUMN OF THIS ARRAY WILL CONTAIN
C     A SET OF EIGENVECTORS
C ERR:ERROR ALLOWED
C N : ACTUAL ORDER OF A
C NX: ROW AND COLUMN DIMENSION OF A
C
C
      DIMENSION A(NX,NX),V(NX,NX)
      ITM=200
      IT=0
C
C PUT A UNIT MATRIX IN ARRAY V
C
      DO 10 I=1,N
      DO 10 J=1,N
      IF(I-J)3,1,3
```

```
      3  V(I,J)=0.
         GO TO 10
      1  V(I,J)=1.
     10  CONTINUE
C
C FIND LARGEST OFF DIAGONAL COEFFICIENT
C
     13  T=0
         M=N-1
         DO 20 I=1,M
         J1=I+1
         DO 20 J=J1,N
         IF(ABS(A(I,J))-T)20,20,2
      2  T=ABS(A(I,J))
         IR=I
         IC=J
     20  CONTINUE
         IF(IT) 5,4,5
C
C TAKE FIRST LARGEST OFF DIAGONAL COEFFICIENT
C TIMES ERR AS COMPARISON VALVE FOR ZERO
C
      4  T1=T*ERR
      5  IF(T-T1)999,999,6
C
C COMPUTE TAN(TA), SIN(S), AND COSINE(C) OF ROTATION ANGLE
C
      6  PS=A(IR,IR)-A(IC,IC)
         TA=(-PS+SQRT(PS*PS+4*T*T))/(2*A(IR,IC))
         C=1./SQRT(1+TA*TA)
         S=C*TA
C
C MULTIPLY ROTATION MATRIX TIMES V AND STORE IN V
C
         DO 50 I=1,N
         P=V(I,IR)
         V(I,IR)=C*P+S*V(I,IC)
     50  V(I,IC)=C*V(I,IC)-S*P
         I=1
    100  IF(I-IR)7,200,7
C
C APPLY ORTHOGONAL TRASFORMATION TO A AND STORE IN A
C
      7  P=A(I,IR)
         A(I,IR)=C*P+S*A(I,IC)
         A(I,IC)=C*A(I,IC)-S*P
         I=I+1
         GO TO 100
    200  I=IR+1
    300  IF(I-IC)8,400,8
      8  P=A(IR,I)
         A(IR,I)=C*P+S*A(I,IC)
         A(I,IC)=C*A(I,IC)-S*P
         I=I+1
         GO TO 300
    400  I=IC+1
    500  IF(I-N)9,9,600
      9  P=A(IR,I)
         A(IR,I)=C*P+S*A(IC,I)
         A(IC,I)=C*A(IC,I)-S*P
         I=I+1
         GO TO 500
```

```
600   P=A(IR,IR)
      A(IR,IR)=C*C*P+2.*C*S*A(IR,IC)+S*S*A(IC,IC)
      A(IC,IC)=C*C*A(IC,IC)+S*S*P-2.*C*S*A(IR,IC)
      A(IR,IC)=0
      IT=IT+1
      IF(IT-ITM)13,13,999
999   RETURN
      END
```

2.4.3 General eigenvalue–eigenvector problem

In the general case, instead of an expression such as Equation (2.80) we will have

$$AX = \lambda BX \tag{2.121}$$

In particular we will consider the case where **B** is a symmetric and positive definite matrix. One approach to compute the eigenvalues and eigenvectors of (2.121) is to reduce it to the form (2.80) and then to apply one of the methods seen previously, such as Jacobi. To be able to do that, however, the reduction should be done without loosing the symmetry of the matrices involved.

Since **B** is a symmetric and positive definite matrix we can decompose it as in (2.56). Then we will have

$$AX = \lambda S^T SX \tag{2.122}$$

$$AIX = \lambda S^T SX \tag{2.123}$$

where the unit matrix can be substituted as follows

$$AS^{-1}SX = \lambda S^T SX \tag{2.124}$$

Noticing that

$$(S^T)^{-1} = (S^{-1})^T \tag{2.125}$$

we can premultiply Equation (2.124) by $(S^T)^{-1}$ obtaining

$$(S^{-1})^T AS^{-1}SX = \lambda SX \tag{2.126}$$

Defining a new vector

$$X' = SX \tag{2.127}$$

Equation (2.126) can be written as

$$(S^{-1})^T A S^{-1} X' = \lambda X' \qquad\qquad (2.128)$$

and calling

$$H = (S^{-1})^T A S^{-1} \qquad\qquad (2.129)$$

Equation (2.128) becomes

$$HX' = \lambda X' \qquad\qquad (2.130)$$

which has the same form as Equation (2.80). Since H is a symmetric matrix, we can use the subroutine JACOB to obtain the eigenvectors X'. Then we can compute the eigenvectors X by inverting Equation (2.127) so that

$$X = S^{-1} X'. \qquad\qquad (2.131)$$

The eigenvalues of Equation (2.121) are the same as those of (2.130).

Program 25   Computation of eigenvalues and eigenvectors, for a system of the type $AX = \lambda BX$ (EIGG)

The following routine can be used to find the eigenvalues and eigenvectors of equation

$$AX = \lambda BX \qquad\qquad (a)$$

The eigenvalues will be written in A and the eigenvectors in B.

```
      SUBROUTINE EIGG(A,B,H,V,ERR,N,NX)
C
C PROGRAM 25
C
C THIS PROGRAM COMPUTES THE EIGENVALUES
C AND EIGENVECTORS OF AN EQUATION OF TYPE
C A * X = LAMBDA * B * X
C
C N : ACTUAL ORDER OF A AND B
C NX: ROW AND COLUMN DIMENSION OF A AND B
C ERR:ERROR LIMIT USED IN SUBROUTINE JACOB
C
C V : AUXILIARY ARRAY
C
      DIMENSION A(NX,NX),B(NX,NX),H(NX,NX),V(NX)
C
C DECOMPOSE MATRIX B USING CHOLESKI'S METHOD
C
      CALL DECOG(B,N,NX)
C
C INVERT MATRIX B
C
      CALL INVCH(B,H,N,NX)
C
C MULTIPLY TRANSPOSE(H) * A * H
```

```
C
      CALL BTAB3(A,H,V,N,NX)
C
C
C COMPUTE THE EIGENVALUES
C
      CALL JACOB(A,B,ERR,N,NX)
C
C COMPUTE THE EIGENVECTORS
C
      CALL MATMB(H,B,V,N,NX)
C
      RETURN
      END
```

2.5   QUADRATIC FORMS

A second degree polynomial of the form

$$F = a_{11} x_1^2 + a_{22} x_2^2 + \ldots a_{nn} x_n^2 + 2a_{12} x_1 x_2 \ldots 2a_{n-1,n} x_{n-1} x_n \quad (2.132)$$

is called a quadratic form in $x_1, x_2, \ldots, x_n$. For instance, a quadratic form in two variables x_1, x_2 is of the form

$$F = a_{11} x_1^2 + a_{22} x_2^2 + 2a_{12} x_1 x_2 \tag{2.133}$$

In many engineering problems there is an associated quadratic form, whose properties are of interest.

A quadratic expression can be organized in matrix form, according to:

$$F = [x_1 \, x_2 \ldots x_n] \begin{bmatrix} a_{11} & a_{12} & \cdots & a_{1n} \\ a_{21} & a_{22} & \cdots & a_{2n} \\ \cdot & \cdot & & \cdot \\ \cdot & \cdot & & \cdot \\ \cdot & \cdot & & \cdot \\ a_{n1} & a_{n2} & & a_{nn} \end{bmatrix} \begin{bmatrix} x_1 \\ x_2 \\ \cdot \\ \cdot \\ \cdot \\ x_n \end{bmatrix} \tag{2.134}$$

or

$$F = X^T A X \tag{2.135}$$

where A is a symmetric matrix.

When a quadratic form is equal to zero only if

$$x_i = 0 \quad i = 1, n \tag{2.136}$$

and is positive for all other values of the variables x_i, it is called a *positive definite* quadratic form. In such case the matrix $\mathbf{A}$ is called a positive definite matrix.

Let us now assume that we perform the change in variables

$$\mathbf{X} = \mathbf{QY} \tag{2.137}$$

such that $\mathbf{Q}$ is an orthogonal matrix. Then

$$F = \mathbf{Y}^T \mathbf{Q}^T \mathbf{AQY} \tag{2.138}$$

or

$$F = \mathbf{Y}^T \mathbf{A'Y} \tag{2.139}$$

where

$$\mathbf{A'} = \mathbf{Q}^T \mathbf{AQ}$$

If we select $\mathbf{Q}$ such that $\mathbf{A'}$ is a diagonal matrix the quadratic form becomes

$$F = a'_{11} y_1^2 + a'_{22} y_2^2 + \ldots a'_{nn} y_n^2 \tag{2.140}$$

which is called the *canonical form* of the quadratic form. Furthermore, according to what was seen previously, the diagonal coefficients of $\mathbf{A'}$ must be the eigenvalues of $\mathbf{A}$. Then

$$F = \lambda_1 y_1^2 + \lambda_2 y_2^2 + \ldots + \lambda_n y_n^2 \tag{2.141}$$

Thus we conclude that for the quadratic form to be positive definite all the eigenvalues of $\mathbf{A}$ must be positive, different than zero, values. In addition we see that the determinant of $\mathbf{A}$ will be positive.

When a quadratic form is non-negative for all permissible values of the variables, but in some cases it is null without all the variables being also null, it is called a *positive semidefinite* quadratic form. For this it is required that some of the eigenvalues of $\mathbf{A}$ be equal to zero. We can then conclude that a quadratic form will be positive semidefinite when the associated real symmetric matrix $\mathbf{A}$ is singular and does not have negative eigenvalues.

2.6   MATRIX REPRESENTATION OF EXTREMUM PROBLEMS

In the calculus of variations we are concerned with the determination of stationary conditions of functions (or functionals). The stationary conditions means that the derivatives of the function with respect to each of the variables are zero. This is a necessary condition for the existence of an extremum but is not sufficient and we have to investigate higher derivatives in order to identify the points.

Before we go into the matrix representation for the general case of a function of many variables, let us review the case of a function of only one variable.

Consider a function $f(x)$ derivable in the (a, b) interval. We are interested in investigating the behaviour of this function in the vicinity of the point $x_0 (a < x_0 < b)$. This can be done by expanding f in terms of a Taylor's series about x_0.

Thus,

$$\Delta f = f(x_0 + \delta x) - f(x_0) = f_x \delta x + \tfrac{1}{2} f_{xx}(\delta x)^2 + \ldots (1/n!) f_{x(n)}(\delta x)^n + \ldots$$

$$(2.142)$$

where δx is an increment of the independent variable, Δf is the total increment of the f function and, we call $\delta f = f_x \delta x =$ first order increment of the function f, $\delta^2 f = f_{xx}(\delta x)^2 =$ second order increment of the function f, etc.

Thus Equation (2.102) can be written as

$$\Delta f = \delta f + \frac{1}{2!} \delta^2 f \ldots \frac{1}{n!} \delta^n f \qquad (2.143)$$

The *necessary* condition for the $f(x)$ function to have an extremum at x_0 is

$$\delta f = \frac{\partial f}{\partial x} \delta x = 0 \qquad \text{or} \qquad \frac{\partial f}{\partial x} = 0 \qquad (2.144)$$

In addition it is *sufficient* that $f_{xx} > 0$ for the point to be a minimum for arbitrary δx (or $f_{xx} < 0$ for a maximum). If $f_{xx} = 0$ we have a neutral point, and we need them to go on differentiating. For instance, if the first non zero derivative is odd the value of Δf is positive or negative according with δx. Thus no extremum exists and the point is called indifferent. If the first non-zero derivative is even we can have a maximum (even derivative is negative) or a minimum (even derivative is positive).

For a function of n variables, the problem of determining a minimum or maximum becomes more complex. Here we have the total increment,

$$\Delta f = \delta f + \frac{1}{2!} \delta^2 f + \ldots \qquad (2.145)$$

where $f = f(x_1 x_2 \ldots)$. Thus

$$\delta f = \frac{\partial f}{\partial x_1} \delta x_1 + \frac{\partial f}{\partial x_2} \delta x_2 + \ldots = \sum_i \frac{\partial f}{\partial x_i} \delta x_i \qquad (2.146)$$

$$\delta^2 f = \sum_i \sum_j \frac{\partial^2 f}{\partial x_i \partial x_j} \delta x_i \, \delta x_j \ldots, \text{etc.} \qquad (2.147)$$

The *necessary* (stationary) condition for an extremum is

$$\delta f = 0 \qquad \text{or} \; \frac{\partial f}{\partial x_1} = \frac{\partial f}{\partial x_2} = \ldots = 0 \qquad (2.148)$$

The *sufficient* conditions for an extremum are more involved. In order to illustrate how they can be obtained and represented in matrix form, we can consider the case of a function of two variables,

$$f(x, y)$$

Expanding in Taylor's series we have

$$\Delta f = f(x_0 + \delta x, y_0 + \delta y) - f(x_0, y_0) =$$

$$= (f_x \delta x + f_y \delta y) + \frac{1}{2!} \{(\delta x)^2 f_{xx} + 2\delta x \delta y \, f_{xy} + (\delta y)^2 f_{yy}\} + \ldots \qquad (2.149)$$

If the point is stationary ($\delta f = 0$) and Equation (2.149) reduces to,

$$\Delta f = \frac{1}{2!} \{(\delta x)^2 f_{xx} + 2\delta x \delta y \, f_{xy} + (\delta y)^2 f_{yy}\} + \ldots \qquad (2.150)$$

Let us now investigate the second variation by writing it as

$$\delta^2 f = f_{xx} \left\{ \left(\delta x + \delta y \, \frac{f_{xy}}{f_{xx}} \right)^2 + \frac{(\delta y)^2 \, (f_{xx} f_{yy} - f_{xy}^2)}{f_{xx}^2} \right\} \qquad (2.151)$$

If $f_{xx} > 0$ and $f_{xx} f_{yy} - f_{xy}^2 > 0$ we said $\delta^2 f$ is positive definite for all values of $\delta x, \delta y$. Thus

$$f(x_0 + \delta x, y_0 + \delta y) > f(x_0, y_0)$$

and $f(x_0 \, y_0)$ is a minimum value of $f(x, y)$. If $f_{xx} < 0$ and

$$f_{xx} f_{yy} - f_{xy}^2 > 0$$

we said $\delta^2 f$ is negative definite for all values of δx, δy. Thus

$$f(x_0 + \delta x, y_0 + \delta y) < f(x_0, y_0)$$

and $f(x_0, y_0)$ is a maximum value of f. If we have

$$f_{xx} f_{yy} - f_{xy}^2 < 0$$

the $\delta^2 f$ is indefinite (neither a maximum nor a minimum). The case $f_{xx} f_{yy} - f_{xy}^2 = 0$ is undecided and is called a neutral case; we need to investigate other terms in Taylor series in order to identify the point.

We could proceed in this way to define the condition for functions of several independent variables, but we will now switch to matrix notation for simplicity.

Thus for the 2-variables system we have for the first variation

$$\delta f = \{\delta x \; \delta y\} \quad \begin{Bmatrix} f_x \\ f_y \end{Bmatrix} = \delta \mathbf{x}^T \mathbf{f}^{(1)} \tag{2.152}$$

and for the second variation

$$\delta^2 f = \{\delta x \; \delta y\} \begin{bmatrix} f_{xx} f_{xy} \\ f_{xy} f_{yy} \end{bmatrix} \begin{Bmatrix} \delta x \\ \delta y \end{Bmatrix}$$

$$= \delta \mathbf{x}^T \mathbf{f}^{(2)} \delta x \tag{2.153}$$

The condition for a stationary value is $\mathbf{f}^{(1)} = 0$ and for defining the point we have,

(1)   If $\mathbf{f}^{(2)}$ is positive definite the point is a relative *minimum*
(2)   If $\mathbf{f}^{(2)}$ is negative definite, the point is a relative *maximum*
(3)   If $\mathbf{f}^{(2)}$ is semi-definite, the point is *neutral*
(4)   If $\mathbf{f}^{(2)}$ has negative and positive eigenvalues, the point is *indefinite*.

See Figure 2.13.

The above results will be valid for functions of n variables. Thus for a matrix,

$$f^{(2)} = \begin{bmatrix} f_{11} & f_{12} & f_{13} & \cdots \\ f_{21} & f_{22} & f_{23} & \\ \cdot & \cdot & \cdot & \end{bmatrix} \qquad (2.154)$$

The condition for minimum is that the following determinants (minors) are all positive:

$$|f_{11}| > 0, \quad \begin{vmatrix} f_{11} & f_{12} \\ f_{21} & f_{22} \end{vmatrix} > 0 \cdots \quad \begin{vmatrix} f_{11} & f_{12} & \cdots \\ f_{21} & f_{22} & \\ \cdot & \cdot & \cdot \end{vmatrix} > 0 \qquad (2.155)$$

Example 2.9
Consider that $\delta^2 f$ is given by the following quadratic function,

$$\delta^2 f = 2\delta x_1^2 - 2\delta x_1 \delta x_2 - 2\delta x_2 \delta x_3 + \delta x_2^2 + 3\delta x_3^2 \qquad \text{(a)}$$

This function can be shown to be positive definite by ordering it as follows:

$$\{\delta x_1 \ \delta x_2 \ \delta x_3\} \begin{bmatrix} 2 & -1 & 0 \\ -1 & 1 & -1 \\ 0 & -1 & 3 \end{bmatrix} \begin{Bmatrix} \delta x_1 \\ \delta x_2 \\ \delta x_3 \end{Bmatrix} \qquad \text{(b)}$$

We have the following principal minors:

$$|2| = 2 > 0$$

$$\begin{vmatrix} 2 & -1 \\ -1 & 1 \end{vmatrix} = 1 > 0$$

$$\begin{vmatrix} 2 & -1 & 0 \\ -1 & 1 & -1 \\ 0 & -1 & 3 \end{vmatrix} = 1 > 0$$

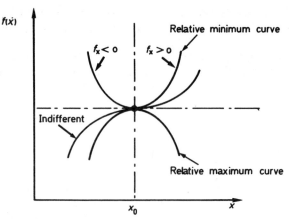

Figure 2.13   Different types of stationary points

Note that a positive definite function is such that for any value of x_1 x_2 x_3 one will have a positive F function.

Exercises

(1)   Show that

$$(AB)^T = B^T A^T$$

(2)   Show that the matrix **C**, given by

$$C = B^T A B$$

for **A** symmetric, is also a symmetric matrix.

(3)   Given the following matrix

$$A = \begin{vmatrix} A_{11} & A_{12} \\ A_{21} & A_{22} \end{vmatrix}$$

find its inverse in partitioned form given that A_{11} and A_{22} are square mat

(4)   Repeat the previous problem for the case in which $A_{22} = 0$.

(5) Modify routine SLSIM for symmetric systems of equations, to allow for zero diagonal coefficients, such that symmetry is not destroyed.

(6) Modify routine SLBSI to allow for the simultaneous solution of several systems of equations, having the same coefficient matrix **A**, but different vectors of independent coefficients.

(7) Any non-singular matrix can be decomposed into a lower **L** and an upper triangular matrix **U**, such that

$$A = LU \tag{a}$$

In general such decomposition can be done for different **L** and **U** matrices. However, if the diagonal coefficients of one of them, say **U**, are fixed, the decomposition will be unique. Let us now consider the system

$$AX = B \tag{b}$$

Introducing (a) we obtain

$$LUX = B$$

such that

$$UX = L^{-1}B$$

This means that the Gauss elimination process is equivalent to premultiplying Equation (b) by L^{-1}

$$L^{-1}AX = UX = L^{-1}B$$

when the diagonal coefficients of **U** are set equal to 1. Find the coefficients of **L** for a system of order 3.

(8) Find an algorithm to compute the inverse of a lower triangular matrix, such as the matrix **L** of the previous problem.

(9) Formula (2.71) shows general equation for the r iteration cycle in the Gauss–Seidel technique. A variation of this is the Gauss scheme in which the values of the unknowns for cycle r are computed using only the values of the unknowns for cycle r^{-1}, such that

$$X_k^{(r)} = \frac{1}{a_{kk}} \left[b_k - \sum_{i=1}^{k-1} a_{ki} x_i^{(r-1)} - \sum_{i=k+1}^{n} a_{ki} x_i^{(r-1)} \right]$$

Implement a program to apply the Gauss iterative technique and compare the number of iterations required when using the Gauss and Gauss–Seidel schemes.

(10) A continuous beam, such as that of Figure 2.5, will always lead to a symmetric tridiagonal system, no matter the number of supports. Find an efficient algorithm to solve symmetric tridiagonal systems of order n.

(11) Implement a program to apply the Gauss–Seidel technique including an over-relaxation factor. Make a numerical study of the effect of using different values for such factor.

(12) Compute the largest frequency and the corresponding mode of vibration for the three degrees of freedom system shown in Figure 2.14.

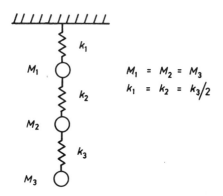

$M_1 = M_2 = M_3$
$k_1 = k_2 = k_3/2$

Figure 2.14

(13) Find the canonical form of the following quadratic form:

$$F = 4x_1^2 + 8x_2^2 + 4x_3^2 + 4x_1 x_2 + 4x_2 x_3$$

(14) Consider the case of a two-spring body under an applied load P at 2 and an applied displacement u at 3 (Figure 2.15). Assuming that the two springs have the same k constant deduce their elongations and forces.

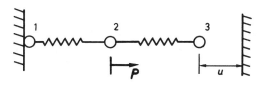

Figure 2.15

Bibliography

Booth, A. D., *Numerical Methods,* Butterworths (1966)

Conte, S. D. and de Boor, Elementary Numerical Analysis, 2nd edn, McGraw Hill (1972)

Hamming, R. W., *Introduction to Applied Numerical Analysis,* McGraw Hill (1971)

Kuo, S. S., *Computer Applications of Numerical Methods,* Addison-Wesley (1972)

Faddeeva, V. N., *Computational Methods of Linear Algebra,* Dover Publications, New York (1959)

Hildebrand, F. B., *Methods of Applied Mathematics,* 2nd edn, Prentice Hall (1965)

Ralston, A. and Wilf, H. S., *Mathematical Methods for Digital Computers,* Vol.1, John Wiley (1967)

Chapter 3

Matrix analysis of simple structural systems

3.1   INTRODUCTION

In many engineering applications we are confronted with network systems
consisting of a collection of different elements, or branches, interconnected at
a series of nodal points, or nodes. In civil engineering this type of system
appears in structural applications (trusses, frames), hydraulics (pipe systems),
transportation, construction, etc.

For identification purposes it is required that different names are assigned
to each of the nodes and elements of the system. For simplicity, numbers are
normally used. The connectivity relations, usually given in tabular form,
depict the topology of the system. Let us consider, for instance, the network
system shown in Figure 3.1. There are a number of nodes, N = 10, and a
number of elements, M = 14. Each of them is identified by a number. The
connectivity table for this network is given in Table 3.1, and defines the way
in which the elements are interconnected. It is built by listing the numbers of
the nodes pertaining to each element.

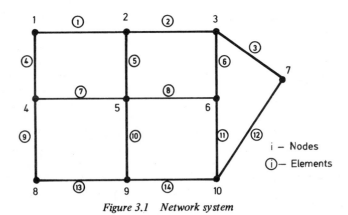

Figure 3.1   Network system

100

Table 3.1 CONNECTIVITY TABLE

Element	Nodes	
1	1	2
2	2	3
3	3	7
4	1	4
5	2	5
6	3	6
7	4	5
8	5	6
9	4	8
10	5	9
11	6	10
12	7	10
13	8	9
14	9	10

The solution of a network problem can be expressed in terms of nodal variables, which can be one or more per node. For some nodes in the system these variables will be unknown. These nodes will be called free nodes. For the other nodes there will be known values for the nodal variables as, for example, in the case of supports in a structural system, or nodes with specified potentials in a hydraulic system, etc. These type of nodes will be called boundary nodes.

The solution approach for studying network systems consists of first analysing the behaviour of each element, unassembled or isolated from the others. The cause—effect relationships between the actions applied to the element and its nodal variables are established, normally in matrix form. Then the study of the interaction of the collection of assembled elements leads to a system of equations, obtained by adding up the contributions of each unassembled element. This system of equations represents the behaviour of the total system, and its solution provides the values for the unknown nodal variables. Having obtained them other results of interest can then be computed.

In this chapter, truss and frame structural systems, using the displacement method, will be studied. Since in this case the elements or branches of the network receive the name of bars, these systems are known as bar systems. The complete description of a bar system includes, in addition to the topology and boundary values, the geometry, bar properties, and loads. The system geometry is normally specified in terms of the nodal coordinates, given with regard to some basic reference frame, arbitrarily selected. They define the length and orientation of each bar. The bar properties are given in terms of the cross-sectional characteristics, such as area, moments of inertia, etc., and the material elastic constants. The loads or actions applied to the system can be concentrated or distributed forces and moments, temperature effects, initial deformations, etc.

In a structural system the basic unknowns are the nodal displacements at the free nodes. At the boundary nodes the nodal displacements are known. They take a zero value at fixed supports or known non-zero values when there are support settlements. The study of the behaviour of each unassembled bar permits the establishment of relationships between the end forces and the nodal variables for the bar. In the displacement method these relationships are defined in terms of the bar stiffness matrix. The governing system of equations is then formed by conveniently adding up the bar stiffness matrices into a total stiffness matrix. The solution of the governing system of equations provides the values of the nodal displacements. From these we can later compute the bar end forces, support reactions, etc.

In the following sections the process outlined above will be discussed in detail, first for truss systems, and then for frame systems.

3.2   DISPLACEMENT METHOD – TRUSS

In structures we work with nodal and element quantities. The nodal quantities will be the nodal displacements, and the externally applied nodal loads. These nodal loads can eventually be totally or partially derived from distributed loads, temperature effects, and other actions on the structure. The element quantities will be the member end forces and the element deformations.

As in any problem of elasticity, the relevant quantities will have to satisfy the following 3 sets of relations
   (1)   Compatibility equations
   (2)   Constitutive equations
   (3)   Equilibrium equations
For small deformations, i.e. geometrically linear problems, the compatibility equations which relate nodal displacements and element deformations will be linear equations. The equilibrium equations can be established with regard to the undeformed configuration of the structure, rather than with regard to its deformed configuration and will only include nodal loads and element forces. If, in addition, the material is linearly elastic, i.e. corresponding to a physically linear problem, the constitutive equations will be linear equations, the same applying for the equilibrium equations. Thus, all the governing equations will be linear, and we will be performing a linear structural analysis.

In the displacement method we first consider the compatibility equations relating element deformations to nodal displacements. Introducing them into the constitutive equations we can express the element forces in terms of the nodal displacements. When these last expressions are substituted into the equilibrium relations, we obtain a set of equations which relate the nodal loads and the nodal displacements. These can be considered as the equilibrium equations expressed in terms of the nodal displacements. The solution of this

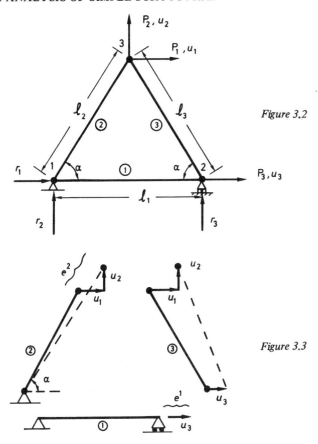

Figure 3.2

Figure 3.3

linear system of equations allows us to determine the values of the unknown nodal displacements, from which other results can be computed.

To illustrate more clearly the development of the displacement method, let us examine the following example.

Example 3.1

Consider the three-bar truss of Figure 3.2. For simplicity, we assume no support movement or temperature change.

The first step consists of establishing the compatibility equations. Since the strains and stresses are constant throughout a bar, we can use the bar elongation $e = \epsilon L$ as the element deformation, and the axial force $F = \sigma A$ as the element force, where L is the bar length. A is the bar cross-sectional area, and σ is the axial stress. Studying the kinematics of each bar, as shown in Figure 3.3, we can easily establish the compatibility equations

$$e^1 = u_3$$

$$e^2 = u_1 \cos \alpha + u_2 \sin \alpha$$

$$e^3 = -u_1 \cos \alpha + u_2 \sin \alpha + u_3 \cos \alpha \qquad \text{(a)}$$

which can be called bar-elongation nodal displacement relations.

In the matrix form we can write

$$
\mathbf{e} = \begin{Bmatrix} e^1 \\ e^2 \\ e^3 \end{Bmatrix} = \begin{bmatrix} 0 & 0 & 1 \\ \cos \alpha & \sin \alpha & 0 \\ -\cos \alpha & \sin \alpha & \cos \alpha \end{bmatrix} \begin{Bmatrix} u_1 \\ u_2 \\ u_3 \end{Bmatrix} = \mathbf{AU} \qquad \text{(b)}
$$

When the material is linearly elastic, the expressions for the bar force for the ith bar, in terms of the bar elongation, have the following form

$$F^i = \frac{A_i E}{L_i} e^i = k_i e^i \qquad \text{(c)}$$

which for this case are 3 equations determining the constitutive relations. In matrix form they can be written as

$$
\mathbf{F} = \begin{Bmatrix} F^1 \\ F^2 \\ F^3 \end{Bmatrix} = \begin{bmatrix} \dfrac{A_1 E}{l_1} & 0 & 0 \\ 0 & \dfrac{A_2 E}{l_2} & 0 \\ 0 & 0 & \dfrac{A_3 E}{l_3} \end{bmatrix} \begin{Bmatrix} e^1 \\ e^2 \\ e^3 \end{Bmatrix} = \begin{bmatrix} k_1 & 0 & 0 \\ 0 & k_2 & 0 \\ 0 & 0 & k_3 \end{bmatrix} \begin{Bmatrix} e^1 \\ e^2 \\ e^3 \end{Bmatrix} =
$$

where $\overline{\mathbf{K}}$ is normally called the unassembled stiffness matrix.

Finally, analysing the nodal equilibrium, as suggested by Figure 3.4, we can write the equilibrium equation

$$P_1 = F^2 \cos \alpha - F^3 \cos \alpha$$

$$P_2 = F^2 \sin \alpha + F^3 \sin \alpha$$

$$P_3 = F^1 + F^3 \cos \alpha \qquad \text{(f)}$$

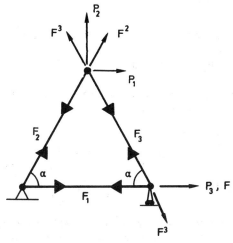

Figure 3.4

which can be written in matrix form as

$$\mathbf{P} = \begin{Bmatrix} P_1 \\ P_2 \\ P_3 \end{Bmatrix} = \begin{bmatrix} 0 & \cos\alpha & -\cos\alpha \\ 0 & \sin\alpha & \sin\alpha \\ 1 & 0 & \cos\alpha \end{bmatrix} \begin{Bmatrix} F^1 \\ F^2 \\ F^3 \end{Bmatrix} = \mathbf{GF} \qquad\text{(g)}$$

By inspection we verify that

$$\mathbf{A}^T = \mathbf{G} \qquad\qquad\text{(h)}$$

a result which can be established theoretically by the use of the principle of virtual displacements, studied in the next chapter.

Combining (b) and (d) we obtain

$$\mathbf{F} = \overline{\mathbf{K}}\mathbf{AU} \qquad\qquad\text{(i)}$$

which, when replaced in (g) gives

$$\mathbf{P} = \mathbf{A}^T\overline{\mathbf{K}}\mathbf{AU} \qquad\qquad\text{(j)}$$

or

$$\mathbf{P} = \mathbf{KU} \qquad\qquad\text{(k)}$$

where $\mathbf{K}$ is the structure stiffness matrix, equal to

$$
\mathbf{K} =
\begin{vmatrix}
0 & \cos\alpha & -\cos\alpha \\
0 & \sin\alpha & \sin\alpha \\
1 & 0 & \cos\alpha
\end{vmatrix}
\begin{vmatrix}
k_1 & 0 & 0 \\
0 & k_2 & 0 \\
0 & 0 & k_3
\end{vmatrix}
\begin{vmatrix}
0 & 0 & 1 \\
\cos\alpha & \sin\alpha & 0 \\
-\cos\alpha & \sin\alpha & \cos\alpha
\end{vmatrix}
$$

$$
\begin{vmatrix}
(k_2 + k_3)\cos^2\alpha & (k_2 - k_3)\sin\alpha\cos\alpha & -k_3\cos^2\alpha \\
(k_2 - k_3)\sin\alpha\cos\alpha & (k_2 + k_3)\sin^2\alpha & k_3\sin\alpha\cos\alpha \\
-k_3\cos^2\alpha & k_3\sin\cos\alpha & k_1 + k_3\cos^2\alpha
\end{vmatrix} \quad (1)
$$

To solve this example numerically we will assume that

$$
\alpha = 45°, \quad k_1 = k_2 = k_3 = 10, \quad P_1 = 1, \quad P_2 = P_3 = 0 \tag{m}
$$

Hence

$$
\begin{Bmatrix} 1 \\ 0 \\ 0 \end{Bmatrix} =
\begin{vmatrix}
10 & 0 & -5 \\
0 & 10 & 5 \\
-5 & 5 & 15
\end{vmatrix}
\begin{Bmatrix} u_1 \\ u_2 \\ u_3 \end{Bmatrix} \tag{n}
$$

whose solution is

$$
\mathbf{U} = \begin{Bmatrix} u_1 \\ u_2 \\ u_3 \end{Bmatrix} = \begin{Bmatrix} 0.125 \\ -0.025 \\ 0.05 \end{Bmatrix} \tag{o}
$$

Finally from (b) we obtain the element deformations

$$
\mathbf{e} = \mathbf{AU} =
\begin{bmatrix}
0 & 0 & 1 \\
0.707 & 0.707 & 0 \\
-0.707 & 0.707 & 0.707
\end{bmatrix}
\begin{Bmatrix} 0.125 \\ -0.025 \\ 0.05 \end{Bmatrix} =
\begin{Bmatrix} 0.05 \\ 0.0707 \\ -0.0707 \end{Bmatrix} \tag{p}
$$

and from (d) we get the element forces

$$
\mathbf{F} = \overline{\mathbf{K}}\mathbf{e} =
\begin{bmatrix}
10 & 0 & 0 \\
0 & 10 & 0 \\
0 & 0 & 10
\end{bmatrix}
\begin{Bmatrix}
0.05 \\
0.0707 \\
-0.0707
\end{Bmatrix}
=
\begin{Bmatrix}
0.5 \\
0.707 \\
-0.707
\end{Bmatrix}
\tag{q}
$$

The previous example, although a very simple statically determinate truss, was useful to illustrate the basic operations required to apply the displacement method. This method, however can be applied to statically indeterminate structures, exactly in the same way.

Usually it is required to take into account external actions such as temperature variation, forces applied along the bars, etc. These actions are converted to equivalent nodal forces, and then the displacement method can be applied the same as above. Let us take, for instance, the case in which the structure undergoes a temperature variation of value T. In that case the forces acting on a bar will be due both to its elastic deformation, and to the temperature dilatation which is prevented because of the bar constraints. Thus we can write

$$
F^i = \frac{A_i E}{l_i} e^i - A_i E\beta T = \frac{A_i E}{l_i} e^i + F_B^i
\tag{3.1}
$$

where β is the thermal expansion coefficient. Note that the term $F_{Bi}^i = -A_i E\beta T$ is the bar force when its end movements are completely prevented. Because of it, these type of forces receive the name of built in forces. These type of forces can derive from a variety of external actions. When these forces are present, the constitutive equations can be written as

$$
\mathbf{F} = \overline{\mathbf{K}}\mathbf{e} + \mathbf{F}_B
\tag{3.2}
$$

where

$$
\mathbf{F}_B = \{F_B^1 \quad F_B^2 \quad F_B^3 \ldots\}
\tag{3.3}
$$

After operating, the governing equations become

$$
\mathbf{P} = \mathbf{A}^T\overline{\mathbf{K}}\mathbf{A}\mathbf{U} + \mathbf{A}^T\mathbf{F}_B = \mathbf{K}\mathbf{U} + \mathbf{A}^T\mathbf{F}_B
\tag{3.4}
$$

and then are transformed into

$$
\mathbf{P} - \mathbf{A}^T\mathbf{F}_B = \mathbf{K}\mathbf{U}
\tag{3.5}
$$

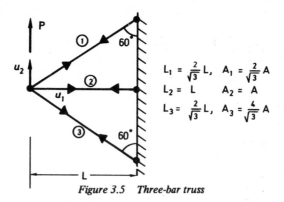

Figure 3.5   Three-bar truss

which is the system to be solved for the unknown displacements. The vector $-\mathbf{A}^T \mathbf{F}_B$ can be considered as a vector of equivalent nodal forces due to the temperature variation.

Example 3.2

To illustrate the application of the displacement method to a statically indeterminate structure, and also the consideration of a temperature variation T, let us analyse the truss system of Figure 3.5.

In this case, we can easily construct the matrices

$$
\mathbf{A} = \begin{bmatrix} -\dfrac{\sqrt{3}}{2} & -\dfrac{1}{2} \\[2ex] -1 & 0 \\[2ex] -\dfrac{\sqrt{3}}{2} & \dfrac{1}{2} \end{bmatrix} \tag{a}
$$

and

$$
\overline{\mathbf{K}} = \begin{bmatrix} \dfrac{AE}{L} & 0 & 0 \\[2ex] 0 & \dfrac{AE}{L} & 0 \\[2ex] 0 & 0 & \dfrac{2EA}{L} \end{bmatrix} \tag{b}
$$

so that

$$\mathbf{K} = \mathbf{A}^T \overline{\mathbf{K}} \mathbf{A} = \frac{EA}{4L} \begin{bmatrix} 13' & -\sqrt{3} \\ -\sqrt{3} & 3 \end{bmatrix} \tag{c}$$

In addition we see that

$$\mathbf{F}_B = AE\beta T \begin{Bmatrix} -2/\sqrt{3} \\ -1 \\ -4/\sqrt{3} \end{Bmatrix} \tag{d}$$

and

$$-\mathbf{A}^T \mathbf{F}_B = AE\beta T \begin{Bmatrix} -4 \\ 1/\sqrt{3} \end{Bmatrix} \tag{e}$$

Finally, the governing system of equations is written as

$$\begin{Bmatrix} -4\,AE\beta T \\ P + \dfrac{AE\beta T}{\sqrt{3}} \end{Bmatrix} = \frac{EA}{4L} \begin{vmatrix} 13 & -\sqrt{3} \\ -\sqrt{3} & 3 \end{vmatrix} \begin{Bmatrix} u_1 \\ u_2 \end{Bmatrix} \tag{f}$$

Once the values of P, A, L, β, and T, are known the system (f) can be solved for the unknown displacements u_1 and u_2.

In order to program the solution scheme (i.e. the displacement method), one has to identify the various computational phases and represent the operations in explicit form. Matrix notation is indispensable. From the previous discussion we can conclude that general procedure for the application of the displacement method consists of the following three basic steps:

(1) The bar elongation e is expressed in terms of the joint displacements at the ends of the bar.

(2) Using the material properties (stress–strain relations) the bar force F is expressed in terms of the elongation, and then, using the result of step 1, F is obtained in terms of the joint displacements.

(3) The force equilibrium condition at the joints are used. The final

equations are joint-force equilibrium equations in terms of joint displacements.

This scheme, however, may be inefficient since to compute the system matrix **K**, we have to operate with large matrices, namely **A** and **K̄**, which in a practical case will have few non-zero coefficients. This can be avoided, as shown in the matrix scheme described in the next paragraph.

3.3   COMPUTER MATRIX DISPLACEMENT METHOD

The characteristic of this computer scheme consists of assembling the system matrix **K** directly, without using the matrices **A** and **K̄**. Hence this method is sometimes called the direct stiffness method. From a systems approach we regard each element as isolated from the structure, to establish a matrix equation defining the individual element behaviour. Then considering the interaction of each element with the remaining elements of the structure, according to its connectivity, we define the total behaviour of the structure, which leads to the problem solution.

The basic steps of this scheme are

(1)   Structural identification
(2)   Evaluation of the element matrix equations
(3)   Assembling of the total system of equations
(4)   Introduction of the boundary conditions
(5)   Solution of the system of equations
(6)   Evaluation of the element forces

The fundamental aspects of each of these steps are discussed in detail below.

3.3.1 Structural identification

This step is normally carried out outside of the computer, and corresponds to number nodes and elements for identification purposes. Then a global reference system is selected to which all the nodal quantities will be referred. For each element or bar, a local reference system is selected, to which all the bar quantities are referred. For some problems we can dispense with the local reference systems, and express all quantities in the global system.

Based on the node and element numbers, and taking into account the reference systems selected, the structural data is prepared. As a minimum the data will include the nodal coordinates, connectivity table, element properties, boundary or support conditions, and a description of the applied loads.

For practical problems of large magnitude, the data preparation, and their checking, can become a very time consuming task. Lately some automatic data generation programs are being introduced, which operate from a few basic

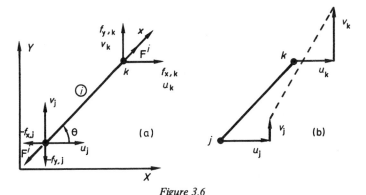

Figure 3.6

parameters to internally generate all the data required. However, in most cases, for bar structures, these programs are not general enough.

3.3.2 Evaluation of the element matrix equations

To evaluate the characteristic matrix equation for a bar of a truss system we consider the bar of Figure 3.6, arbitrarily oriented with regard to the global axes of reference X, Y. We will work in terms of the ith bar, having end nodes generically called j and k. The positive sense for the local x axis is assumed to be from node j to node k. The bar axial force F^i is positive when in tension. The bar elongation e^i is defined as the increase in length due to the nodal displacements u_j, v_j at node j, and u_k, v_k at node k, in the direction of the axes x, y respectively. For the geometrically linear case, e is equal to the projections of the translations on the initial positive direction of the bar, that is:

$$e^i = -(u_j - u_k)\cos\theta - (v_j - v_k)\sin\theta \tag{3.6}$$

which can be written as

$$e^i = [-\cos\theta \quad -\sin\theta \quad \cos\theta \quad \sin\theta] \begin{Bmatrix} u_j \\ v_j \\ u_k \\ v_k \end{Bmatrix} \tag{3.7}$$

Defining the direction cosines matrix

$$\mathbf{r}^i = [\cos\theta \quad \sin\theta] \tag{3.8}$$

we can write

$$e^i = [-\mathbf{r}^i \quad \mathbf{r}^i] \quad \begin{Bmatrix} \mathbf{u}_j \\ \mathbf{u}_k \end{Bmatrix} = \mathbf{R}^i \mathbf{U}^i \tag{3.9}$$

where $\mathbf{u}_j = \{u_j \; v_j\}$ and $\mathbf{u}_k = \{u_k \; v_k\}$ are the nodal displacement vectors for nodes j and k respectively. Also

$$\mathbf{U}^i = \begin{Bmatrix} \mathbf{u}_j \\ \mathbf{u}_k \end{Bmatrix} \tag{3.10}$$

is the elemental nodal displacement vector, and

$$\mathbf{R}^i = [-\mathbf{r}^i \quad \mathbf{r}^i] \tag{3.11}$$

is the element rotation matrix.

The axial force F^i can be written in terms of the bar elongation. For elastic linear materials it will be

$$F^i = k_i(e^i - \beta T L_i) \tag{3.12}$$

where β is the thermal expansion coefficient, T is the temperature variation, and L_i is the bar length. The bar end forces, with reference to the global axes, are:

$$\mathbf{F}_j^i = \begin{Bmatrix} f_{x,j} \\ f_{y,j} \end{Bmatrix}, \qquad \mathbf{F}_k^i = \begin{Bmatrix} f_{x,k} \\ f_{y,k} \end{Bmatrix} \tag{3.13}$$

We can easily verify that

$$f_{x,j} = -F^i \cos \theta$$

$$f_{y,j} = -F^i \sin \theta$$

$$f_{x,k} = F^i \cos \theta$$

$$f_{y,k} = F^i \sin \theta \tag{3.14}$$

which in matrix form become

$$\mathbf{F}_j^i = \begin{bmatrix} -\cos \theta \\ -\sin \theta \end{bmatrix} F^i = -\mathbf{r}^{i,T} F^i \tag{3.15}$$

and

$$\mathbf{F}_k^i = {}^\bullet \begin{bmatrix} \cos\theta \\ \sin\theta \end{bmatrix} \qquad F^i = \mathbf{r}^{i,T} F^i \tag{3.16}$$

Then we can write

$$\mathbf{F}^i = \begin{Bmatrix} \mathbf{F}_j^i \\ \mathbf{F}_k^i \end{Bmatrix} = \begin{Bmatrix} -\mathbf{r}^{i,T} \\ \mathbf{r}^{i,T} \end{Bmatrix} \qquad F^i = \mathbf{R}^{i,T} F^i \tag{3.17}$$

where $\mathbf{F}^i$ is the element end forces vector.

Finally, from Equations (3.9), (3.12) and (3.17), we get

$$\mathbf{F}^i = \mathbf{R}^{i,T} F^i = \mathbf{R}^{i,T} k_i(e^i - \beta T L_i) = \mathbf{R}^{i,T} k_i \mathbf{R}^i \mathbf{U}^i - \mathbf{R}^{i,T} k_i \beta T L_i \tag{3.18}$$

or

$$\mathbf{F}^i = \mathbf{K}^i \mathbf{U}^i + \mathbf{F}_B^i \tag{3.19}$$

where $\mathbf{K}^i$ is the bar stiffness matrix with regard to the global reference system, equal to

$$\mathbf{K}^i = \mathbf{R}^{i,T} k_i \mathbf{R}^i = \begin{Bmatrix} -\mathbf{r}^{i,T} \\ \mathbf{r}^{i,T} \end{Bmatrix} k_i \begin{bmatrix} -\mathbf{r}^i & \mathbf{r}^i \end{bmatrix} = \begin{bmatrix} \mathbf{r}^{i,T} k_i \mathbf{r}^i & -\mathbf{r}^{i,T} k_i \mathbf{r}^i \\ -\mathbf{r}^{i,T} k_i \mathbf{r}^i & \mathbf{r}^{i,T} k_i \mathbf{r}^i \end{bmatrix} =$$

$$= \begin{bmatrix} \mathbf{k}^i & -\mathbf{k}^i \\ -\mathbf{k}^i & \mathbf{k}^i \end{bmatrix} \tag{3.20}$$

with

$$\mathbf{k}^i = \mathbf{r}^{i,T} k_i \mathbf{r}^i = \begin{bmatrix} k_i \cos^2\theta & k_i \sin\theta \cos\theta \\ k_i \sin\theta \cos\theta & k_i \sin^2\theta \end{bmatrix} \tag{3.21}$$

which is commonly called the nodal stiffness matrix for the truss bar i.

We also see that

$$\mathbf{F}_B^i = -\mathbf{R}^{i,T} k_i \beta T L_i = - \begin{Bmatrix} -\mathbf{r}^{i,T} \\ \\ \mathbf{r}^{i,T} \end{Bmatrix} k_i \beta T L_i = k_i \beta T L_i \begin{Bmatrix} \cos\theta \\ \sin\theta \\ -\cos\theta \\ -\sin\theta \end{Bmatrix} \qquad (3.22)$$

Expression (3.19) is the characteristic matrix equation for the ith bar. Normally we visualize it in partitioned form, organized according to nodal partitions as

$$\begin{Bmatrix} \mathbf{F}_j^i \\ \\ \mathbf{F}_k^i \end{Bmatrix} = \begin{bmatrix} \mathbf{k}_{jj}^i & \mathbf{k}_{jk}^i \\ \\ \mathbf{k}_{kj}^i & \mathbf{k}_{kk}^i \end{bmatrix} \begin{Bmatrix} \mathbf{u}_j \\ \\ \mathbf{u}_k \end{Bmatrix} + \begin{Bmatrix} \mathbf{F}_{B,j}^i \\ \\ \mathbf{F}_{B,k}^i \end{Bmatrix} \qquad (3.23)$$

which corresponds to the equations

$$\mathbf{F}_j^i = \mathbf{k}_{jj}^i \mathbf{u}_j + \mathbf{k}_{jk}^i \mathbf{u}_k + \mathbf{F}_{B,j}^i$$
$$\mathbf{F}_k^i = \mathbf{k}_{kj}^i \mathbf{u}_j + \mathbf{k}_{kk}^i \mathbf{u}_k + \mathbf{F}_{B,j}^i \qquad (3.24)$$

(See Program 29 for extended version of **K** matrix.)

3.3.3 Assembling of the total system of equations

We use the force equilibrium equations for the nodes, representing the equality between the external forces applied at a node and the forces of the bars connected to that node. For each node we will have an equation of the type

$$\mathbf{P}_j = \Sigma \, \mathbf{F}_j^i \qquad (3.25)$$

where the summation covers all the bars connected to node j. A typical equation will be of the form

$$\mathbf{P}_j = \ldots + \mathbf{k}_{jj}^i \mathbf{u}_j + \mathbf{k}_{jk}^i \mathbf{u}_k + \mathbf{F}_{B,j}^i + \mathbf{k}_{jj}^l \mathbf{u}_j + \mathbf{k}_{jn}^l \mathbf{u}_n + \mathbf{F}_{B,j}^l + \ldots \qquad (3.26)$$

which can also be written as

$$\mathbf{P}_j \ldots - \mathbf{F}_{B,j}^i - \mathbf{F}_{B,j}^l \ldots = \ldots + \mathbf{k}_{jj}^i \mathbf{u}_j + \mathbf{k}_{jk}^i \mathbf{u}_k + \mathbf{k}_{jj}^l \mathbf{u}_j + \mathbf{k}_{jn}^l \mathbf{u}_n + \ldots \quad (3.27)$$

defining a row of the total system of Equations (3.5). We have previously indicated that the formation of the total system using (3.5) is inefficient because it requires multiplying large sparse matrices such as **A** and $\overline{\mathbf{K}}$. How-

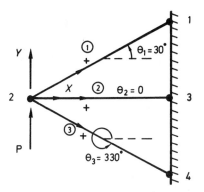

Figure 3.7   Three-bar truss

ever, in Equations (3.26) or (3.27), we see that in each row of the system we only have terms corresponding to the bars connected to the node being considered. The equation corresponding to the row under consideration could be formed directly from the bar stiffness matrices, using the connectivity table. Considering the bar stiffness matrix partitioned as in (3.23), we assemble the total system of equations, by systematically performing the following operations.

Add k_{st}^i to matrix row s and matrix column t of $\mathbf{K}$, for $s = j, k$ and $t = j, k$.

Add $\mathbf{F}_{B,s}^i$ to matrix row s of vector $\mathbf{P}$, for $s = j, k$.

Once these operations are carried out for all bars, the total system of equations is complete, and can be written, organized by nodal partitions, as:

$$\mathbf{P} = \left\{ \begin{array}{c} \mathbf{P}_1 \\ \hline \mathbf{P}_2 \\ \hline \cdot \\ \cdot \\ \cdot \\ \hline \mathbf{P}_n \end{array} \right\} = \left[\begin{array}{c|c|c|c} \mathbf{K}_{11} & \mathbf{K}_{12} & \ldots & \mathbf{K}_{1n} \\ \hline \mathbf{K}_{21} & \mathbf{K}_{22} & \ldots & \mathbf{K}_{2n} \\ \hline \cdot & \cdot & & \cdot \\ \cdot & \cdot & & \cdot \\ \cdot & \cdot & & \cdot \\ \hline \mathbf{K}_{n1} & \mathbf{K}_{n2} & \ldots & \mathbf{K}_{nn} \end{array} \right] \left\{ \begin{array}{c} \mathbf{u}_1 \\ \hline \mathbf{u}_2 \\ \hline \cdot \\ \cdot \\ \cdot \\ \hline \mathbf{u}_n \end{array} \right\} = \mathbf{K}\mathbf{U} \qquad (3.28)$$

where the vector $\mathbf{P}$ includes both forces directly applied to the nodes and equivalent nodal forces due to temperature variation, distributed loads, etc.

Example 3.3

We can now consider again the truss of Example 3.2, shown in Figure 3.7, which indicates the positive sense selected for each bar. According to the

Table 3.2

Bar	First joint	Second joint
1	2	1
2	2	3
3	2	4

node and element numbers chosen, the connectivity table will be as shown in Table 3.1.

The load and displacement vectors of the structure will be

$$
\mathbf{P} = \left\{ \begin{array}{c} \mathbf{P_1} \\ \mathbf{P_2} \\ \mathbf{P_3} \\ \mathbf{P_4} \end{array} \right\} = \left\{ \begin{array}{c} P_{x,1} \\ P_{y,1} \\ -- \\ P_{x,2} \\ P_{y,2} \\ -- \\ P_{x,3} \\ P_{y,3} \\ -- \\ P_{x,4} \\ P_{y,4} \end{array} \right\}
\qquad
\mathbf{U} = \left\{ \begin{array}{c} \mathbf{u_1} \\ \mathbf{u_2} \\ \mathbf{u_3} \\ \mathbf{u_4} \end{array} \right\} = \left\{ \begin{array}{c} u_1 \\ v_1 \\ -- \\ u_2 \\ v_2 \\ -- \\ u_3 \\ v_3 \\ -- \\ u_4 \\ v_4 \end{array} \right\}
\qquad (3.29)
$$

To compute the bar stiffness matrices, we use Equation (3.21) to obtain

$$
\mathbf{K}^1 = \left[\begin{array}{c|c} \mathbf{K_{22}^1} & \mathbf{K_{21}^1} \\ \hline \mathbf{K_{12}^1} & \mathbf{K_{11}^1} \end{array} \right] = \frac{AE}{L} \left[\begin{array}{cc|cc} \dfrac{3}{4} & \dfrac{\sqrt{3}}{4} & -\dfrac{3}{4} & -\dfrac{\sqrt{3}}{4} \\[2mm] \dfrac{\sqrt{3}}{4} & \dfrac{1}{4} & -\dfrac{\sqrt{3}}{4} & -\dfrac{1}{4} \\[2mm] \hline -\dfrac{3}{4} & -\dfrac{\sqrt{3}}{4} & \dfrac{3}{4} & \dfrac{\sqrt{3}}{4} \\[2mm] -\dfrac{\sqrt{3}}{4} & -\dfrac{1}{4} & \dfrac{\sqrt{3}}{4} & \dfrac{1}{4} \end{array} \right]
\qquad (3.30)
$$

$$K^2 = \begin{bmatrix} K_{22}^2 & | & K_{23}^2 \\ -- & | & -- \\ K_{32}^2 & | & K_{33}^2 \end{bmatrix} = \frac{AE}{L} \begin{bmatrix} 1 & 0 & | & -1 & 0 \\ 0 & 0 & | & 0 & 0 \\ -- & -- & | & -- & -- \\ -1 & 0 & | & 1 & 0 \\ 0 & 0 & | & 0 & 0 \end{bmatrix} \qquad (3.31)$$

$$K^3 = \begin{bmatrix} K_{22}^3 & | & K_{24}^3 \\ -- & | & -- \\ K_{42}^3 & | & K_{44}^3 \end{bmatrix} = \frac{AE}{L} \begin{bmatrix} \dfrac{6}{4} & -\dfrac{2\sqrt{3}}{4} & | & -\dfrac{6}{4} & \dfrac{2\sqrt{3}}{4} \\[2mm] -\dfrac{2\sqrt{3}}{4} & \dfrac{2}{4} & | & \dfrac{2\sqrt{3}}{4} & -\dfrac{2}{4} \\[2mm] -- & -- & | & -- & -- \\[2mm] -\dfrac{6}{4} & \dfrac{2\sqrt{3}}{4} & | & \dfrac{6}{4} & -\dfrac{2\sqrt{3}}{4} \\[2mm] \dfrac{2\sqrt{3}}{4} & -\dfrac{2}{4} & | & -\dfrac{2\sqrt{3}}{4} & \dfrac{2}{4} \end{bmatrix} \qquad (3.32)$$

From these bar stiffness matrices we can form the total stiffness matrix $\mathbf{K}$. Using the systematic scheme explained above, we can easily obtain

$$\begin{Bmatrix} P_1 \\ P_2 \\ P_3 \\ P_4 \end{Bmatrix} = \begin{bmatrix} K_{11}^1 & K_{12}^1 & 0 & 0 \\ K_{21}^1 & K_{22}^1 + K_{22}^2 + K_{22}^3 & K_{23}^2 & K_{24}^3 \\ 0 & K_{32}^2 & K_{33}^2 & 0 \\ 0 & K_{42}^3 & 0 & K_{44}^3 \end{bmatrix} \begin{Bmatrix} u_1 \\ u_2 \\ u_3 \\ u_4 \end{Bmatrix} + \begin{Bmatrix} F_{B,1}^1 \\ F_{B,2}^1 + F_{B,2}^2 + F_{B,2}^3 \\ F_{B,3}^2 \\ F_{B,4}^3 \end{Bmatrix} \qquad (3.33)$$

which corresponds to the equilibrium equations

$$P_1 = K_{11}^1 u_1 + K_{12}^1 u_2 + F_{B,1}^1$$
$$P_2 = K_{21}^1 u_1 + K_{22}^1 u_2 + K_{22}^2 u_2 + K_{22}^3 u_2 + K_{23}^2 u_3 + K_{24}^3 u_4 + F_{B,2}^1 + F_{B,2}^2 + F_{B,3}^3$$

$$P_3 = K_{32}^2 u_2 + K_{33}^2 u_3 + F_{B,3}^2$$

$$P_4 = K_{42}^3 u_2 + K_{44}^3 u_4 + F_{B,4}^3 \tag{3.34}$$

For our particular structure, considering a load P, in the y direction, applied at joint 2, and a temperature variation T, (3.33) becomes:

$$
\begin{Bmatrix}
R_{x,1} \\
R_{y,1} \\
-4AE\beta T \\
P + \dfrac{AE\beta T}{\sqrt{3}} \\
R_{x,3} \\
R_{y,3} \\
R_{x,4} \\
R_{y,4}
\end{Bmatrix}
= \frac{AE}{L}
\begin{bmatrix}
\dfrac{3}{4} & \dfrac{\sqrt{3}}{4} & -\dfrac{3}{4} & -\dfrac{\sqrt{3}}{4} & 0 & 0 & 0 & 0 \\
 & \dfrac{1}{4} & -\dfrac{\sqrt{3}}{4} & -\dfrac{1}{4} & 0 & 0 & 0 & 0 \\
 & & \dfrac{13}{4} & -\dfrac{\sqrt{3}}{4} & -1 & 0 & -\dfrac{3}{2} & \dfrac{\sqrt{3}}{2} \\
 & & & \dfrac{3}{4} & 0 & 0 & \dfrac{\sqrt{3}}{2} & -\dfrac{1}{2} \\
 & & & & 1 & 0 & 0 & 0 \\
 & \text{SYMM.} & & & & 0 & 0 & 0 \\
 & & & & & & \dfrac{3}{2} & -\dfrac{\sqrt{3}}{2} \\
 & & & & & & & \dfrac{1}{2}
\end{bmatrix}
\begin{Bmatrix}
u_1 \\
v_1 \\
u_2 \\
v_2 \\
u_3 \\
v_3 \\
u_4 \\
v_4
\end{Bmatrix}
\tag{3.35}
$$

Note that the forces at nodes 1, 3, and 4, correspond to support reactions, and therefore are so far unknowns. As a consequence the displacements for those nodes are prescribed. Before proceeding to the solution of Equation (3.35) it is necessary to introduce the boundary conditions.

As we have seen, the assembling of the system matrix $\mathbf{K}$ can be easily carried out using the systematic approach of the direct stiffness matrix. For structural systems $\mathbf{K}$ will be a positive definite symmetric matrix. It is also a sparse banded matrix, in the sense that it will have few non-zero coefficients, which will be grouped around the main diagonal. The bandwidth will depend on the numbering scheme adopted for the nodes. To see this let us take the

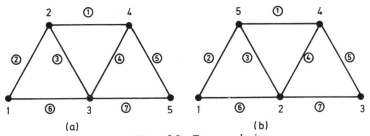

Figure 3.8   Truss numbering

Table 3.3

Bar	First node	Second node
1	2	4
2	1	2
3	2	3
4	3	4
5	4	5
6	1	3
7	3	5

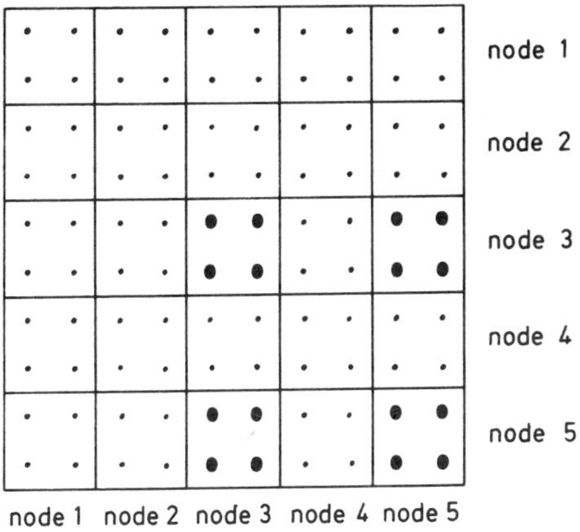

Figure 3.9   Element coefficients

truss of Figure 3.8(a). The connectivity table corresponding to this truss is Table 3.3.

A typical element like (7) will have 16 coefficients which will be distributed to the *global* matrix (10 × 10) as shown in Figure 3.9.

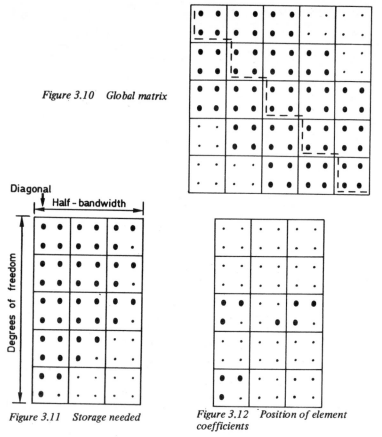

Figure 3.10   Global matrix

Figure 3.11   Storage needed

Figure 3.12   Position of element coefficients

Once all the elements are superimposed we will have a global matrix as in Figure 3.10, where the dots represent zero elements. This matrix is symmetric and banded, the band being equal to 6. Note that the band is proportional to the larger difference between nodes in the same element. We can write only the diagonal and the upper diagonal elements, organizing the storage according to the scheme shown in Figure 3.11 for symmetric banded matrices.

The coefficients of element 7, for instance, are now stored as shown in Figure 3.12.

The half-bandwidth is equal to 3 nodes or, since there are 2 degrees of freedom per node, 6 coefficients.

When we consider the same truss, but numbered according to Figure 3.8(b), we have the connectivity Table 3.4. The total stiffness matrix will, therefore, have the structure shown in Figure 3.13, in which case the half-bandwidth is equal to 5 nodes or 10 coefficients.

Table 3.4

Bar	First node	Second node
1	5	4
2	1	5
3	5	2
4	2	4
5	4	3
6	1	2
7	2	3

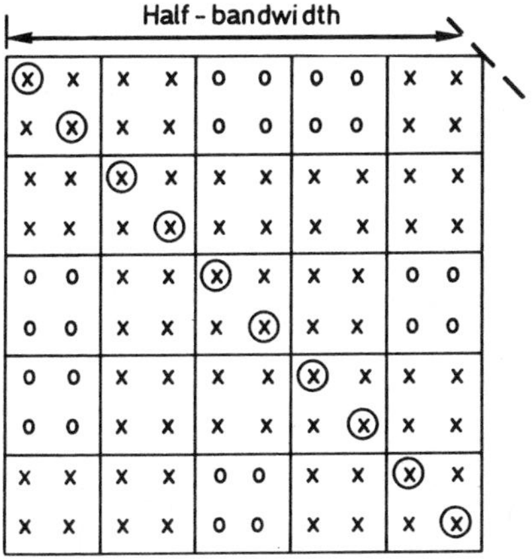

Figure 3.13   Global matrix for structure of Figure 3.8(b)

We see then that the half-bandwidth depends on the nodal numbering. Of all the possible nodal numberings which can be adopted, there will be one, or eventually several, which give a minimum half-bandwidth. The half-bandwidth, counted in nodes, is obtained by adding 1 to the greatest difference in number for the two nodes associated with a bar. Thus for the truss of Figure 3.8(a), the largest difference in node numbers is equal to 2, for bars 1, 6 and 7, and the half-bandwidth will be 3. For the truss of Figure 3.8(b) the largest difference is 4, for bar 1, and the half-bandwidth will be 5. To obtain the half-bandwidth counted in coefficients, we simply multiply by the number of nodal degrees of freedom.

Since the half-bandwidth has a great effect on the efficiency with which the system of equations can be solved, the nodal numbering should be selected such as to minimise the differences in number, of the two nodes connected to each bar.

3.3.4 Introduction of the boundary conditions

Once all the element matrices have been assembled into the global stiffness matrix we have to apply the displacement boundary conditions.

Our essential boundary conditions can be of two types, the first type is when some of the unknowns are zero (homogeneous) and t' e second type when they are equal to given values (non-homogeneous).

To better analyse the process of introducing the boundary conditions, we can reorganize the system of equations as

$$\left\{ \begin{array}{c} \mathbf{P}_f \\ -- \\ \mathbf{P}_s \end{array} \right\} = \left[\begin{array}{c|c} \mathbf{K}_{ff} & \mathbf{K}_{fs} \\ \hline \mathbf{K}_{sf} & \mathbf{K}_{ss} \end{array} \right] \left\{ \begin{array}{c} \mathbf{u}_f \\ -- \\ \mathbf{u}_s \end{array} \right\} \tag{3.36}$$

where $\mathbf{u}_f$ are the unknown displacements, and $\mathbf{u}_s$ are the prescribed displacements at the support or boundary nodes. Correspondingly, $\mathbf{P}_f$ are the known applied loads, and $\mathbf{P}_s$ are the unknown support reactions.

Equation (3.36) can be written as

$$\mathbf{P}_f = \mathbf{K}_{ff}\mathbf{u}_f + \mathbf{K}_{fs}\mathbf{u}_s \tag{3.37a}$$

$$\mathbf{P}_s = \mathbf{K}_{sf}\mathbf{u}_f + \mathbf{K}_{ss}\mathbf{u}_s \tag{3.37b}$$

Since $\mathbf{P}_f$ and $\mathbf{u}_s$ are known, from (3.37a) we obtain the system of equations

$$(\mathbf{P}_f - \mathbf{K}_{fs}\mathbf{u}_s) = \mathbf{K}_{ff}\mathbf{u}_f \tag{3.38}$$

which can be solved for the unknown displacements $\mathbf{u}_f$. Once these are known, we can compute the support reactions from (3.37b).

In practice this process is inefficient since it might require a large reorganization of the total system of equations, which is very time consuming in a computer. Also, the nodal partition organization will be destroyed. To avoid this, normally another process is used, which can best be explained by considering the following system of equations

$$a_{11}x_1 + a_{12}x_2 + a_{13}x_3 = b_1$$

$$a_{21}x_1 + a_{22}x_2 + a_{23}x_3 = b_2$$

$$a_{31}x_1 + a_{32}x_2 + a_{33}x_3 = b_3 \tag{3.39}$$

If the second unknown x_2 has a prescribed value

$$x_2 = \bar{x}_2 \tag{3.40}$$

the system (3.39) becomes

$$a_{11}x_1 + 0 + a_{13}x_3 = b_1 - a_{12}\overline{x}_2$$

$$a_{21}x_1 + 0 + a_{23}x_3 = b_2 - a_{22}\overline{x}_2$$

$$a_{31}x_1 + 0 + a_{33}x_3 = b_3 - a_{32}\overline{x}_2 \qquad (3.41)$$

We now have 3 equations for 2 unknowns. To solve this system we could use the first and third equations, but if the matrix coefficients are already stored in a computer array, this would have to be reorganized. Rather, we prefer to replace the second equation by condition (3.40), obtaining the system

$$a_{11}x_1 + 0 + a_{13}x_3 = b_1 - a_{12}\overline{x}_2$$

$$0 + x_2 + 0 = \overline{x}_2$$

$$a_{31}x_1 + 0 + a_{33}x_3 = b_3 - a_{32}\overline{x}_2 \qquad (3.42)$$

which can be written in matrix from as

$$\begin{bmatrix} a_{11} & 0 & a_{13} \\ 0 & 1 & 0 \\ a_{31} & 0 & a_{33} \end{bmatrix} \begin{Bmatrix} x_1 \\ x_2 \\ x_3 \end{Bmatrix} = \begin{Bmatrix} b_1 - a_{12}\overline{x}_2 \\ \overline{x}_2 \\ b_3 - a_{32}\overline{x}_2 \end{Bmatrix} \qquad (3.43)$$

The solution of this system will provide the values of x_1 and x_3, while x_2 will maintain its prescribed value.

The use of this scheme to introduce the boundary conditions in our structural systems of equations has several advantages. It does not require row and column reorganization, is simple to apply, and can very easily be implemented in a computer program. When the ith coefficient of U is prescribed, we carry out the following operations.

(1) Subtract from the jth coefficient of P the product of the (j, i) coefficient of K times the known ith coefficient of U. Do this for all $j \neq i$.
(2) Zero the ith row and the ith column of K.
(3) Give a unit value to the (i, i) coefficient of K.
(4) Make the ith coefficient of P equal to the prescribed displacement value.

Example 3.4
As an illustration let us consider again the truss in Figure 3.7. The total system

of equations for this truss, before the introduction of the boundary conditions, is given by Equation (3.35). We assume that

$$v_1 = u_3 = v_3 = u_4 = v_4 = 0 \tag{a}$$

and

$$u_1 = \bar{u} \tag{b}$$

where $\bar{u}$ is a known non zero value. The displacement components u_2 and v_2 are unknowns.

Applying the systematic scheme described above Equation (3.35) becomes

$$
\left\{
\begin{array}{c}
\bar{u} \\[4pt]
0 \\[4pt]
\dfrac{3}{4}\dfrac{AE}{L}\bar{u} - 4AE\beta T \\[8pt]
P + \dfrac{\sqrt{3}}{4}\dfrac{AE}{L}\bar{u} + \dfrac{AE\beta T}{\sqrt{3}} \\[8pt]
0 \\[4pt]
0 \\[4pt]
0 \\[4pt]
0
\end{array}
\right\}
= \dfrac{AE}{L}
\left[
\begin{array}{cc|cc|cc|cc}
1 & 0 & 0 & 0 & 0 & 0 & 0 & 0 \\
0 & 1 & 0 & 0 & 0 & 0 & 0 & 0 \\ \hline
 & & \dfrac{13}{4} & -\dfrac{\sqrt{3}}{4} & 0 & 0 & 0 & 0 \\
 & & -\dfrac{\sqrt{3}}{4} & \dfrac{3}{4} & 0 & 0 & 0 & 0 \\ \hline
 & & & & 1 & 0 & 0 & 0 \\
 & & & & 0 & 1 & 0 & 0 \\ \hline
 & & & & & & 1 & 0 \\
 & & & & & & 0 & 1
\end{array}
\right]
\left\{
\begin{array}{c}
u_1 \\ v_1 \\ u_2 \\ v_2 \\ u_3 \\ v_3 \\ u_4 \\ v_4
\end{array}
\right\}
\tag{c}
$$

which can now be solved for u_2 and v_2. Note that in this case we have destroyed the coefficients of **K** which could be used to compute the support reactions, according to Equation (3.37b). Therefore, they will have to be computed by other means.

3.3.5 Solution of a system of equations

The solution of a system of equations can be carried out by any of the methods explained in Chapter 2. A large majority of the structural analysis computer programs implemented use a direct method, which can be Gauss elimination, Choleski's, or some variation of these.

The solution of a system of equations, excluding very simple problems, takes a large proportion of the total analysis time. Thus, it is very important to use an efficient solution scheme. For this, the special characteristics of the stiffness matrix should be taken into account. An efficient program should at least consider the symmetric and banded characteristics. The more sophisticated programs also take advantage of the sparseness of the stiffness matrix.

It should be mentioned that for the analysis of medium and, especially, large size structures, the computer storage organization selected for the stiffness matrix becomes a crucial factor. This is so because K is not normally wholly held in the core storage but is in secondary storage. The time spent in transferring information to and from secondary storage becomes very important. The scheme adopted should attempt to minimise that time.

3.3.6 Evaluation of member forces

Once the nodal displacements are known, we can compute the axial forces for each bar. For this we use Equation (3.18) or (3.24).

Normally, after the axial forces have been obtained, the nodal force resultants in the x and y directions, are computed. For each node we add the x and y projections of the axial forces of the bars connected to it. For support nodes the resultants will be the support reactions, while for free nodes they must be equal to the loads applied to them. These last values provide a useful check on the correctness of the solution.

3.4   A COMPUTER PROGRAM FOR TRUSS ANALYSIS

In this section we present a computer program for the structural analysis of plane trusses, as a means of illustrating the details of the computer implementation of the analysis method previously described.

When implementing a structural analysis program it is necessary to achieve a balance between efficiency and compactness, and clarity and generality.

In our case we put the emphasis on the last two aspects, both to provide a better understanding of the problem, and to have a program which can be easily adapted for other applications.

The program for static analysis of plane trusses is subdivided according to the analysis steps given in Section 3.3, which can be organized in flow-chart form as shown in Figure 3.14.

For each of the steps indicated in the flow-chart there will be a specific subprogram, linked together by a main program.

The geometry, connectivity, properties, boundary conditions, and loading data will be stored in one-dimensional arrays, which will be dimensioned in the main program. The stiffness matrix will be stored in a two-dimensional array, according to the symmetric banded storage scheme given in Figure 2.7(b). The

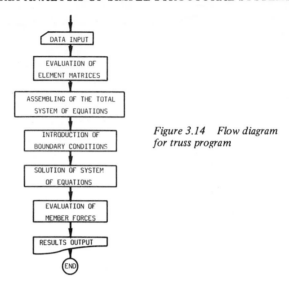

Figure 3.14   Flow diagram for truss program

absolute dimensioning of this array will be done in the main program, while in the subprograms variable dimensioning will be used. In summary, whenever the array dimensions need to be changed only the main program is altered, while the subprograms remain unchanged.

3.4.1 Data structure

The general integer variables used by the program, together with their meaning, are given below.

NN   = total number of nodes

NE   = total number of elements or bars

NLN   = total number of loaded nodes

NBN   = total number of support nodes

NNE   = number of nodes per element (which for bar structures is equal to 2)

NDF   = number of degrees of freedom per node (which for plane trusses is equal to 2)

NDFEL = number of degrees of freedom per element, given by NDF*NNE (which for plane trusses is equal to 4)

NRMX = row dimension of the array TK, to store the total stiffness matrix

NCMX = column dimension of the array TK

N   = total number of unknowns for a problem, given by NDF*NN. It is also equal to the actual number of rows in the total stiffness matrix

MS   = actual half-bandwidth of the total stiffness matrix

The only real variable used is

E = modulus of elasticity

The program uses several integer and real arrays to store both data and results. In most cases these are one dimensional arrays so that, excluding the main program, all programs can be made independent of array dimensioning.

The integer arrays used are

CON: One dimensional array of element connectivities.
        CON(NNE*(L − 1) + 1) contains the first node and
        CON(NNE*(L − 1) + 2) the second node of element L.

IB: One dimensional array of boundary condition data.
        IB((NDF + 1) * (J − 1) + 1) contains the node number,
        IB((NDF + 1) * (J − 1) + 2) the status for the x displacement
        component, and IB(NDF + 1) * (J − 1) + 3) the status for the y
        displacement component, for the Jth support node.

The real arrays used are

X: One dimensional array of x coordinates. X(J) contains the x co-
        ordinate of node J.

Y: One dimensional array of y coordinates. Y(J) contains the y co-
        ordinate of node J.

PROP: One dimensional array of bar cross-sectional areas. PROP(L) con-
        tains the cross-sectional area of element L.

AL: One dimensional array of nodal loads. AL(NDF*(J − 1) + 1) con-
        tains the x component and AL(NDF*(J − 1) + 2) the y component
        of the load applied to node J. It will have zero values for nodes
        with no loads. After solution of the system of equations, AL is
        used to store the nodal displacements, such that
        AL(NDF*(J − 1) + 1) contains the x component and
        AL(NDF*(J − 1) + 2) the y component of the displacement of
        node J.

TK: Bidimensional array containing the total stiffness matrix, accord-
        ing to the symmetric banded storage scheme explained in Chapter
        2. NRMX is its row dimension and NCMX is its column dimension.

ELST: Bidimensional array storing the element stiffness matrix for the
        element being processed. Its row and column dimensions are
        equal to NDFEL.

V: One dimensional auxiliary array used by the program performing
        the solution of the system of equations. Its dimension is equal to
        NCMX.

FORC: One dimensional array of bar axial forces. FORC(L) contains the
        axial force of element L.

REAC: One dimensional array initially containing prescribed nodal un-
        knowns. REAC(NDF*(J − 1) + 1) contains the x component
        and REAC(NDF*(J − 1) + 2) the y component of the prescribed

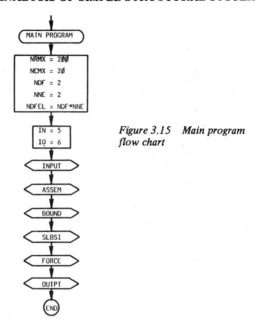

Figure 3.15　Main program flow chart

displacement for node J. It will contain zero values for nodes without prescribed displacements. After the bar axial forces are computed it is used to store the nodal resultants, obtained from the summation of the global projections of the axial forces of the bars connected to each node. REAC(NDF*(J − 1) + 1) contains the x component and REAC(NDF*(J − 1) + 2) the y component of the nodal resultant for node J. For support nodes these are the support reactions, while for free nodes they must equal the applied nodal loads.

Note that within some of the subprograms some auxiliary arrays may be used. Their meanings will be evident analyzing the flow of operations of those subprograms. In most cases they are also described in the subprogram commentaries.

Program 26　Main program

The main program will initialize the basic program parameters, and then will call on a set of 6 different subprograms, each performing the operations relative to the analysis steps described before. The sequence of operations is shown in Figure 3.15, corresponding to the following FORTRAN code.

```
C
C
C
C                   PROGRAM 26 - MAIN PROGRAM
C      STATIC ANALYSIS OF PLANE TRUSSES
C
       COMMON NRMX,NCMX,NDFEL,NN,NE,NLN,NBN,NDF,NNE,N,MS,IN,IO,E,G
       DIMENSION X(100),Y(100),CON(200),PROP(100),IB(60),TK(200,20),
      *AL(200),FORC(100),REAC(200),ELST(4,4),V(20)
C
C      INITIALIZATION OF PROGRAM PARAMETERS
C
C      NRMX  = ROW DIMENSION FOR THE STIFFNESS
C              MATRIX
C      NCMX  = COLUMN DIMENSION FOR THE STIFFNESS
C              MATRIX, OR MAXIMUM HALF BAND WIDTH
C              ALLOWED
C      NDF   = NUMBER OF DEGREES OF FREEDOM PER
C              NODE
C      NNE   = NUMBER OF NODES PER ELEMENT, EQUAL
C              TO 2 FOR BAR SYSTEMS
C      NDFEL = TOTAL NUMBER OF DEGREES OF FREEDOM
C              FOR ONE ELEMENT
C
       NRMX=200
       NCMX=20
       NDF=2
       NNE=2
       NDFEL=NDF*NNE
C
C      ASSIGN DATA SET NUMBERS TO IN, FOR INPUT,
C      AND IO, FOR OUTPUT
C
       IN=5
       IO=6
C
C      APPLY THE ANALYSIS STEPS
C
C      INPUT
C
       CALL INPUT(X,Y,CON,PROP,AL,IB,REAC)
C
C      ASSEMBLING OF TOTAL STIFFNESS MATRIX
C
       CALL ASSEM(X,Y,CON,PROP,TK,ELST,AL)
C
C      INTRODUCTION OF BOUNDARY CONDITIONS
C
       CALL BOUND(TK,AL,REAC,IB)
C
C      SOLUTION OF THE SYSTEM OF EQUATIONS
C
       CALL SLBSI(TK,AL,V,N,MS,NRMX,NCMX)
C
C      COMPUTATION OF MEMBER FORCES
C
       CALL FORCE(CON,PROP,FORC,REAC,X,Y,AL)
C
C      OUTPUT
C
       CALL OUTPT(AL,FORC,REAC)
       CALL EXIT
       END
```

The dimensions declared in this main program allow a maximum of 100 nodes, 100 elements or bars, and 20 support nodes. Accordingly the integer variable NRMX, indicating the maximum number of rows in the total stiffness matrix is set equal to 200 (total number of degrees of freedom, or unknowns, for a plane truss of 100 nodes). The maximum half-bandwidth, NCMX, is set equal to 20. All these values can be changed according to the memory capacity of the computer available.

The program also initializes the nodal degrees of freedom equal to 2, in NDF; the number of nodes per element equal to 2, in NNE; and the number of degrees of freedom per element equal to NDF*NNE, in NDFEL. The integer variables IN and IO, which are the data set numbers for the input and output devices are set equal to 5 and 6, respectively. Note that these numbers should be set according to the conventions of the computer available.

The analysis steps are applied calling the subprograms INPUT, for data input; ASSEM, to compute the element stiffness matrices and to assemble the total system of equations; BOUND, to introduce the boundary conditions; SLBSI to solve the system of equations (this program is the same given in Chapter 2); FORCE, to compute the bar axial forces and nodal resultants, and OUTPT, for result output. Note, in particular, that for efficiency reasons the subprogram ASSEM performs two of the analysis steps: the evaluation of the element matrices and the assembling of the total system of equations. This eliminates the eventual need to have an array to store all the element matrices, as required in the case in which both steps are carried out separately.

Finally, it should be mentioned that all basic integer and real variables are placed in COMMON. The COMMON variable G, not used by this program, was included in provision for its extension to other applications, where a shear modulus, a Poisson coefficient, or another such variable, might be needed.

Program 27   Data input (INPUT)

All the input required by the program for static analysis of plane trusses will be read by the subprogram INPUT. The input of data will consist of the following group of cards;

(1) *Basic parameters card.* One card containing the number of nodes, number of elements, number of loaded nodes, number of support nodes, and Young's modulus, with format 4I10, F10.0.

(2) *Nodal coordinate cards.* As many cards as there are nodes in the truss. Each card will contain the node number, and its x and y coordinates, with format I10, 2F10.2.

(3) *Element connectivity and properties cards.* As many cards as bars there are in the structure. Each card will contain the bar number, the number of its first and second node, and its cross-sectional area, with format 3I10, F10.5.

(4) *Nodal loads cards.* As many cards as there are loaded nodes in the

structure. Each card will contain the node number, and the load
components in the x and y directions, with format I10, 2F10.2.

(5)   *Support data cards.* As many cards as there are support nodes in the
structure. Each card will contain the node number, and indicator of
the support conditions for the displacement component in the x and
y directions, and the values of the prescribed displacement components, with format 3I10, 2F10.4. The indicator for a displacement
component will be given equal to 1 when unknown or equal to 0 when
prescribed. For unknown displacement components, the columns
assigned to the prescribed values are left blank.

At the end of this chapter there is an example of the input data for a practical case (Example 3.5).

The FORTRAN code for the subprogram INPUT is as follows:

```
      SUBROUTINE INPUT(X,Y,CON,PROP,AL,IB,REAC)
C
C          PROGRAM 27 - INPUT PROGRAM
C
      COMMON NRMX,NCMX,NDFEL,NN,NE,NLN,NBN,NDF,NNE,N,MS,IN,IO,E,G
      DIMENSION X(1),Y(1),CON(1),PROP(1),AL(1),IB(1),REAC(1),W(3),IC(2)
C
C   W = AN AUXILIARY VECTOR TO TEMPORARELY STORE A SET OF
C          NODAL LOADS AND PRESCRIBED UNKNOWN VALUES
C   IC = AUXILIARY ARRAY TO STORE TEMPORARELY THE CONNECTIVITY
C          OF AN ELEMENT, AND THE BOUNDARY UNKNOWNS STATUS
C          INDICATORS
C
C   READ BASIC PARAMETERS
C
C   NN  = NUMBER OF NODES
C   NE  = NUMBER OF ELEMENTS
C   NLN = NUMBER OF LOADED NODES
C   NBN = NUMBER OF BOUNDARY NODES
C   E   = MODULUS OF ELASTICITY
C
      WRITE(IO,20)
   20 FORMAT(' ',130('*'))
      READ(IN,1) NN,NE,NLN,NBN,E
      WRITE(IO,21) NN,NE,NLN,NBN,E
   21 FORMAT(//' INTERNAL DATA'//' NUMBER OF NODES          :',I5/' NUMBE
     *R OF ELEMENTS       :',I5/' NUMBER OF LOADED NODES  :',I5/' NUMBER
     *OF SUPPORT NODES :',I5/' MODULUS OF ELASTICITY :',F15.0//' NODAL C
     *OORDINATES'/7X,'NODE',6X,'X', 9X,'Y')
    1 FORMAT(4I10,F10.0)
C
C   READ NODAL COORDINATES IN ARRAY X AND Y
C
      READ(IN,2) (I,X(I),Y(I),J=1,NN)
      WRITE(IO,2) (I,X(I),Y(I),I=1,NN)
    2 FORMAT(I10,2F10.2)
C
C   READ ELEMENT CONNECTIVITY IN ARRAY CON
C   AND ELEMENT PROPERTIES IN ARRAY PROP
C
```

```
      WRITE(IO,22)
   22 FORMAT(/' ELEMENT CONNECTIVITY AND PROPERTIES'/4X,'ELEMENT',3X,
     *ART NODE  END NODE',5X,'AREA')
      DO 3 J=1,NE
      READ(IN,4) I,IC(1),IC(2),PROP(I)
      WRITE(IO,34) I,IC(1),IC(2),PROP(I)
      N1=NNE*(I-1)
      CON(N1+1)=IC(1)
    3 CON(N1+2)=IC(2)
    4 FORMAT(3I10,F10.5)
   34 FORMAT(3I10,F15.5)
C
C  COMPUTE N, ACTUAL NUMBER OF UNKNOWNS, AND CLEAR THE LOAD
C  VECTOR
C
      N=NN*NDF
      DO 5 I=1,N
    5 AL(I)=0.
C
C  READ THE NODAL LOADS AND STORE THEM IN ARRAY AL
C
      WRITE(IO,23)
   23 FORMAT(/' NODAL LOADS'/7X,'NODE',5X,'PX',8X,'PY')
      DO 6 I=1,NLN
      READ (IN,2) J,(W(K),K=1,NDF)
      WRITE(IO,2) J,(W(K),K=1,NDF)
      DO 6 K=1,NDF
      L=NDF*(J-1)+K
    6 AL(L)=W(K)
C
C  READ BOUNDARY NODES DATA. STORE UNKNOWNS STATUS
C  INDICATORS IN ARRAY IB, AND PRESCRIBED UNKNOWNS
C  VALUES IN ARRAY REAC
C
      WRITE(IO,24)
   24 FORMAT(/' BOUNDARY CONDITION DATA'/23X,'STATUS',14X,'PRESCRIBED
     *LUES'/15X,'(0:PRESCRIBED, 1:FREE)'/7X,'NODE',8X,'U',9X,'V',16X,
     *,9X,'V')
      DO 7 I=1,NBN
      READ(IN,8) J,(IC(K),K=1,NDF),(W(K),K=1,NDF)
      WRITE(IO,9) J,(IC(K),K=1,NDF),(W(K),K=1,NDF)
      L1=(NDF+1)*(I-1)+1
      L2=NDF*(J-1)
      IB(L1)=J
      DO 7 K=1,NDF
      N1=L1+K
      N2=L2+K
      IB(N1)=IC(K)
    7 REAC(N2)=W(K)
    8 FORMAT(3I10,2F10.4)
    9 FORMAT(3I10,10X,2F10.4)
      RETURN
      END
```

This subprogram can be easily understood simply by reading it. Therefore it does not require a detailed explanation. All data read is also printed, for checking purposes. The output formats are not identical to the input formats, but give a clearer and more detailed picture of the input data.

Before reading the nodal loads the total number of system unknowns (N)

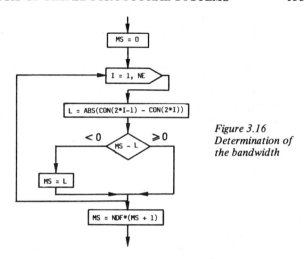

Figure 3.16
Determination of
the bandwidth

is computed. Then the vector AL is cleared. This is done because in some computers unused memory positions cannot be assumed to be zero. Otherwise that DO can be eliminated. The nodal loads are read and placed into the array AL. Thus after the reading of the nodal loads is completed the vector of independent coefficients, of the total system of equations, is already formed.

Program 28   Assembling of the total stiffness matrix (ASSEM)

The subprogram ASSEM is the main program to assemble the total stiffness matrix. It first computes the half-bandwidth, then clears the array TK and, finally, looping on the elements calls the subprogram STIFF, to compute the element stiffness matrix, and ELASS, to add it to the array TK where the total stiffness matrix must be stored.

In the case of bar structures the computation of the half-bandwidth is straightforward. This can be done by checking all bars and registering the maximum difference existing between the node numbers of the two nodes connected to each bar, then adding 1 to that number, and multiplying by the number of nodal degrees freedom. These operations can be organized according to the flow chart in Figure 3.16, where the half-bandwidth is stored in MS.

If, however, we consider the possibility that an element can have more than 2 nodes, as will be the case for finite element applications, the computation of the half-bandwidth becomes a little more involved. In particular, for each element, rather than considering the difference between just two nodes, each node will have to be compared with all remaining ones to find the maximum nodal number difference, as shown by the flow chart in Figure 3.17. In order for the subprogram ASSEM to be as general as possible the second alternative is adopted.

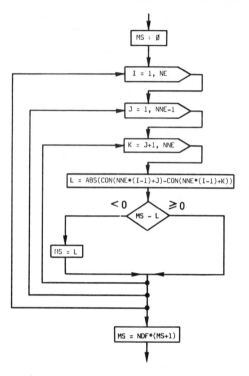

Figure 3.17
General bandwidth
flow-chart

The FORTRAN code for the subprogram ASSEM is as follows:

```
      SUBROUTINE ASSEM(X,Y,CON,PROP,TK,ELST,AL)
C
C           PROGRAM 28
C  ASSEMBLING OF THE TOTAL MATRIX FOR THE PROBLEM
C
      COMMON NRMX,NCMX,NDFEL,NN,NE,NLN,NBN,NDF,NNE,N,MS,IN,IO,E,G
      DIMENSION X(1),Y(1),CON(1),TK(NRMX,NCMX),ELST(NDFEL,NDFEL),
     *PROP(1),AL(1)
C
C  COMPUTE HALF BAND WIDTH AND STORE IN MS
C
      N1=NNE-1
      MS=0
      DO 7 I=1,NE
      L1=NNE*(I-1)
      DO 7 J=1,N1
      L2=L1+J
      J1=J+1
      DO 7 K=J1,NNE
      L3=L1+K
      L=ABS(CON(L2)-CON(L3))
      IF(MS-L)6,7,7
    6 MS=L
```

```
    7 CONTINUE
      MS=NDF*(MS+1)
C
C   CLEAR THE TOTAL STIFFNESS MATRIX
C
      DO 10 I=1,N
      DO 10 J=1,MS
   10 TK(I,J)=0
C
C   LOOP ON THE ELEMENTS AND ASSEMBLE THE TOTAL STIFFNESS MATRIX
C
      DO 20 NEL=1,NE
C
C   STIFF COMPUTES THE STIFFNESS MATRIX FOR ELEMENT NEL
C
      CALL STIFF(NEL,X,Y,PROP,CON,ELST,AL)
C
C   ELASS PLACES THE STIFFNESS MATRIX OF ELEMENT NEL IN THE TOTAL
C   STIFFNESS MATRIX
C
   20 CALL ELASS(NEL,CON,TK,ELST)
      RETURN
      END
```

After computing the half-bandwidth according to the flow chart in Figure 3.17, the array TK which will contain the total stiffness matrix is cleared. Then, looping on the elements, the element stiffness matrix is computed in array ELST by the subprogram STIFF, and added to the array TK by subprogram ELASS.* The subprograms STIFF and ELASS are explained below.

Program 29   Computation of the plane truss bar stiffness matrix (STIFF)

The element stiffness matrix for the plane truss bar is computed by the subprogram STIFF, according to formulas (3.20) and (3.21). The FORTRAN code for this subprogram is as follows:

```
      SUBROUTINE STIFF(NEL,X,Y,PROP,CON,ELST,AL)
C
C                   PROGRAM 29
C   COMPUTATION OF ELEMENT STIFFNESS MATRIX FOR CURRENT ELEMENT
C
      COMMON NRMX,NCMX,NDFEL,NN,NE,NLN,NBN,NDF,NNE,N,MS,IN,IO,E,G
      DIMENSION X(1),Y(1),CON(1),PROP(1),ELST(NDFEL,NDFEL),AL(1)
C
C   NEL  = CURRENT ELEMENT NUMBER
C   N1   = NUMBER OF START NODE
C   N2   = NUMBER OF END NODE
C
      L=NNE*(NEL-1)
      N1=CON(L+1)
      N2=CON(L+2)
```

* Note that the parameter list includes the load vector array AL, which is not needed in this case. It was included to facilitate a future modification of this program, to consider external actions applied to the element or bar level, such as distributed loads, temperature effects, etc. (see Chapter 6).

```
C
C   COMPUTE LENGTH OF ELEMENT,AND SIN AND COSINE OF ITS LOCAL
C   X AXIS, AND STORE IN D,SI, AND CO RESPECTIVELY.
C
      D=SQRT((X(N2)-X(N1))**2+(Y(N2)-Y(N1))**2)
      CO=(X(N2)-X(N1))/D
      SI=(Y(N2)-Y(N1))/D
C
C   COMPUTE ELEMENT STIFFNESS MATRIX
C
      COEF=E*PROP(NEL)/D
      ELST(1,1)=COEF*CO*CO
      ELST(1,2)=COEF*CO*SI
      ELST(2,2)=COEF*SI*SI
      DO 10 I=1,2
      DO 10 J=I,2
      K1=I+NDF
      K2=J+NDF
      ELST(K1,K2)=ELST(I,J)
   10 ELST(I,K2)=-ELST(I,J)
      ELST(2,3)=-ELST(1,2)
      RETURN
      END
```

This subprogram computes the element stiffness matrix given by

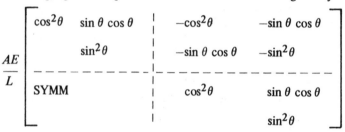

$$\frac{AE}{L} \begin{bmatrix} \cos^2\theta & \sin\theta\cos\theta & -\cos^2\theta & -\sin\theta\cos\theta \\ & \sin^2\theta & -\sin\theta\cos\theta & -\sin^2\theta \\ \hline \text{SYMM} & & \cos^2\theta & \sin\theta\cos\theta \\ & & & \sin^2\theta \end{bmatrix}$$

for the current element whose number is indicated by the integer variable NEL (which is passed through the argument list). Because of symmetry only the upper triangular part of the element stiffness matrix needs to be computed.

The subprogram first sets the integre variables N1 and N2 equal to the start and end node numbers of element NEL. The length of the bar and the cosine and sine of the angle β (see Figure 3.6) are computed and stored in the variables D, CO and SI, respectively. The stiffness coefficient, equal to the Young's modulus times the cross-sectional area divided by the length of the bar, is stored in the variable COEF. Finally the upper triangular part of the element stiffness matrix is computed and stored in array ELST. The array AL was included in the argument list for the same reasons given when explaining the previous program.

Program 30   Addition of an element stiffness matrix to the total stiffness matrix (ELASS)

The subprogram ELASS takes the element matrix for the current element

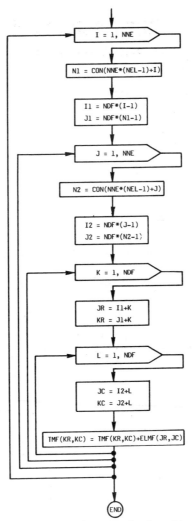

Figure 3.18   Flow-chart for Assembler

NEL, stored in array ELMAT, and adds it to the total matrix in array TM. Both the arrays ELMAT and TM are passed through the argument list. These matrices may be stiffness matrices, mass matrices, or any other type of similar matrices. In the present case, and according to the calling sequence for ELASS in subprogram ASSEM, the array ELMAT is equivalent to array ELST, and the array TM is equivalent to array TK.

To better explain the operations to be carried out by the subprogram

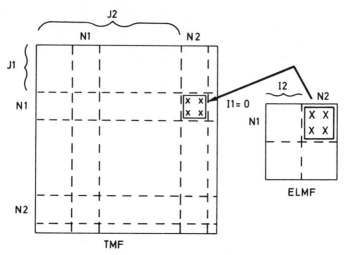

Figure 3.19   Assembling the element coefficients

ELASS let us first assume that the total matrix would be stored in a full square array, of name TMF, not according to the symmetric banded storage scheme but as conventionally in matrix notation. We also assume that the element matrix is stored in full in array ELMF. In such case, the operations required for an element of NNE nodes, eventually greater than 2, are indicated in the flow chart (Figure 3.18), prepared according to the scheme presented in Section 3.3.3.

This operation requires to add the nodal submatrices of ELMF into the corresponding hyper-rows and hyper-columns of TMF, as suggested by Figure 3.19, for an element of 2 nodes.

The integer variable N1 assumes the number of different element nodes, the same applying for the integer variable N2. Each combination of N1 and N2 identifies a nodal submatrix in ELMF. The integer variables J1 and J2 are the number of rows and columns, respectively, before the first coefficient of the nodal submatrix corresponding to nodes N1 and N2, in the array TMF. The integer variables I1 and I2 are similar, but for the matrix ELMF. Then for each coefficient of ELMF, of row and column positions JR and JC it is possible to compute the corresponding row and column positions KR and KC in TMF, allowing adding up all coefficients of ELMF into the proper positions of TMF.

As explained before, however, it is more efficient to take advantage of the symmetry and banded characteristics of the total matrix. Using a symmetric banded storage scheme, only the upper triangular part of the total matrix, and up to the half-bandwidth, is stored in array TM. Consistently, only the upper triangular part of the element matrix is stored in array ELMAT.

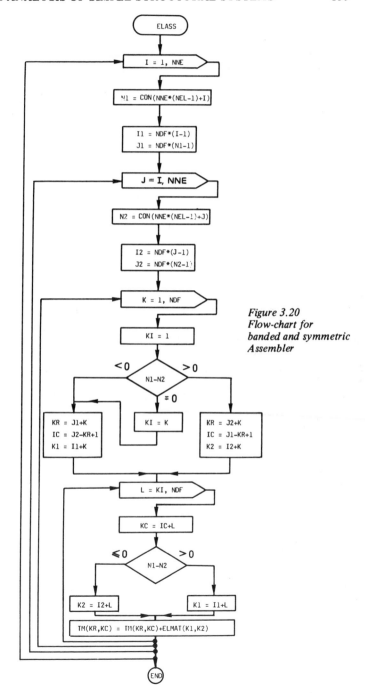

*Figure 3.20
Flow-chart for
banded and symmetric
Assembler*

Based on these considerations the operations to be carried out by subprogram ELASS can be organized as shown by the flow-chart given in Figure 3.20.

First, we notice that the DO loop on J starts in I, rather than in 1. This is to avoid considering both the off-diagonal submatrices representing the coupling between nodes N1 and N2, and between N2 and N1. We know that these submatrices are each the transpose of the other. On the other hand, since we are registering in TK only the upper triangular part of the total stiffness matrix only one of those submatrices is needed. In particular, if N1 $<$ N2 we need $K_{N1,N2}^{NEL}$, while if N1 $>$ N2 we need $K_{N2,N1}^{NEL}$. Correspondingly, before adding a coefficient into TK, N1 and N2 are compared. If N1 $<$ N2 the coefficients of $K_{N1,N2}^{NEL}$ are directly added to TK, properly computing the row and column subscripts for the symmetric banded storage scheme. When N1 = N2 we can only store in TK the upper triangular part of $K_{N1,N1}^{NEL}$. Correspondingly the DO loop on L starts with KI = K, rather than with KI = 1. Finally, when N1 $>$ N2 we interchange the corresponding row and column subscripts, so as to register in TK the transpose of $K_{N1,N2}^{NEL}$.

Based on the flow chart in Figure 3.20 the FORTRAN code for the subprogram ELASS is the following:

```
      SUBROUTINE ELASS(NEL,CON,TM,ELMAT)
C
C              PROGRAM 30
C
C   THIS PROGRAM STORES THE ELEMENT MATRIX FOR ELEMENT NEL IN
C   THE TOTAL MATRIX FOR THE PROBLEM
C
      COMMON NRMX,NCMX,NDFEL,NN,NE,NLN,NBN,NDF,NNE,N,MS,IN,IO,E,G
      DIMENSION CON(1),TM(NRMX,NCMX),ELMAT(NDFEL,NDFEL)
C
C   NEL  = CURRENT ELEMENT NUMBER
C   N1   = NUMBER OF START NODE
C   N2   = NUMBER OF END NODE
C
      L1=NNE*(NEL-1)
      DO  50  I=1,NNE
      L2=L1+I
      N1=CON(L2)
      I1=NDF*(I-1)
      J1=NDF*(N1-1)
      DO  50  J=I,NNE
      L2=L1+J
      N2=CON(L2)
      I2=NDF*(J-1)
      J2=NDF*(N2-1)
      DO 50 K=1,NDF
      KI=1
      IF(N1-N2)20,10,30
C
C   STORE A DIAGONAL SUBMATRIX
C
   10 KI=K
C
C   STORE AN OFF DIAGONAL SUBMATRIX
C
```

```
   20 KR=J1+K
      IC=J2-KR+1
      K1=I1+K
      GO TO 40
C
C  STORE THE TRANSPOSE OF AN OFF DIAGONAL MATRIX
C
   30 KR=J2+K
      IC=J1-KR+1
      K2=I2+K
   40 DO 50 L=KI,NDF
      KC=IC+L
      IF(N1-N2)45,45,46
   45 K2=I2+L
      GO TO 50
   46 K1=I1+L
   50 TM(KR,KC)=TM(KR,KC)+ELMAT(K1,K2)
      RETURN
      END
```

Program 31   Introduction of boundary conditions (BOUND)

The problem boundary conditions will be introduced according to the scheme presented in Section 3.3.4. As in the previous case let us consider, for simplicity, the case in which the nodal stiffness matrix is fully stored in a square array TMF. The corresponding operations can be organized in flow chart form as in Figure 3.21.

The number of the support node being processed is stored in the integer NO. The integer K1 contains the number of rows in TK before the first row for node NO. When the indicator of the status of the Ith displacements component of node NO is equal to zero, in which case it is prescribed, the vector AL is modified to take into account the prescribed component, and the corresponding rows and columns of TK are zeroed. Then, the diagonal coefficient is made equal to 1, and the prescribed component is stored in $AL(K1 + I)$.

In the case of storing the full matrix all the coefficients of the corresponding row and column are made zero, as indicated in Figure 3.22(a)., and the introduction of the boundary conditions is straightforward. If we store only the upper triangular part of the symmetric banded stiffness matrix, we have the case of Figure 3.22(b) where we have to zero the part of the column above the diagonal, and the part of the row to the right of the diagonal. When we use the symmetric banded storage the column coefficients are diagonally stored, as shown in Figure 3.22(c) and that is the case for the program being implemented. Then, when zeroing the row it must be checked that the column subscript does not exceed the total number of equations, as it will happen when the row is one of the last rows of the matrix. Similarly, when zeroing the column it must be checked that the row subscript does not become zero, as it will happen for one of the first columns of the matrix.

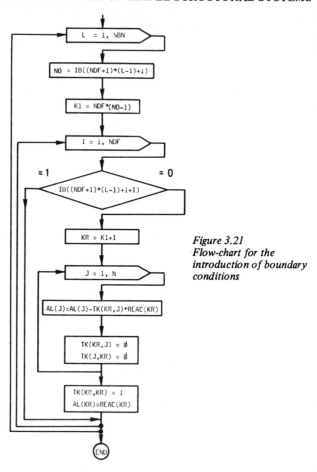

Figure 3.21
Flow-chart for the introduction of boundary conditions

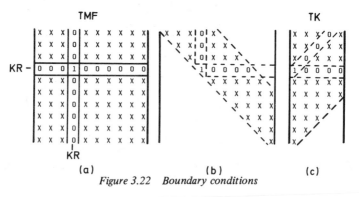

Figure 3.22   Boundary conditions

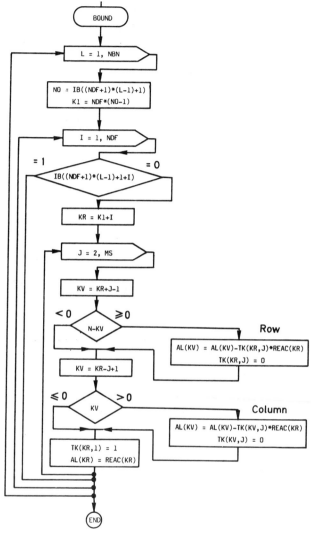

Figure 3.23   Flow-chart for the introduction of boundary conditions (banded and symmetric matrix)

Taking into account these considerations, the introduction of the boundary conditions can be carried out according to the flow chart in Figure 3.23.

The FORTRAN code for the subprogram BOUND, programmed according to the last flow chart, is the following:

```
      SUBROUTINE BOUND(TK,AL,REAC,IB)
C
C              PROGRAM 31
C
C  INTRODUCTION OF THE BOUNDARY CONDITIONS
C
      COMMON NRMX,NCMX,NDFEL,NN,NE,NLN,NBN,NDF,NNE,N,MS,IN,IO,E,G
      DIMENSION AL(1),IB(1),REAC(1),TK(NRMX,NCMX)
      DO 100 L=1,NBN
C
C  NO =  NUMBER OF THE CURRENT BOUNDARY NODE
C
      L1=(NDF+1)*(L-1)+1
      NO=IB(L1)
      K1=NDF*(NO-1)
      DO 100 I=1,NDF
      L2=L1+I
      IF(IB(L2))100,10,100
C
C  PRESCRIBED UNKNOWN TO BE CONSIDERED
C
C  SET DIAGONAL COEFFICIENT OF TK EQUAL TO 1
C  PLACE PRESCRIBED UNKNOWN VALUE IN AL
C
   10 KR=K1+I
      DO 50 J=2,MS
      KV=KR+J-1
      IF(N-KV)30,20,20
C
C  MODIFY ROW OF TK AND CORRESPONDING ELEMENTS IN AL
C
   20 AL(KV)=AL(KV)-TK(KR,J)*REAC(KR)
      TK(KR,J)=0
   30 KV=KR-J+1
      IF(KV)50,50,40
C
C  MODIFY COLUMN IN TK AND CORRESPONDING ELEMENT  IN AL
C
   40 AL(KV)=AL(KV)-TK(KV,J)*REAC(KR)
      TK(KV,J)=0
   50 CONTINUE
      TK(KR,1)=1
      AL(KR)=REAC(KR)
  100 CONTINUE
      RETURN
      END
```

Program 32   Evaluation of bar axial forces (FORCE)

The bar axial forces can be easily computed by multiplying the stiffness coefficient AE/L times the element deformation e_i, given by formula (3.6). Once the bar axial forces are available, by adding up, for each node, the global projections of the axial forces pertaining to all bars connected to that node, the nodal resultants are computed. These operations are performed by the subprogram FORCE, having the following FORTRAN code:

```
      SUBROUTINE FORCE(CON,PROP,FORC,REAC,X,Y,AL)
C
C               PROGRAM 32
C  COMPUTATION OF ELEMENT FORCES
C
      COMMON NRMX,NCMX,NDFEL,NN,NE,NLN,NBN,NDF,NNE,N,MS,IN,IO,E,G
      DIMENSION CON(1),PROP(1),FORC(1),REAC(1),X(1),Y(1),AL(1)
C
C CLEAR THE REACTIONS ARRAY
C
      DO 1 I=1,N
    1 REAC(I)=0
C
C  NEL= = NUMBER OF CURRENT ELEMENT
C  N1   = NUMBER OF START NODE FOR CURRENT ELEMENT
C  N2   = NUMBER OF END NODE FOR CURRENT ELEMENT
C
      DO 100 NEL=1,NE
      L=NNE*(NEL-1)
      N1=CON(L+1)
      N2=CON(L+2)
      K1=NDF*(N1-1)
      K2=NDF*(N2-1)
C
C  COMPUTE LENGTH OF ELEMENT,AND SIN AND COSINE OF ITS LOCAL
C  X AXIS, AND STORE IN D,SI, AND CO RESPECTIVELY.
C
      D=SQRT((X(N2)-X(N1))**2+(Y(N2)-Y(N1))**2)
      CO=(X(N2)-X(N1))/D
      SI=(Y(N2)-Y(N1))/D
      COEF=E*PROP(NEL)/D
C
C  COMPUTE MEMBER AXIAL FORCE AND STORE IN ARRAY FORC
C
      FORC(NEL)=COEF*((AL(K2+1)-AL(K1+1))*CO+(AL(K2+2)-AL(K1+2))*SI)
C
C  COMPUTE GLOBAL PROJECTIONS OF MEMBER AXIAL FORCE AND
C  ADD TO THE REACTION ARRAY REAC
C
      REAC(K1+1)=REAC(K1+1)-FORC(NEL)*CO
      REAC(K1+2)=REAC(K1+2)-FORC(NEL)*SI
      REAC(K2+1)=REAC(K2+1)+FORC(NEL)*CO
  100 REAC(K2+2)=REAC(K2+2)+FORC(NEL)*SI
      RETURN
      END
```

First the array REAC, to store the nodal resultants, is cleared. The integer N1 and N2 are set equal to the numbers of the start and end node of element NEL. The integers K1 and K2 are indexes to retrieve the nodal displacements from AL. D is the bar length, and SI and CO are the sine and cosine, respectively, of the angle formed by the bar axis with the global x-axis. COEF is the coefficient $k_{NEL} = AE/L$. The axial force is computed and stored in

FORC(NEL). Finally, the projections of the bar axial force on the x and y directions are added, with the proper sign, to the array REAC.

Program 33   Result output (OUTPT)

The FORTRAN code of the subprogram OUTPT, to print the nodal displacements, nodal reactions, and bar axial forces, is the following:

```
       SUBROUTINE OUTPT(AL,FORC,REAC)
C
C             PROGRAM 33 - OUTPUT PROGRAM
C
       COMMON NRMX,NCMX,NDFEL,NN,NE,NLN,NBN,NDF,NNE,N,MS,IN,IO,E,G
       DIMENSION AL(1),REAC(1),FORC(1)
C
C   WRITE NODAL DISPLACEMENTS
C
       WRITE(IO,1)
     1 FORMAT(//1X,130('*')//' RESULTS'//' NODAL DISPLACEMENTS'/7X,'NOD
      *,11X,'U',14X,'V')
       DO 10 I=1,NN
       K1=NDF*(I-1)+1
       K2=K1+NDF-1
    10 WRITE(IO,2) I,(AL(J),J=K1,K2)
     2 FORMAT(I10,6F15.4)
C
C   WRITE NODAL REACTIONS
C
       WRITE(IO,3)
     3 FORMAT(/' NODAL REACTIONS'/7X,'NODE',10X,'PX',13X,'PY')
       DO 20 I=1,NN
       K1=NDF*(I-1)+1
       K2=K1+NDF-1
    20 WRITE(IO,2) I,(REAC(J),J=K1,K2)
C
C   WRITE MEMBER AXIAL FORCES
C
       WRITE(IO,4)
     4 FORMAT(/' MEMBER FORCES'/6X,'MEMBER      AXIAL FORCE')
       DO 30 I=1,NE
    30 WRITE(IO,2) I,FORC(I)
       WRITE(IO,5)
     5 FORMAT(//1X,130('*'))
       RETURN
       END
```

Example 3.5

As an illustration of the application of the program for static analysis of plane trusses, we will consider the truss shown by Figure 3.24, which indicates the overall dimensions, the node and bar numbering, and the applied loads. The

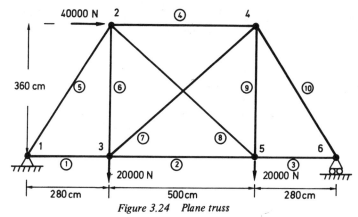

Figure 3.24   Plane truss

bar areas are

$$A_1 = A_2 = A_3 = 25 \text{ cm}^2$$

$$A_4 = A_5 = A_{10} = 30 \text{ cm}^2$$

$$A_6 = A_7 = A_8 = A_9 = 10 \text{ cm}^2$$

Young's modulus is taken equal to 21 000 000 N/cm^2.
    The input cards for this problem are the following:

```
****** INPUT FOR PLANE TRUSS ANALYSIS PROGRAM

    6           10          3        2 21000000.
    1
    2   280.        360.
    3   280.
    4   780.        360.
    5   780.
    6  1060.
    1           1           3   25.
    2           3           5   25.
    3           5           6   25.
    4           2           4   30.
    5           1           2   30.
    6           3           2   10.
    7           3           4   10.
    8           5           2   10.
    9           5           4   10.
   10           4           6   30.
    2   40000.
    3               -20000.
    5               -20000.
    1
    6           1
```

giving the following printout:

****
INTERNAL DATA

```
NUMBER OF NODES          :        6
NUMBER OF ELEMENTS       :       10
NUMBER OF LOADED NODES   :        3
NUMBER OF SUPPORT NODES  :        2
MODULUS OF ELASTICITY    :        21000000.
```

NODAL COORDINATES

NODE	X	Y
1	0.00	0.00
2	280.00	360.00
3	280.00	0.00
4	780.00	360.00
5	780.00	0.00
6	1060.00	0.00

ELEMENT CONNECTIVITY AND PROPERTIES

ELEMENT	START NODE	END NODE	AREA
1	1	3	25.00000
2	3	5	25.00000
3	5	6	25.00000
4	2	4	30.00000
5	1	2	30.00000
6	3	2	10.00000
7	3	4	10.00000
8	5	2	10.00000
9	5	4	10.00000
10	4	6	30.00000

NODAL LOADS

NODE	PX	PY
2	40000.00	0.00
3	0.00	−20000.00
5	0.00	−20000.00

BOUNDARY CONDITION DATA

	STATUS (0: PRESCRIBED, 1: FREE)		PRESCRIBED VALUES	
NODE	U	V	U	V
1	0	0	0.0000	0.0000
6	1	0	0.0000	0.0000

****
RESULTS

NODAL DISPLACEMENTS

NODE	U	V
1	0.0000	0.0000
2	0.0865	−0.0748
3	0.0240	−0.0919
4	0.0548	−0.0490
5	0.0536	−0.0894
6	0.0678	0.0000

NODAL REACTIONS

NODE	PX	PY
1	−40000.0000	6415.0943
2	40000.0000	0.0000
3	−0.0000	−20000.0000
4	0.0000	−0.0000
5	−0.0000	−20000.0000
6	−0.0000	33584.9057

MEMBER FORCES

MEMBER	AXIAL FORCE
1	44989.5178
2	31130.6387
3	26121.5933
4	−39980.4724
5	−8127.0366
6	10021.6071
7	17077.3785
8	−6172.3148
9	23606.5127
10	−42547.4268

****

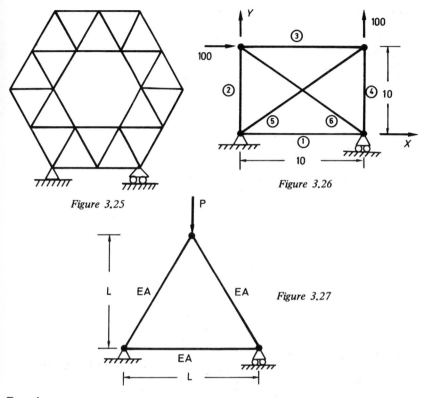

Figure 3.25

Figure 3.26

Figure 3.27

Exercises

(1) Find the minimum bandwidth nodal numbering for the structure shown in Figure 3.25.

(2) Given the plane truss shown in Figure 3.26, apply the displacement method to find the nodal displacements and bar forces for the loads shown.

(3) Modify the program for plane truss analysis to consider the case of a uniform temperature variation, applied to one or more bars.

(4) Modify the program for plane truss analysis to consider more than one loading condition.

(5) Given the truss shown in Figure 3.27, subjected to a load P and to a temperature increase T, find the value of P, such that the upper node does not move in the vertical direction.

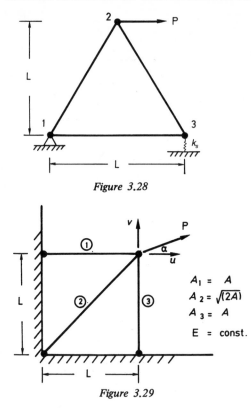

Figure 3.28

Figure 3.29

(6) Considering Example 3.4, compute $\bar{u}$ such that $v_2 = 0$.

(7) Consider the truss of Example 3.3 (Figure 3.7). For the case where the modulus of elasticity of bars 1 and 2 is equal to E_1, and for bar 3 is equal to E_2, find the relationship between E_1 and E_2 such that $u_2 = -v_2$, for any vertical load.

(8) Given the truss of Figure 3.28, where all bars have the same cross-sectional properties, find the displacements for the load shown. Notice that there is an elastic connection at node 3, such that its vertical reaction is equal to $k_s v_3$.

(9) Find the relationship between u and v displacements and the load $\bar{P}$ for the pin-jointed structure shown in Figure 3.29. Determine how this relation varies according with the α angle and find the value of $\tan \alpha$ for which there is no force in bar 3.

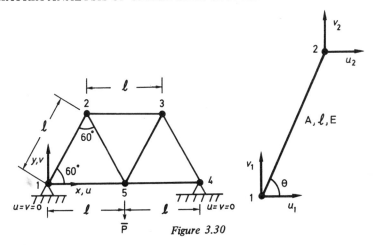

Figure 3.30

(10) The stiffness matrix of each individual member of the truss, shown in
Figure 3.30, is

$$K = \frac{AE}{l} \begin{bmatrix} c^2 & \text{symm.} & & \\ sc & s^2 & & \\ -c^2 & -sc & c^2 & \\ -sc & -s^2 & sc & s^2 \end{bmatrix}$$

where $c = \cos\theta$, $s = \sin\theta$, and $AE/l =$ constant and equal for all the bars.
(Note that only axial forces act on the members.)

   Form the general stiffness matrix of the structure and the force matrix
and solve applying symmetry considerations. Find the vertical deflection
due to a vertical P load applied at node 5 and compute the force in each
bar.

Chapter 4

Solid mechanics

4.1 INTRODUCTION

Within certain limits the behaviour of solids can be considered as linear. In general, however, materials are such that their properties are a function of time and the state of stress. Certain materials, for instance, creep with time and others present plastic modes of deformation at certain loads. In addition a body can crack, which produces a redistribution of stresses.

Even if its material properties are linear the deformation of a body can be such that its behaviour cannot be assumed linear, i.e. the new geometry of the body will have to be taken into consideration.

Such problems are too complex to be analysed here and we will restrict this chapter to the case of linear-elastic systems.

The linear theory of elasticity for a solid body is based on the following assumptions:

(1) Linear material behaviour, i.e. linear stress–strain relations.
(2) The change in orientation of a body due to displacements is negligible. This assumption leads to linear strain displacement relations and also allows us to refer the equilibrium relations to the undeformed geometry.

In what follows we will consider a two-dimensional body for simplicity. The treatment of general, three-dimensional bodies does not involve any additional concepts but merely requires matrices with more terms in the formulae.

4.1.1 State of stress

The state of stress at an internal point of the body is defined by σ_x, σ_y and τ-components. The σ_x, σ_y are called direct stresses and τ is a shear stress. They can be written as an array,

$$\begin{bmatrix} \sigma_x & \tau \\ \tau & \sigma_y \end{bmatrix} \qquad (4.1)$$

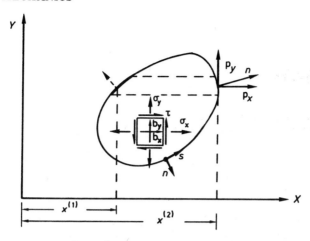

Figure 4.1   Two-dimensional body

The stress components must satisfy the following equilibrium equations throughout the interior of the body:

$$\frac{\partial \sigma_x}{\partial x} + \frac{\partial \tau}{\partial y} + b_x = 0 \qquad (4.2)$$

$$\frac{\partial \tau}{\partial x} + \frac{\partial \sigma_y}{\partial y} + b_y = 0 \qquad (4.3)$$

where b_x, b_y are body forces per unit volume.

4.1.2 Stress boundary conditions

The prescribed surface force intensities on the surface S_σ (where forces are prescribed) are denoted by $\overline{p}_x$, $\overline{p}_y$ (Figure 4.1). Force-equilibrium at the boundary requires that

$$p_x = \sigma_x \alpha_{nx} + \tau \alpha_{ny} = \overline{p}_x$$

$$p_y = \tau \alpha_{nx} + \sigma_y \alpha_{ny} = \overline{p}_y \qquad (4.4)$$

where α_n is the direction cosine for the outward normal, n, given by

$$\alpha_{nx} = \cos(n, x), \quad \alpha_{ny} = \cos(n, y) \qquad (4.5)$$

4.1.3 State of strain

The state of strain at a point is defined by the strain array,

$$
\begin{bmatrix} \epsilon_x & \epsilon_{xy} \\ \epsilon_{yx} & \epsilon_y \end{bmatrix}
\tag{4.6}
$$

The strain-displacement relations for the linear theory are extensional strains ϵ_x, ϵ_y and shear strain ϵ_{xy}, such that

$$
\epsilon_x = \frac{\partial u}{\partial x}, \qquad\qquad \epsilon_y = \frac{\partial v}{\partial y}
$$

$$
\epsilon_{xy} = \frac{1}{2} \left(\frac{\partial u}{\partial y} + \frac{\partial v}{\partial x} \right) = \epsilon_{yx}
\tag{4.7}
$$

where u, v are displacements in the x and y directions.

In what follows we will use the 'engineering' shear strain γ defined as

$$
\gamma = 2\epsilon_{xy}
$$

4.1.4 Displacement boundary conditions

Let S_u denote the portion of the boundary on which displacements are prescribed. The displacement constraints are,

$$
u = \overline{u}, \quad v = \overline{v}
\tag{4.8}
$$

where $\overline{u}$, $\overline{v}$ are the prescribed values. Note that the surface S of the body is equal to S_u plus S_σ.

4.1.5 Constitutive relations

The stress–strain relations for a linearly elastic material are written as,

$$
\begin{Bmatrix} \epsilon_x \\ \epsilon_y \\ \gamma \end{Bmatrix} = \begin{bmatrix} c_{11} & c_{12} & c_{13} \\ & c_{22} & c_{23} \\ \text{symm} & & c_{33} \end{bmatrix} \begin{Bmatrix} \sigma_x \\ \sigma_y \\ \tau \end{Bmatrix}
\tag{4.9}
$$

where c_{ij} are called elastic compliances. We express the inverted form of (4.9) as

$$\left\{ \begin{matrix} \sigma_x \\ \sigma_y \\ \tau \end{matrix} \right\} = \begin{bmatrix} d_{11} & d_{12} & d_{13} \\ & d_{22} & d_{23} \\ \text{symm} & & d_{33} \end{bmatrix} \left\{ \begin{matrix} \epsilon_x \\ \epsilon_y \\ \gamma \end{matrix} \right\} \tag{4.10}$$

where d_{ij} are the rigidity coefficients.

If the body has two orthogonal planes of symmetry, it is said to be orthotropic, and (4.9) reduces to

$$\left\{ \begin{matrix} \epsilon_x \\ \epsilon_y \\ \gamma \end{matrix} \right\} = \begin{bmatrix} c_{11} & c_{12} & 0 \\ & c_{22} & 0 \\ \text{symm} & & c_{33} \end{bmatrix} \left\{ \begin{matrix} \sigma_x \\ \sigma_y \\ \tau \end{matrix} \right\} \tag{4.11}$$

where the $x-z$ and $y-z$ are the material symmetry planes. The inverted relations are

$$\left\{ \begin{matrix} \sigma_x \\ \sigma_y \\ \tau \end{matrix} \right\} = \begin{bmatrix} d_{11} & d_{12} & 0 \\ & d_{22} & 0 \\ \text{symm} & & d_{33} \end{bmatrix} \left\{ \begin{matrix} \epsilon_x \\ \epsilon_y \\ \gamma \end{matrix} \right\} \tag{4.12}$$

An isotropic material has only *two* independent constants. By definition the form of the stress–strain relations is invariant, i.e. independent of the choice of reference frame

The form of Equation (4.11) for a *plane strain* ($\epsilon_z = 0$) isotropic case, i.e.

$$\left\{ \begin{matrix} \epsilon_x \\ \epsilon_y \\ \gamma \end{matrix} \right\} = \frac{1}{E} \begin{bmatrix} (1-v^2) & -v(1+v) & 0 \\ -v(1+v) & (1-v^2) & 0 \\ 0 & 0 & 2(1+v) \end{bmatrix} \left\{ \begin{matrix} \sigma_x \\ \sigma_y \\ \tau \end{matrix} \right\} \tag{4.13}$$

where E and v are Young's modulus and Poisson's ratio respectively. The inverted relations are written as,

$$
\begin{Bmatrix} \sigma_x \\ \sigma_y \\ \tau \end{Bmatrix} = \frac{E}{(1+v)(1-2v)} \begin{bmatrix} (1-v) & v & 0 \\ v & (1-v) & 0 \\ 0 & 0 & \dfrac{1-2v}{2} \end{bmatrix} \begin{Bmatrix} \epsilon_x \\ \epsilon_y \\ \gamma \end{Bmatrix} \quad (4.14)
$$

For the *plane stress* case ($\sigma_z = 0$) we can write,

$$
\begin{Bmatrix} \epsilon_x \\ \epsilon_y \\ \gamma \end{Bmatrix} = \frac{1}{E} \begin{bmatrix} 1 & -v & 0 \\ -v & 1 & 0 \\ 0 & 0 & 2(1+v) \end{bmatrix} \begin{Bmatrix} \sigma_x \\ \sigma_y \\ \tau \end{Bmatrix} \quad (4.15)
$$

and

$$
\begin{Bmatrix} \sigma_x \\ \sigma_y \\ \tau \end{Bmatrix} = \frac{E}{1-v^2} \begin{bmatrix} 1 & v & 0 \\ v & 1 & 0 \\ 0 & 0 & \dfrac{1-v}{2} \end{bmatrix} \begin{Bmatrix} \epsilon_x \\ \epsilon_y \\ \gamma \end{Bmatrix} \quad (4.16)
$$

The above relations can also be expressed in terms of Lame's constants (λ, μ) which are related to E and v in the following way:

$$ \lambda = vE/[(1+v)(1-2v)] $$

$$ \mu = E/[2(1+v)] $$

The complete set of equations for an elastic solid consists of two equations of equilibrium, the three strain-displacement relations, the three stress–strain relations and the stress and geometrical boundary conditions. The unknowns are the two displacements, the three stresses and the three strains. The strains can be eliminated by combining the stress–strain and strain-displacement relations.

Example 4.1
The volumetric strain ϵ_v is defined as the sum of the three extensional strains

and given by

$$\epsilon_v = \epsilon_x + \epsilon_y + \epsilon_z \tag{a}$$

In the case of *plane strain* $\epsilon_z = 0$ and we have the following constitutive relations:

$$\begin{Bmatrix} \sigma_x \\ \sigma_y \\ \tau \end{Bmatrix} = \frac{E}{(1+v)(1-2v)} \begin{bmatrix} (1-v) & v & 0 \\ v & (1-v) & 0 \\ 0 & 0 & \dfrac{1-2v}{2} \end{bmatrix} \begin{Bmatrix} \epsilon_x \\ \epsilon_y \\ \gamma \end{Bmatrix} \tag{b}$$

plus

$$\sigma_z = \frac{E}{(1+v)(1-2v)} [v\,\epsilon_x + v\,\epsilon_y]$$

We can write Equations (b) in function of Lame parameters which are

$$\sigma_x = \lambda\,\epsilon_v + 2\mu\,\epsilon_x$$

$$\sigma_y = \lambda\,\epsilon_v + 2\mu\,\epsilon_y$$

$$\tau = \mu\,\gamma$$

$$\sigma_z = \lambda\,\epsilon_v \tag{c}$$

The mean volumetric stress is defined as

$$\sigma_m = \frac{1}{3}(\sigma_x + \sigma_y + \sigma_z) \tag{d}$$

which becomes a function of ϵ_v after substituting (c),

$$\sigma_m = \lambda\,\epsilon_v + 2\,\frac{\mu}{3}\,\epsilon_v \tag{e}$$

Note that if $v \to \frac{1}{2}$, $\lambda \to \infty$, which implies that $\epsilon_v \to 0$ for the σ_m to be defined. This means that for $v = \frac{1}{2}$ the material is incompressible.

4.2   PRINCIPLE OF VIRTUAL DISPLACEMENT

Consider a body *in equilibrium* under loading b_x, b_y, $\overline{p}_x$, $\overline{p}_y$ and internal forces σ_x, σ_y, τ. Now, visualize the body displaced from the equilibrium position and let δu, δv define the virtual displacements, which satisfy the displacement boundary conditions on S_u (i.e. δu, δv vanish on S_u).

The first order work, δW_E, done by the external forces is equal to

$$\delta W_E = \int_{S_\sigma} (\overline{p}_x \, \delta u + \overline{p}_y \, \delta v) dS + \iint (b_x \, \delta u + b_y \, \delta v) dA \tag{4.17}$$

Taking into account the equilibrium relations in Equations (4.2) and (4.3), and stress boundary conditions in Equation (4.4), we can write Equation (4.17) as

$$\delta W_E = \int_{S_\sigma} \{(\sigma_x \, \alpha_{nx} + \tau \, \alpha_{ny})\delta u + (\tau \, \alpha_{nx} + \sigma_y \, \alpha_{ny})\delta v\} \, dS$$

$$- \iint \left\{ \left(\frac{\partial \sigma_x}{\partial x} + \frac{\partial \tau}{\partial y} \right) \delta u + \left(\frac{\partial \tau}{\partial x} + \frac{\partial \sigma_y}{\partial y} \right) \delta v \right\} \, dA \tag{4.18}$$

Integrating by parts the terms in the second integral using Gauss' theorem, we have, for the first term

$$\iint \frac{\partial \sigma_x}{\partial x} \, \delta u \, dA = \int [\sigma_x \, \delta u]_{x(1)}^{x(2)} \, dx - \iint \sigma_x \, \delta \, \epsilon_x \, dA$$

$$= \int_{S_\sigma} \sigma_x \, \alpha_{nx} \, \delta u \, dS - \iint \sigma_x \, \delta \, \epsilon_x \, dA$$

and similarly for term in σ_y. For the shear terms we have

$$\iint \left\{ \frac{\partial \tau}{\partial x} \, \delta v + \frac{\partial \tau}{\partial y} \, \delta u \right\} \, dA =$$

$$= \int_{S_\sigma} \tau \, \{\delta v \, \alpha_{nx} + \delta u \, \alpha_{ny}\} \, dS - \iint \tau \, \delta \gamma \, dA$$

Substituting these relationships into Equation (4.18) we find that

$$\delta W_E = \iint \{\sigma_x \, \delta\epsilon_x + \sigma_y \, \delta\epsilon_y + \tau \, \delta\gamma\} \, dA \tag{4.19}$$

The right hand side of Equation (4.19) represents the first order work done by the internal forces (stresses) during the virtual displacements and will be called δW_D. We can now write the Principle of Virtual displacements in two forms.

(1)   The increment of internal work equals the increment of external work, i.e.

$$\delta W_D = \delta W_E \tag{4.20}$$

$$\iint \{\sigma_x \, \delta\epsilon_x + \sigma_y \, \delta\epsilon_y + \tau \, \delta\gamma\} \, dA$$

$$= \int_{S_\sigma} \{\overline{p}_x \, \delta u + \overline{p}_y \, \delta v \; dS + \iint \{b_x \, \delta u + b_y \, \delta v\} \, dA$$

(2)   From Equations (4.17) and (4.18), the Principle of Virtual Displacements can be written in an equilibrium equation form

$$\iint \left\{ \left(\frac{\partial\sigma_x}{\partial x} + \frac{\partial\tau}{\partial y} + b_x\right) \delta u + \left(\frac{\partial\tau}{\partial x} + \frac{\partial\sigma_y}{\partial y} + b_y\right) \delta v \right\} \, dA =$$

$$= \int \{(\sigma_x \, \alpha_{nx} + \tau \, \alpha_{ny} - \overline{p}_x)\delta u + (\tau \, \alpha_{nx} + \sigma_y \, \alpha_{ny} - \overline{p}_y)\delta v\} \, dS \tag{4.21}$$

Note that passing from Equation (4.20) to (4.21) requires the displacement and stresses to be continuous up to their first derivatives and in addition, the stresses must be in equilibrium in the interior and the surface of the body.

Equation (4.20) is the most used expression of the Principle of Virtual Displacements and is independent of material behaviour and magnitude of displacements, (i.e. is still valid for non-linear geometry and arbitrary material behaviour).

Equation (4.21) assumes instead that the strain-displacement relationships are of the type shown in Equation (4.7), i.e. the geometrically linear case. In

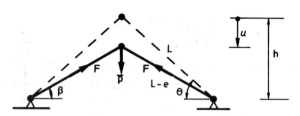

Figure 4.2 Two-bar truss. β deformed angle, θ initial angle

addition the equilibrium equations are assumed to be given by Equations (4.2) and (4.3).

When the strain-displacement relations and geometrically boundary conditions are specified, the equations of equilibrium and the mechanical boundary conditions can be deduced from Equation (4.20). This approach is illustrated in the following example.

Example 4.2
Consider the symmetrical two-bar truss shown in Figure 4.2. The initial and deformed (loaded) positions are defined by the angle θ and β respectively, and e is the shortening of the bar. Assuming small strain,

$$\sin \theta = \frac{h}{L}$$

$$\sin \beta = \frac{h - u}{L - e} \simeq \frac{h - u}{L} = \sin \theta - \frac{u}{L} \tag{a}$$

Note that

$$de = \sin \beta \, du$$

Integrating this expression one finds that

$$e \simeq u \sin \theta - \frac{1}{2L} u^2$$

The first two variations of e are

$$\delta e = \left\{ \sin \theta - \frac{u}{L} \right\} \delta u = \sin \beta \, \delta u$$

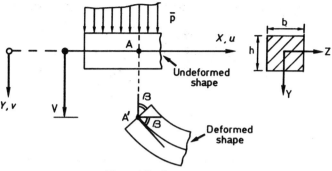

Figure 4.3   Prismatic beam

$$\delta^2 e = -\frac{(\delta u)^2}{L} \tag{b}$$

Applying the Principle of Virtual Displacements,

$$\delta W_E = \delta W_D \tag{c}$$

gives

$$\overline{P}\,\delta u = 2F\,\delta e$$

for arbitrary u, resulting in the equilibrium equation

$$\overline{P} = 2F\sin\beta \tag{d}$$

Example 4.3

Let us consider a uniform beam under a distributed load $\overline{p}$. We will consider the beam is undergoing only bending deformation and neglect any shear effects (Figure 4.3).

We will assume that the beam deforms in such a way that normal plane sections remain plane after deformation, i.e.

$$u = -\beta y \tag{a}$$

where β is the rotation of the middle section.

The vertical deflection v is considered to be constant through the sections. Hence the strain-displacement equations become,

$$\epsilon_x = \frac{\partial u}{\partial x} = -y\,\frac{\partial \beta}{\partial x}$$

$$\epsilon_y = \frac{\partial v}{\partial y} = 0$$

$$\gamma = \frac{\partial u}{\partial y} + \frac{\partial v}{\partial x} = -\beta + \frac{\partial v}{\partial x} \tag{b}$$

Note that if normal sections remain normal after deformation (Figure 4.3) we have

$$\beta = \frac{\partial v}{\partial x} \tag{c}$$

Hence $\gamma = 0$ and

$$\epsilon_x = -y \frac{\partial^2 v}{\partial x^2} \,.$$

This hypothesis implies that shear effects have been neglected.

The Principle of Virtual Displacements can now be written as

$$\delta W_D = \delta W_E$$

or

$$b \iint \sigma_x \, \delta \epsilon_x \, dx \, dy = \int \overline{p} \, \delta v \, dx \tag{d}$$

where b is the width of the beam. Substituting the strain-displacement relation and the stress–strain equation ($\sigma = E(\partial u/\partial x)$ for this case), we have

$$b \int\limits_{-h/2}^{h/2} \int\limits_{0}^{l} E \left(\frac{d^2 v}{dx^2} \right) y^2 \left(\frac{d^2 \, \delta v}{dx^2} \right) \, dx \, dy$$

$$= \int\limits_{0}^{l} EI \left(\frac{d^2 v}{dx^2} \right) \left(\frac{d^2 \, \delta v}{dx^2} \right) \, dx \tag{e}$$

where

$$I = \frac{bh^3}{12}$$

is the moment of inertia. We can write the Principle of Virtual Displacements for the beam as

$$\int_0^l M \left(\frac{d^2 \delta v}{dx^2} \right) dx = \int_0^l \bar{p} \, \delta v \, dx \tag{f}$$

where

$$M = EI \frac{d^2 v}{dx^2}$$

In order to obtain the equilibrium equation of the beam we can integrate twice by parts the left hand side of formulae (f). This gives

$$\int_0^l EI \frac{d^2 v}{dx^2} \frac{d^2 \delta v}{dx^2} \, dx = -EI \int_0^l \frac{d^3 v}{dx^3} \frac{d \delta v}{dx} \, dx + \left. EI \frac{d^2 v}{dx^2} \frac{d \delta v}{dx} \right|_0^l$$

$$= EI \int_0^l \frac{d^4 v}{dx^4} \delta v \, dx + \left. EI \frac{d^2 v}{dx^2} \frac{d \delta v}{dx} \right|_0^l - \left. EI \frac{d^3 v}{dx^3} \delta v \right|_0^l \tag{g}$$

Note that the shear force can be written as

$$Q = -EI \frac{d^3 v}{dx^3}$$

and that at the $x = 0$ and $x = L$ ends of the beam we know the displacement or stress boundary conditions, i.e. for $x = 0$, l either M or $d\delta v/dx$, and Q or δv are known.

The Principle of Virtual Displacements for the beam can then be written

$$\int_0^l \left\{ EI \frac{d^4 v}{dx^4} - \bar{p} \right\} \delta v \, dx = 0 \tag{h}$$

The expression between brackets in the above integral is the equilibrium equation for a uniform beam.

4.3   PRINCIPLE OF MINIMUM POTENTIAL ENERGY

Let us now restrict the Principle of Virtual Displacements to elastic behaviour and static loading. The object is to express the relationship

$$\delta W_D - \delta W_E = 0$$

as the stationary requirement for some functional (or function of functions) of the displacements.

When the material is elastic, and the deformation process is adiabatic, the work done by the external static forces is equal to the change in internal energy — which is called strain energy, since it is a function of the deformation measures. Let $\mathscr{V}$ be the strain energy per unit volume, i.e. the strain energy density. By definition, $\delta \mathscr{V}$ is equal to the first order work per unit volume done by the stresses during the incremental displacements δu, δv from an equilibrium position.

One can show that,

$$\delta \mathscr{V} = \sigma_x \, \delta \epsilon_x + \sigma_y \, \delta \epsilon_y + \tau \, \delta \gamma$$

$$= [\sigma_x \, \sigma_y \, \tau] \left\{ \begin{array}{c} \delta \epsilon_x \\ \delta \epsilon_y \\ \delta \gamma \end{array} \right\} = \boldsymbol{\sigma}^T \delta \boldsymbol{\epsilon} \tag{4.22}$$

Expanding $\delta \mathscr{V}$

$$\delta \mathscr{V} = \frac{\partial \mathscr{V}}{\partial \epsilon_x} \delta \epsilon_x + \frac{\partial \mathscr{V}}{\partial \epsilon_y} \delta \epsilon_y + \frac{\partial \mathscr{V}}{\partial \gamma} \delta \gamma \tag{4.23}$$

leads to an alternate form of the stress—strain relations

$$\sigma_x = \frac{\partial \mathscr{V}}{\partial \epsilon_x}, \quad \sigma_y = \frac{\partial \mathscr{V}}{\partial \epsilon_y}, \quad \tau = \frac{\partial \mathscr{V}}{\partial \gamma} \tag{4.24}$$

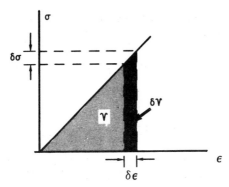

Figure 4.4   Stress-strain curve

For a plane strain case ($\epsilon_z = 0$) and isotropic materials we can write

$$\begin{Bmatrix} \sigma_x \\ \sigma_y \\ \tau \end{Bmatrix} = \frac{E}{(1+v)(1-2v)} \begin{bmatrix} 1-v & v & 0 \\ v & 1-v & 0 \\ 0 & 0 & \dfrac{1-v}{2} \end{bmatrix} \begin{Bmatrix} \epsilon_x \\ \epsilon_y \\ \gamma \end{Bmatrix}$$

or

$$\sigma = D \,\epsilon \qquad\qquad\qquad\qquad\qquad\qquad\qquad\qquad\qquad (4.25)$$

$\mathscr{V}$ is then a *quadratic* function of ϵ,

$$\mathscr{V} = \tfrac{1}{2}\epsilon^T D\,\epsilon \qquad\qquad\qquad\qquad\qquad\qquad\qquad\qquad (4.26)$$

Lastly we can interpret $\mathscr{V}$ as the area under the $\sigma-\epsilon$ line for a one dimensional stress case (Figure 4.4).

The total strain energy is

$$V = \iiint \mathscr{V}\,\mathrm{d}\,(\text{volume}) \qquad\qquad\qquad\qquad\qquad\qquad (4.27)$$

With this definition,

$$\delta W_D = \delta V$$

If the surface and body forces are independent of the displacements, one can express the external work terms in terms of a force potential, Ω, defined as,

$$\delta\Omega = -\iint (b_x\,\delta u + b_y\,\delta v)\mathrm{d}x\,\mathrm{d}y -- \int_{S_\sigma} (\bar{p}_x\delta u + \bar{p}_y\delta v)\mathrm{d}S$$

$$= -\iint \mathbf{b}^T\,\delta\mathbf{u}\,\mathrm{d}x\,\mathrm{d}y - \int_{S_\sigma} \bar{\mathbf{p}}^T\,\delta\mathbf{u}\,\mathrm{d}S$$

$$\equiv -\delta W_E \tag{4.28}$$

where,

$$\mathbf{b} = \{b_x, b_y\}, \quad \bar{\mathbf{p}} = \{\bar{p}_x, \bar{p}_y\}, \quad \mathbf{u} = \{u, v\}$$

We obtain Ω by integrating Equation (4.28) with respect to the displacement components

$$\Omega = -\iint (b_x\,u + b_y\,v)\mathrm{d}x\,\mathrm{d}y - \int_{S_\sigma} (\bar{p}_x\,u + \bar{p}_y\,v)\mathrm{d}S \tag{4.29}$$

Note that the boundary integral involves only S_σ, since $\delta\mathbf{u} = 0$ on S_u. Finally, let

$$\Pi_p(\mathbf{u}) = V + \Omega \tag{4.30}$$

and call Π_p the 'total potential energy'. The equilibrium requirement can be stated as

$$\delta\Pi_p = \delta V + \delta\Omega = 0 \tag{4.31}$$

for arbitrary $\delta\mathbf{u}$, which satisfies the displacement boundary conditions. It follows that the displacements defining an equilibrium position correspond to a stationary value of the total potential energy.

When the surface forces are independent of displacement and the geometry is linear, the second variation of Π_p reduces to

$$\delta^2\Pi_p = \delta^2 V \tag{4.32}$$

$$\delta^2 V = \iint \delta\boldsymbol{\sigma}^T\delta\boldsymbol{\epsilon}\,\mathrm{d}x\,\mathrm{d}y = \iint \delta\boldsymbol{\epsilon}^T\,\mathbf{D}_t\,\delta\boldsymbol{\epsilon}\,\mathrm{d}x\,\mathrm{d}y$$

where $\mathbf{D}_t$ is the tangent material rigidity matrix.

Since $\mathbf{D}_t$ must be positive definite for a stable material, we see that the true displacements correspond to a relative minimum value of Π_p. Then, the potential energy for an approximate displacement field $u^{(a)}$ which satisfies the displacement boundary conditions is greater than the true value

$$\Pi_{P_{\text{approx}}} \geqslant \Pi_{P_{\text{true}}} \tag{4.33}$$

We use this result to establish bounds on *approximate* displacement solutions.

4.3.1 Matrix formulation

The governing equations for a linear discrete system have the following general form,

$$P = KU \tag{4.34}$$

where P contains the prescribed loads, U contains the discrete displacement unknowns and K is the system stiffness matrix. Noting that these equations follow from (4.31), we can write,

$$\Pi_p = \tfrac{1}{2}U^T K U - U^T P \tag{4.35}$$

The value of Π_p at an equilibrium position is obtained by substituting for U using (4.34)

$$\Pi_p\big|_{\text{equilib. position}} = \tfrac{1}{2}U^T P - U^T P = -\tfrac{1}{2}U^T P \tag{4.36}$$

Suppose there is only a single force, say P_c, acting on the system. Let u_c be the exact value and $u_c^{(a)}$ an approximation for the displacement in the direction of P_c. The exact and approximate energies are

$$2\Pi_p = -u_c P_c$$

$$2\Pi_{P_{\text{approx}}} = -u_c^{(a)} P_c$$

Noting Equation (4.33) we obtain a bound on u_c,

$$(u_c)_{\text{approx}} \leqslant u_c \tag{4.37}$$

This result shows that an approximate compatible displacement field corresponds to a structure which is stiffer than the actual structure and therefore will give a *lower bound* on displacement.

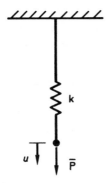

Figure 4.5   Linear spring under concentrated load

Example 4.4
Consider a linear spring of stiffness k, subjected to a load $\overline{P}$ (Figure 4.5). The various energy terms are,

$$V = \tfrac{1}{2}ku^2$$

$$\Omega = -\overline{P}u$$

$$\Pi_p = \tfrac{1}{2}ku^2 - \overline{P}u \tag{a}$$

The equilibrium equation is obtained by requiring $\delta\Pi_p$ to be stationary for arbitrary u,

$$\delta\Pi_p = ku\,\delta u - \overline{P}\,\delta u = 0 \tag{b}$$

therefore

$$ku_e - \overline{P} = 0 \tag{c}$$

To show that the potential energy is a minimum at the exact value of u (i.e. u_e) we consider another state $u = u_e + \delta u$ and evaluate Π_p.

$$\Pi_p' = \tfrac{1}{2}k(u_e + \delta u)^2 - \overline{P}(u_e + \delta u)$$

$$= \Pi_p + \delta\Pi_p + \tfrac{1}{2}\delta^2\Pi_p$$

$$= \Pi_p + \tfrac{1}{2}k(\delta u)^2 \tag{d}$$

Since $k > 0$, it follows that

$$\Pi'_p \geqslant \Pi_p$$

for arbitrary δu.

Example 4.5
Let us consider the case of a uniform elastic beam. The strain energy of the beam is obtained by integrating

$$V = \tfrac{1}{2} \iiint \sigma\epsilon \, d(\text{volume}) \tag{a}$$

which gives

$$V = \tfrac{1}{2} \int_0^l M \frac{d^2 v}{dx^2} \, dx = \frac{EI}{2} \int_0^l \left(\frac{d^2 v}{dx^2}\right)^2 dx \tag{b}$$

The potential energy of the distributed load $\bar{p}$ is,

$$\Omega = - \int_0^l \bar{p} v \, dx \tag{c}$$

Hence the total potential energy, in the absence of any other external forces, can be written as

$$\Pi_p = \frac{EI}{2} \int_0^l \left(\frac{d^2 v}{dx^2}\right)^2 dx - \int_0^l \bar{p} v \, dx \tag{d}$$

If we think of this energy function Π_p as a function of the displacement, we have the stationarity condition,

$$\delta\Pi_p = EI \int_0^l \left(\frac{d^2 v}{dx^2}\right) \left(\frac{d^2 \delta v}{dx^2}\right) dx - \int_0^l \bar{p} \delta v \, dx = 0 \tag{e}$$

which is the principle of virtual displacements.

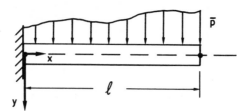

Figure 4.6   Fixed-free end beam

We can have some applied moment or shear at the end of the beam, such as

$$\bar{M} = EI \frac{d^2 v}{dx^2}, \quad \bar{Q} = -EI \frac{d^3 v}{dx^3} \tag{f}$$

The virtual work produced by these forces will give a new term in Equation (e) which becomes

$$EI \int_0^l \left(\frac{d^2 v}{dx^2}\right)\left(\frac{d^2 \delta v}{dx^2}\right) dx = \int_0^l \bar{p}\delta v \, dx$$

$$+ \left[\bar{M} \frac{d\delta v}{dx}\right]_{x=0}^{x=l} + [\bar{Q}\delta v]_{x=0}^{x=l} \tag{g}$$

If we integrate the right hand side term in (g) twice by parts, we obtain the equilibrium equation and force boundary conditions, i.e.

$$\delta\Pi_p = \int_0^l \left(EI \frac{d^4 v}{dx^4} - \bar{p}\right) \delta v \, dx + \left[(M - \bar{M}) \frac{d\delta v}{dx}\right]_{x=0}^{x=l}$$

$$+ [(Q - \bar{Q})\delta v]_{x=0}^{x=l} \tag{h}$$

In order to see the type of function v which satisfies the equilibrium conditions let us consider the case of a fixed-free beam as the one shown in Figure 4.6.

We can solve this problem by integrating the equilibrium equation,

$$EI \frac{d^4 v}{dx^4} - \bar{p} = 0 \tag{i}$$

Thus

$$\frac{d^3v}{dx^3} = \frac{\bar{p}}{EI} x + C_1$$

where C_1 is an arbitrary constant which can be determined from the condition

$$\left.\frac{d^3v}{dx^3}\right|_{x=l} = 0 \therefore C_1 = -\frac{\bar{p}l}{EI} \qquad\qquad\text{(j)}$$

Hence

$$\frac{d^3v}{dx^3} = \frac{\bar{p}}{EI} (x - l) \qquad\qquad\text{(k)}$$

Integrating again

$$\frac{d^2v}{dx^2} = \frac{\bar{p}}{EI} \left(\frac{x^2}{2} - xl + C_2\right)$$

The value of C_2 is defined by the moment equal zero condition

$$\left.\frac{d^2v}{dx^2}\right|_{x=l} = 0 \therefore C_2 = \frac{l^2}{2}$$

Thus

$$\frac{d^2v}{dx^2} = \frac{\bar{p}}{2EI} (x^2 - 2lx + l^2) \qquad\qquad\text{(l)}$$

We can integrate again

$$\frac{dv}{dx} = \frac{\bar{p}}{2EI} \left(\frac{x^3}{3} - lx^2 + l^2x + C_3\right) \qquad\qquad\text{(m)}$$

and determine C_3 from

$$\left.\frac{dv}{dx}\right|_{x=0} = 0, \quad \therefore C_3 = 0$$

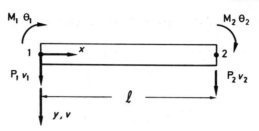

Figure 4.7   Beam element

Finally we integrate (m)

$$v = \frac{\bar{p}}{2EI}\left(\frac{x^4}{12} - \frac{lx^3}{3} + \frac{l^2x^2}{2} + C_4\right)$$

As $v|_{x=0} = 0$, $C_4 = 0$. The deflection of the beam is given by

$$v = \frac{\bar{p}}{2EI}\left(\frac{x^4}{12} - \frac{lx^3}{3} + \frac{l^2x^2}{2}\right) \tag{n}$$

Note that if $\bar{p} = 0$, the deflection of the beam is a cubic. We will use this result in the following example.

Example 4.6
Let us find the stiffness coefficients of the beam shown in Figure 4.7 using the Principle of Minimum Potential Energy.

The bending strain energy of the beam is

$$V = \frac{EI}{2}\int_0^l \left(\frac{d^2v}{dx^2}\right)^2 dx \tag{a}$$

The forces at nodes 1 and 2 give the following potential

$$\Omega = -(P_1v_1 + P_2v_2 + M_1\theta_1 + M_2\theta_2) \tag{b}$$

Note that the exact function for displacements is a cubic, as the moment varies linearly from 1 to 2 (Example 4.5). We can start with

$$v = \alpha_1 + \alpha_2 x + \alpha_3 x^2 + \alpha_4 x^3$$

$$= [1 \quad x \quad x^2 \quad x^3] \begin{Bmatrix} \alpha_1 \\ \alpha_2 \\ \alpha_3 \\ \alpha_4 \end{Bmatrix} = A\alpha \qquad (c)$$

and specialize v and its derivative ($\theta = dv/dx$) for nodes 1 and 2,

$$\begin{Bmatrix} v_1 \\ \theta_1 \\ v_2 \\ \theta_2 \end{Bmatrix} = \begin{bmatrix} 1 & 0 & 0 & 0 \\ 0 & 1 & 0 & 0 \\ 1 & l & l^2 & l^3 \\ 0 & 1 & 2l & 3l^2 \end{bmatrix} \begin{Bmatrix} \alpha_1 \\ \alpha_2 \\ \alpha_3 \\ \alpha_4 \end{Bmatrix} \qquad (d)$$

or

$$U^e = C\alpha$$

Inverting Equation (d) we find the values of α which can then be substituted in (c). This gives,

$$v = (AC^{-1})U^e = \Phi U^e \qquad (e)$$

$$= v_1 \phi_1 + \theta_1 \phi_2 + v_2 \phi_3 + \theta_2 \phi_4$$

The ϕ_i are called *interpolation functions* and Φ is the interpolation function vector, such that

$$\phi_1 = [1 - 3(x/l)^2 + 2(x/l)^3]$$

$$\phi_2 = \left[x - 2\frac{x^2}{l} + \frac{x^3}{l^2} \right]$$

$$\phi_3 = [3(x/l)^2 - 2(x/l)^3]$$

$$\phi_4 = \left[-\frac{x^2}{l} + \frac{x^3}{l^2} \right]$$

Now we can write

$$\frac{d^2 v}{dx^2} = \left[-\frac{6}{l^2} + \frac{12x}{l^3}, \quad -\frac{4}{l} + \frac{6x}{l^2}, \quad \frac{6}{l^2} - \frac{12x}{l^3}, \quad -\frac{2}{l} + \frac{6}{l^2} \right] \begin{Bmatrix} v_1 \\ \theta_1 \\ v_2 \\ \theta_2 \end{Bmatrix}$$

$$= \mathbf{B} \mathbf{U}^e \qquad\qquad\qquad\qquad (f)$$

Let us now substitute Equation (f) into the increment of strain energy

$$\delta V = EI \int_0^l \left(\frac{d^2 v}{dx^2} \right) \left(\frac{d^2 \delta v}{dx^2} \right) \, dx$$

$$= [\delta v_1 \, \delta \theta_1 \, \delta v_2 \, \delta \theta_2] \left\{ \int_0^l \mathbf{B}^T EI \, \mathbf{B} \, dx \right\} \begin{Bmatrix} v_1 \\ \theta_1 \\ v_2 \\ \theta_2 \end{Bmatrix}$$

$$= [\delta v_1 \, \delta \theta_1 \, \delta v_2 \, \delta \theta_2] \mathbf{K}^e \begin{Bmatrix} v_1 \\ \theta_1 \\ v_2 \\ \theta_2 \end{Bmatrix} = \delta \mathbf{U}^{e,T} \mathbf{K}^e \mathbf{U}^e \qquad (g)$$

$\mathbf{K}^e$ is the stiffness matrix for the beam element

$$\mathbf{K}^e = EI \int_0^l \mathbf{B}^T \mathbf{B} \, dx = \frac{EI}{l^3} \begin{bmatrix} 12 & 6l & -12 & 6l \\ & 4l^2 & -6l & 2l^2 \\ & & 12 & -6l \\ \text{symm} & & & 4l^2 \end{bmatrix} \qquad (h)$$

The variation of the potential of the nodal loads is

$$-\delta\Omega = [\delta v_1 \ \delta\theta_1 \ \delta v_2 \ \delta\theta_2] \begin{Bmatrix} P_1 \\ M_1 \\ P_2 \\ M_2 \end{Bmatrix} = \delta \, U^{e,T} \mathbf{P}^e \tag{i}$$

Thus the following relationship will be valid for a beam element

$$\mathbf{K}^e \mathbf{U}^e = \mathbf{P}^e \text{ (for arbitrary } \delta \mathbf{U}^e) \tag{j}$$

where $\mathbf{U}^e$ are the nodal unknowns.

The stiffness matrix (h) considers the bending behaviour of the beam, but neglects shear deformation. That stiffness matrix is given with reference to the local member coordinate system x, y shown in Figure 4.7.

Excluding the case of continuous beams, which can be considered as a special case, a plane frame system will include beam members with different relative orientation. Then, if we want to apply the matrix displacement method of analysis to such system, all the member stiffness matrices will have to be rotated to a common reference coordinate system, before proceeding to the assembling of the total system of equations. In order to do this, we will have to include to the stiffness matrix the stiffness coefficients due to the axial deformation, which are the same as those of the truss bar, studied in Chapter 3. Thus, the force and displacement quantities considered will be those of Figure 4.8, so that

$$\mathbf{U}^e = \begin{Bmatrix} u_1 \\ v_1 \\ \theta_1 \\ u_2 \\ v_2 \\ \theta_2 \end{Bmatrix}, \qquad \mathbf{P}^e = \begin{Bmatrix} F_1 \\ P_1 \\ M_1 \\ F_2 \\ P_2 \\ M_2 \end{Bmatrix} \tag{k}$$

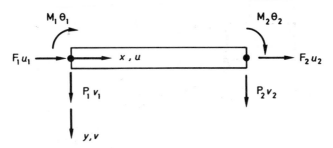

Figure 4.8   Beam element with plane displacements

$$\mathbf{K}^e = \begin{bmatrix} \dfrac{AE}{l} & 0 & 0 & -\dfrac{AE}{l} & 0 & 0 \\[2ex] & \dfrac{12EI}{l^3} & \dfrac{6EI}{l^2} & 0 & -\dfrac{12EI}{l^3} & \dfrac{6EI}{l^2} \\[2ex] & & \dfrac{4EI}{l} & 0 & -\dfrac{6EI}{l^2} & \dfrac{2EI}{l} \\[2ex] & & & \dfrac{AE}{l} & 0 & 0 \\[2ex] & \text{Symm} & & & \dfrac{12EI}{l^3} & -\dfrac{6EI}{l^2} \\[2ex] & & & & & \dfrac{4EI}{l} \end{bmatrix} \tag{l}$$

The components of the vectors shown in Equation (k) are referred to the local reference system x, y of Figure 4.9, and will have to be rotated to the global reference system X, Y, according to the following expressions

$$\mathbf{U}^e = \mathbf{R}\mathbf{U}^{g,e}; \qquad \mathbf{U}^{g,e} = \mathbf{R}^T\mathbf{U}^e$$

$$\mathbf{P}^e = \mathbf{R}\mathbf{P}^{g,e}; \qquad \mathbf{P}^{g,e} = \mathbf{R}^T\mathbf{P}^e \tag{m}$$

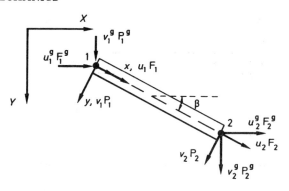

Figure 4.9 Member local and global reference systems

where

$$\mathbf{U}^{g,e} = \begin{Bmatrix} u_1^g \\ v_1^g \\ \theta_1 \\ u_2^g \\ v_2^g \\ \theta_2 \end{Bmatrix} \qquad \mathbf{P}^{g,e} = \begin{Bmatrix} F_1^g \\ P_1^g \\ M_1 \\ F_2^g \\ P_2^g \\ M_2 \end{Bmatrix} \tag{n}$$

and

$$\mathbf{R} = \begin{bmatrix} \cos\beta & \sin\beta & 0 & 0 & 0 & 0 \\ -\sin\beta & \cos\beta & 0 & 0 & 0 & 0 \\ 0 & 0 & 1 & 0 & 0 & 0 \\ 0 & 0 & 0 & \cos\beta & \sin\beta & 0 \\ 0 & 0 & 0 & -\sin\beta & \cos\beta & 0 \\ 0 & 0 & 0 & 0 & 0 & 1 \end{bmatrix} \tag{o}$$

Note that $\mathbf{R}$ is an orthogonal matrix, such that $\mathbf{R}^{-1} = \mathbf{R}^T$. Then taking into account Equations (j) and (m), we see that

$$K^e R U^{g,e} = R P^{g,e} \tag{p}$$

or

$$(R^T K^e R) U^{g,e} = K^{g,e} U^{g,e} = P^{g,e} \tag{q}$$

where

$$K^{g,e} = R^T K^e R \tag{r}$$

is the member global stiffness matrix.

If distributed or concentrated forces acting along the member, temperature variation, initial deformations, etc., are to be considered, the member matrix Equation (j) becomes

$$K^e U^e + F^e_b = P^e \tag{s}$$

where F^e_b takes into account those types of actions. Note that by making $U^e = 0$ we can conclude that F^e_b contains the forces generated in the member, due to the actions mentioned above, when the member is considered with built-in ends. Because of this F^e_b is normally called the fixed-end forces vector. The coefficients of F^e_b, for different types of actions, are tabulated in many structural analysis books.

4.4   PROGRAM FOR STATIC ANALYSIS OF PLANE FRAME SYSTEMS

We can now easily modify the previous program, for static analysis of plane trusses, such that a program for static analysis of plane frames is obtained. The main changes will occur in the subprograms to compute the member stiffness matrix (STIFF), and the member end forces (FORCE), which must take into account the expressions presented above. The other subprograms will require very small modifications or none at all, as it is described in what follows.

4.4.1 Data structure

The data structure requires only minor changes affecting some of the arrays used, to take into account the additional data needed.

Considering the array definitions given for the plane truss analysis program, the changes to be introduced for the case of plane frame analysis are as follows:

CON: Unchanged.

IB: Expanded, so that there are four positions for each boundary node. The first, IB[(NDF + 1)*(J − 1) + 1], contains the node number for the jth boundary node, while the following three, IB((NDF + 1)*(J − 1) + I) with I = 2, 3, 4, contain the status indicators for the nodal unknowns, u, v, and θ, of that node (0: prescribed,; 1: unknown).

X: Unchanged.

Y: Unchanged.

PROP: Expanded to include two cross-sectional properties per element. PROP(2*L − 1) contains the cross-sectional area, and PROP(2*L) contains the moment of inertia, for element L.

AL: Expanded to consider three nodal degrees of freedom rather than two. AL[NDF*(J − 1) + I) with I = 1, 2, 3, will contain the loads F, P, and M for node J, before solution, and the values of the nodal unknowns, u, v, and θ, for node J, after solution of the system of equations.

TK: Unchanged.

ELST: Unchanged.

V: Unchanged.

FORC: It has six positions for each element, storing the member end forces. FORC[6*(L − 1) + I] with I = 1, 2, 3 will contain the end forces F_x, F_y, and M_z, for the first node, and FORC[6*(L − 1) + I] with I = 4, 5, 6, will contain the end forces F_x, F_y, and M_z, for the end node, of element L.

REAC: Expanded similarly to AL, to consider three nodal degrees of freedom rather than two.

In what follows all programs which need to be modified, or are new, are described. The programs ASSEM, ELASS, BOUND, and SLBSI do not require any change. Therefore they will not be included again, since they have already been described in the previous section.

Program 34   Main program

The dimensions of some of the arrays used are modified as required by the data structure changes explained above, in such a way that the program can treat plane frames with a maximum of 100 nodes, 100 bars, and 20 boundary nodes. The basic program parameters are initialized accordingly. Note that the nodal degrees of freedom indicator, NDF, is now set equal to 3.

The FORTRAN code of the main program is the following:

```
C
C          PROGRAM 34 - MAIN PROGRAM
C
C    STATIC ANALYSIS FOR   PLANE FRAME SYSTEMS
C
```

```
      COMMON NRMX,NCMX,NDFEL,NN,NE,NLN,NBN,NDF,NNE,N,MS,IN,IO,E,G
      DIMENSION X(100),Y(100),CON(200),PROP(200),IB(80),TK(300,30),
     *AL(300),REAC(300),FORC(600),ELST(6,6),V(30)
C
C     INITIALIZATION OF PROGRAM PARAMETERS
C
C     NRMX  = ROW DIMENSION FOR THE STIFFNESS
C             MATRIX
C     NCMX  = COLUMN DIMENSION FOR THE STIFFNESS
C             MATRIX, OR MAXIMUM HALF BAND WIDTH
C             ALLOWED
C     NDF   = NUMBER OF DEGREES OF FREEDOM PER
C             NODE
C     NNE   = NUMBER OF NODES PER ELEMENT, EQUAL
C             TO 2 FOR BAR SYSTEMS
C     NDFEL = TOTAL NUMBER OF DEGREES OF FREEDOM
C             FOR ONE ELEMENT
C
      NRMX=300
      NCMX=30
      NDF=3
      NNE=2
      NDFEL=NDF*NNE
C
C     ASSIGN DATA SET NUMBERS TO IN, FOR INPUT,
C     AND IO, FOR OUTPUT
C
      IN=5
      IO=6
C
C APPLY THE ANALYSIS STEPS
C
C     INPUT
C
      CALL INPUT(X,Y,CON,PROP,AL,IB,REAC)
C
C     ASSEMBLING OF TOTAL STIFFNESS MATRIX
C
      CALL ASSEM(X,Y,CON,PROP,TK,ELST,AL)
C
C     INTRODUCTION OF BOUNDARY CONDITIONS
C
      CALL BOUND(TK,AL,REAC,IB)
C
C     SOLUTION OF THE SYSTEM OF EQUATIONS
C
      CALL SLBSI(TK,AL,V,N,MS,NRMX,NCMX)
C
C     COMPUTATION OF MEMBER FORCES
C
      CALL FORCE(CON,PROP,FORC,REAC,X,Y,AL)
C
C     OUTPUT
C
      CALL OUTPT(CON,AL,FORC,REAC)
C
      CALL EXIT
      END
```

Program 35   Input program (INPUT)

The new subprogram for input data reflects the need to take into account som additional data, not needed for plane trusses. Considering the previous case the input cards are as follows.

(1) *Basic Parameters Card.* Unchanged.

(2) *Nodal Coordinates Card.* Unchanged.

(3) *Element Connectivity and Properties Cards.* The moment of inertia of each bar is now also needed. With format 3I10, 2F10.2 there will be a card for each bar, containing the member number, the numbers of the start and end nodes of the member, the cross-sectional area, and the moment of inertia.

(4) *Nodal Loads Cards.* The possibility of applying a bending moment as a nodal load is now considered. For each loaded node there will be one card for the input of the node number, and the loads F, P, and M, with format I10, 3F10.2, corresponding to that node.

(5) *Support Data Cards.* It is considered that the nodal degrees of freedom are three rather than two. For each boundary node three will be a card containing the node number, the indicators of the status of the unknowns u, v, and θ, and the prescribed values, when applicable, for those three unknowns, with format 4I10, 3F10.4.

Considering the modifications in the input of data explained above, the following FORTRAN code for the input program can be easily understood.

```
      SUBROUTINE INPUT(X,Y,CON,PROP,AL,IB,REAC)
C
C            PROGRAM 35 - INPUT PROGRAM
C
      COMMON NRMX,NCMX,NDFEL,NN,NE,NLN,NBN,NDF,NNE,N,MS,IN,IO,E,G
      DIMENSION X(1),Y(1),CON(1),PROP(1),AL(1),IB(1),REAC(1),W(3),IC(3)
C
C         ELEMENT PROPERTIES, NODAL LOADS, AND PRESCRIBED
C         UNKNOWN VALUES
C  IC = AUXILIARY ARRAY TO STORE TEMPORARELY THE CONNECTIVITY
C        OF AN ELEMENT, AND THE BOUNDARY UNKNOWNS STATUS
C        INDICATORS
C
C  READ BASIC PARAMETERS
C
C NN  = NUMBER OF NODES
C NE  = NUMBER OF ELEMENTS
C NLN = NUMBER OF LOADED NODES
C NBN = NUMBER OF BOUNDARY NODES
C E   = MODULUS OF ELASTICITY
C
      WRITE(IO,20)
   20 FORMAT(' ',130('*'))
      READ(IN,1) NN,NE,NLN,NBN,E
      WRITE(IO,21) NN,NE,NLN,NBN,E
   21 FORMAT(//' INTERNAL DATA'//' NUMBER OF NODES          :',I5/' NUMBE
     *R OF ELEMENTS       :',I5/' NUMBER OF LOADED NODES   :',I5/' NUMBER
     *OF SUPPORT NODES :',I5/' MODULUS OF ELASTICITY :',F15.0//' NODAL C
     *OORDINATES'/7X,'NODE',6X,'X', 9X,'Y')
    1 FORMAT(4I10,2F10.0)
C
C  READ NODAL COORDINATES IN ARRAY X ANO Y
C
```

```
      READ(IN,2) (I,X(I),Y(I),J=1,NN)
      WRITE(IO,2) (I,X(I),Y(I),I=1,NN)
    2 FORMAT(I10,2F10.2)
C
C   READ ELEMENT CONNECTIVITY IN ARRAY CON
C   AND ELEMENT PROPERTIES IN ARRAY PROP
C
      WRITE(IO,22)
   22 FORMAT(/' ELEMENT CONNECTIVITY AND PROPERTIES'/4X,'ELEMENT',3X
     *ART NODE   END NODE',5X,'AREA', 5X,'M. OF INERTIA')
      DO 3 J=1,NE
      READ(IN,4) I,IC(1),IC(2),W(1),W(2)
      WRITE(IO,34) I,IC(1),IC(2),W(1),W(2)
   34 FORMAT(3I10,2F15.5)
    4 FORMAT(3I10,2F10.2)
      N1=NNE*(I-1)
      PROP(N1+1)=W(1)
      PROP(N1+2)=W(2)
      CON(N1+1)=IC(1)
    3 CON(N1+2)=IC(2)
C
C   COMPUTE N, ACTUAL NUMBER OF UNKNOWNS, AND CLEAR THE LOAD
C   VECTOR
C
      WRITE(IO,23)
   23 FORMAT(/' NODAL LOADS'/7X,'NODE',5X,'PX',8X,'PY',8X,'MZ')
      N=NN*NDF
      DO 5 I=1,N
    5 AL(I)=0.
C
C   READ THE NODAL LOADS AND STORE THEM IN ARRAY AL
C
      DO 6 I=1,NLN
      READ (IN,8) J,(W(K),K=1,NDF)
      WRITE(IO,8) J,(W(K),K=1,NDF)
    8 FORMAT(I10,3F10.2)
      DO 6 K=1,NDF
      L=NDF*(J-1)+K
    6 AL(L)=W(K)
C
C   READ BOUNDARY NODES DATA. STORE UNKNOWNS STATUS
C   INDICATORS IN ARRAY IB, AND PRESCRIBED UNKNOWNS
C   VALUES IN ARRAY REAC
C
      WRITE(IO,24)
   24 FORMAT(/' BOUNDARY CONDITION DATA'/27X,'STATUS',24X,'PRESCRIB
     *LUES'/19X,'(0:PRESCRIBED, 1:FREE)'/7X,'NODE',8X,'U',9X,'V',8X
     *,16X,'U',9X,'V',8X,'RZ')
      DO 7 I=1,NBN
      READ(IN,9) J,(IC(K),K=1,NDF),(W(K),K=1,NDF)
      WRITE(IO,10) J,(IC(K),K=1,NDF),(W(K),K=1,NDF)
      L1=(NDF+1)*(I-1)+1
      L2=NDF*(J-1)
      IB(L1)=J
      DO 7 K=1,NDF
      N1=L1+K
      N2=L2+K
      IB(N1)=IC(K)
    7 REAC(N2)=W(K)
    9 FORMAT(4I10,3F10.4)
   10 FORMAT(4I10,10X,3F10.4)
      RETURN
      END
```

Program 36   Evaluation of the plane frame member stiffness matrix (STIFF)

The original STIFF subprogram should be replaced by the following:

```
      SUBROUTINE STIFF(NEL,X,Y,PROP,CON,ELST,AL)
C
C               PROGRAM 36
C   COMPUTATION OF ELEMENT STIFFNESS MATRIX FOR THE CURRENT ELEMENT
C
      COMMON NRMX,NCMX,NDFEL,NN,NE,NLN,NBN,NDF,NNE,N,MS,IN,IO,E,G
      DIMENSION X(1),Y(1),CON(1),PROP(1),ELST(NDFEL,NDFEL),AL(1),
     *ROT(6,6),V(6)
C
C   NEL  = CURRENT ELEMENT NUMBER
C   N1   = NUMBER OF START NODE
C   N2   = NUMBER OF END NODE
C   AX   = AREA OF ELEMENT CROSS-SECTION
C   YZ   = MOMENT OF INERTIA OF ELEMENT CROSS-SECTION
C
      L=NNE*(NEL-1)
      N1=CON(L+1)
      N2=CON(L+2)
      AX=PROP(L+1)
      YZ=PROP(L+2)
C
C   COMPUTE LENGTH OF ELEMENT, AND SINE AND COSINE OF ITS LOCAL
C   X AXIS, AND STORE  IN D, SI, AND CO, RESPECTIVELY
C
      DX=X(N2)-X(N1)
      DY=Y(N2)-Y(N1)
      D=SQRT(DX**2+DY**2)
      CO=DX/D
      SI=DY/D
      DO 1 I=1,6
      DO 1 J=1,6
      ELST(I,J)=0.
C
C   FORM ELEMENT ROTATION MATRIX
C
    1 ROT(I,J)=0.
      ROT(1,1)=CO
      ROT(1,2)=SI
      ROT(2,1)=-SI
      ROT(2,2)=CO
      ROT(3,3)=1.
      DO 2 I=1,3
      DO 2 J=1,3
    2 ROT(I+3,J+3)=ROT(I,J)
C
C   COMPUTE ELEMENT LOCAL STIFFNESS MATRIX
C
      ELST(1,1)=E*AX/D
      ELST(1,4)=-ELST(1,1)
      ELST(2,2)=12*E*YZ/(D**3)
      ELST(2,3)=6*E*YZ/(D*D)
      ELST(2,5)=-ELST(2,2)
      ELST(2,6)=ELST(2,3)
      ELST(3,2)=ELST(2,3)
      ELST(3,3)=4*E*YZ/D
      ELST(3,5)=-ELST(2,3)
      ELST(3,6)=2*E*YZ/D
```

```
      ELST(4,1)=ELST(1,4)
      ELST(4,4)=ELST(1,1)
      ELST(5,2)=ELST(2,5)
      ELST(5,3)=ELST(3,5)
      ELST(5,5)=ELST(2,2)
      ELST(5,6)=ELST(3,5)
      ELST(6,2)=ELST(2,6)
      ELST(6,3)=ELST(3,6)
      ELST(6,5)=ELST(5,6)
      ELST(6,6)=ELST(3,3)
C
C  ROTATE ELEMENT STIFFNESS MATRIX TO GLOBAL COORDINATES
C
      CALL BTAB3(ELST,ROT,V,NDFEL,NDFEL)
      RETURN
      END
```

In the above program N1 and N2 are the numbers of the start and end nodes for the current element NEL. AX is the cross-sectional area, and YZ is the moment of inertia. D is the length of the member, while DX and DY are the projections of D over the local x and y axes, respectively. CO and SI are the cosine and sine of the β angle (see Figure 4.9).

The element stiffness matrix array, ELST, is cleared because in this case not all its coefficients are assigned a value, when computing the terms corresponding to the local element stiffness matrix. When this matrix is rotated to global axes new non zero terms will appear, and will remain in array ELST. Thus, when computing the local stiffness matrix for a member all the terms in ELST corresponding to the global stiffness matrix of the previous member must be eliminated.

The element rotation matrix **R** is computed according to Equation (o) of Example 4.6, and is placed in array ROT. Then the terms of the local member stiffness matrix are computed according to Equation (l) of Example 4.6, and are placed in array ELST. Finally this matrix is rotated to the global reference system according to Equation (r). For this the subprogram BTAB3, explained in Chapter 2, is used. Notice that the global member stiffness matrix is also stored in array ELST.

Program 37   Evaluation of member end forces (FORCE)

The FORTRAN code corresponding to the subprogram FORCE, written anew, is the following:

```
      SUBROUTINE FORCE(CON,PROP,FORC,REAC,X,Y,AL)
C
C           PROGRAM 37
C  COMPUTATION OF ELEMENT FORCES
C
      COMMON NRMX,NCMX,NDFEL,NN,NE,NLN,NBN,NDF,NNE,N,MS,IN,IO,E,G
      DIMENSION CON(1),PROP(1),FORC(1),REAC(1),X(1),Y(1),AL(1),ROT(5
     *U(6),F(6),UL(6),FG(6)
      DO 1 I=1,N
    1 REAC(I)=0.
      DO 100 NEL=1,NE
```

```
C
C   NEL   = CURRENT ELEMENT NUMBER
C   N1    = NUMBER OF START NODE
C   N2    = NUMBER OF END NODE
C   AX    = AREA OF ELEMENT CROSS-SECTION
C   YZ    = MOMENT OF INERTIA OF ELEMENT CROSS-SECTION
C
      L=NNE*(NEL-1)
      N1=CON(L+1)
      N2=CON(L+2)
      AX=PROP(L+1)
      YZ=PROP(L+2)
C
C   COMPUTE LENGTH OF ELEMENT, AND SINE AND COSINE OF ITS LOCAL
C   X AXIS, AND STORE  IN D, SI, AND CO, RESPECTIVELY
C
      DX=X(N2)-X(N1)
      DY=Y(N2)-Y(N1)
      D=SQRT(DX**2+DY**2)
      CO=DX/D
      SI=DY/D
C
C   FORM ELEMENT ROTATION MATRIX
C
      ROT(1,1)=CO
      ROT(1,2)=SI
      ROT(1,3)=0.
      ROT(2,1)=-SI
      ROT(2,2)=CO
      ROT(2,3)=0.
      ROT(3,1)=0.
      ROT(3,2)=0.
      ROT(3,3)=1.
C
C   ROTATE ELEMENT NODAL DISPLACEMENTS TO ELEMENT LOCAL
C   REFERENCE FRAME, AND STORE IN ARRAY UL
C
      K1=NDF*(N1-1)
      K2=NDF*(N2-1)
      DO 2 I=1,3
      J1=K1+I
      J2=K2+I
      U(I)=AL(J1)
    2 U(I+3)=AL(J2)
      DO 3 I=1,3
      UL(I)=0.
      UL(I+3)=0.
      DO 3 J=1,3
      UL(I)=UL(I)+ROT(I,J)*U(J)
    3 UL(I+3)=UL(I+3)+ROT(I,J)*U(J+3)
C
C   COMPUTE MEMBER END FORCES IN LOCAL COORDINATES
C
      F(1)=E*AX/D*(UL(1)-UL(4))
      F(2)=12*E*YZ/(D**3)*(UL(2)-UL(5))+6*E*YZ/(D*D)*(UL(3)+UL(6))
      F(3)=6*E*YZ/(D*D)*(UL(2)-UL(5))+2*E*YZ/D*(2*UL(3)+UL(6))
      F(6)=6*E*YZ/(D*D)*(UL(2)-UL(5))+2*E*YZ/D*(UL(3)+2*UL(6))
      F(4)=-F(1)
      F(5)=-F(2)
      I1=6*(NEL-1)
C
C   STORE MEMBER END FORCES IN ARRAY FORC
```

```
C
      DO 4 I=1,6
      I2=I1+I
    4 FORC(I2)=F(I)
C
C  ROTATE MEMBER END FORCES TO THE GLOBAL REFERENCE FRAME
C  AND STORE IN ARRAY FG
C
      DO 5 I=1,3
      FG(I)=0.
      FG(I+3)=0.
      DO 5 J=1,3
      FG(I)=FG(I)+ROT(J,I)*F(J)
    5 FG(I+3)=FG(I+3)+ROT(J,I)*F(J+3)
C
C  ADD ELEMENT CONTRIBUTION TO NODAL RESULTANTS,
C  IN ARRAY REAC
C
      DO 6 I=1,3
      J1=K1+I
      J2=K2+I
      REAC(J1)=REAC(J1)+FG(I)
    6 REAC(J2)=REAC(J2)+FG(I+3)
  100 CONTINUE
      RETURN
      END
```

where the meaning of the variables N1, N2, AX, YZ, DX, DY, D, CO and SI, are the same as for the subprogram STIFF.

This program will compute, for each member, the member end forces and their contribution to the nodal resultants, which will be the reactions for support nodes, and the applied loads for free nodes. These contributions will be added to array REAC, which up to this point contained values of prescribed unknowns. Correspondingly the array REAC is cleared before adding any contribution to it.

The computation of the member end forces is done within a loop from 1 to the total number of elements. For the current element NEL the member parameters are first computed, and then the nodal rotation matrix **r** is evaluated and stored in array ROT. K1 and K2 are the numbers of coefficients in AL before those containing the displacements corresponding to nodes N1 and N2, respectively. These displacements are retrieved from array AL and placed in array U, first those for node N1, and then those for node N2. Those nodal displacements referred to the global reference system are next rotated to the local reference system and placed in array UL.

The member end forces are computed according to expression (j) of Example 4.6 and are placed in array F and copied into the total member end forces array FORC. These member end forces are referred to the local reference system. They are next rotated to the global reference system, and stored in array FG. Finally these member contributions to the nodal resultants are added to the array REAC.

Program 38   Output program (OUTPT)

The output program OUTPT prints the nodal displacements, the nodal resultants or reactions, and the member end forces. The following FORTRAN code for OUTPT is self-explanatory.

```
      SUBROUTINE OUTPT(CON,AL,FORC,REAC)
C
C            PROGRAM 38 - OUTPUT PROGRAM
C
      COMMON NRMX,NCMX,NDFEL,NN,NE,NLN,NBN,NDF,NNE,N,MS,IN,IO,E,G
      DIMENSION CON(1),AL(1),FORC(1),REAC(1)
C
C  WRITE NODAL DISPLACEMENTS
C
      WRITE(IO,1)
    1 FORMAT(//1X,130('*')//' RESULTS'//' NODAL DISPLACEMENTS'/7X,'NODE'
     *,11X,'U',14X,'V',13X,'RZ')
      DO 10 I=1,NN
      K1=NDF*(I-1)+1
      K2=K1+NDF-1
   10 WRITE(IO,2) I,(AL(J),J=K1,K2)
    2 FORMAT(I10,6F15.4)
C
C  WRITE NODAL REACTIONS
C
      WRITE(IO,3)
    3 FORMAT(/' NODAL REACTIONS'/7X,'NODE',10X,'PX',13X,'PY',13X,'MZ')
      DO 20 I=1,NN
      K1=NDF*(I-1)+1
      K2=K1+NDF-1
   20 WRITE(IO,2) I,(REAC(J),J=K1,K2)
C
C  WRITE MEMBER END FORCES
C
      WRITE(IO,4)
    4 FORMAT(/' MEMBER FORCES'/6X,'MEMBER',5X,'NODE', 9X,'FX',13X,'FY',
     *13X,'MZ')
      DO 30 I=1,NE
      K1=6*(I-1)+1
      K2=K1+2
      N1=NNE*(I-1)
      WRITE(IO,6) I,CON(N1+1),(FORC(J),J=K1,K2)
      K1=K2+1
      K2=K1+2
   30 WRITE(IO,7) CON(N1+2),(FORC(J),J=K1,K2)
    6 FORMAT(2I10,3F15.4)
    7 FORMAT(I20,3F15.4)
      WRITE(IO,5)
    5 FORMAT(//1X,130('*'))
      RETURN
      END
```

Example 4.7
As an illustration of the application of this set of programs we consider the case of the plane frame shown in Figure 4.10, which shows the overall dimen-

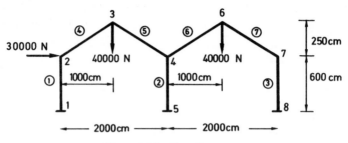

Figure 4.10   Plane frame

sions loads, and node and member numbering. The modulus of elasticity is equal
to $21\,000\,000$ N/cm^2. For members 2, 4, 5, 6, and 7 the cross-sectional area and
moment of inertia are $A_x = 200$ cm^2 and $I_z = 6666$ cm^4, while for members
1 and 3, $A_x = 400$ cm^2 and $I_z = 13\,333$ cm^4.

The input cards for this problem are

```
8           7           3           3  21000000
1
2                       600.
3  1000.             850.
4  2000.             600.
5  2000.
6  3000.             850.
7  4000.             600.
8  4000.
1           1           2  400.              13333.
2           5           4  200.              6666.
3           8           7  400.              13333.
4           2           3  200.              6666.
5           3           4  200.              6666.
6           4           6  200.              6666.
7           6           7  200.              6666.
2  30000.
3                      -40000.
6                      -40000.
1
5
8
```

The output produced by the program is

INTERNAL DATA

NUMBER OF NODES   : 8
NUMBER OF ELEMENTS   : 7
NUMBER OF LOADED NODES   : 3
NUMBER OF SUPPORT NODES : 3
MODULUS OF ELASTICITY   : 21000000.

NODAL COORDINATES

NODE	X	Y
1	0.00	0.00
2	0.00	600.00
3	1000.00	850.00
4	2000.00	600.00
5	2000.00	0.00
6	3000.00	850.00
7	4000.00	600.00
8	4000.00	0.00

ELEMENT CONNECTIVITY AND PROPERTIES

ELEMENT	START NODE	END NODE	AREA	M. OF INERTIA
1	1	2	400.00000	13333.00000
2	5	4	200.00000	6666.00000
3	8	7	400.00000	13333.00000
4	2	3	200.00000	6666.00000
5	3	4	200.00000	6666.00000
6	4	6	200.00000	6666.00000
7	6	7	200.00000	6666.00000

NODAL LOADS

NODE	PX	PY	MZ
2	30000.00	0.00	0.00
3	0.00	-40000.00	0.00
6	0.00	-40000.00	0.00

BOUNDARY CONDITION DATA

	STATUS (0: PRESCRIBED, 1: FREE)			PRESCRIBED VALUES		
NODE	U	V	RZ	U	V	RZ
1	0	0	0	0.0000	0.0000	0.0000
5	0	0	0	0.0000	0.0000	0.0000
8	0	0	0	0.0000	0.0000	0.0000

RESULTS

NODAL DISPLACEMENTS

NODE	U	V	RZ
1	0.0000	0.0000	0.0000
2	-0.8044	-0.0014	-0.0002
3	0.4364	-5.0122	0.0007
4	1.6762	-0.0055	-0.0026
5	0.0000	0.0000	0.0000
6	3.0591	-5.5776	0.0024
7	4.4428	-0.0016	-0.0070
8	0.0000	0.0000	0.0000

NODAL REACTIONS

NODE	PX	PY	MZ
1	13511.5568	19431.6355	-3953576.9840
2	30000.0000	-0.0000	0.0000
3	0.0000	-40000.0000	-0.0002
4	-0.0001	0.0000	-0.0002
5	-7015.7920	38618.3103	2706715.4284
6	-0.0001	-40000.0000	0.0000
7	0.0001	0.0000	-0.0000
8	-36495.7647	21950.0542	14210023.9600

MEMBER FORCES

MEMBER	NODE	FX	FY	MZ
1	1	19431.6355	-13511.5568	-3953576.9840
	2	-19431.6355	13511.5568	-4153357.0949
2	5	38618.3103	7015.7920	2706715.4284
	4	-38618.3103	-7015.7920	1502759.7990
3	8	21950.0542	36495.7647	14210023.9600
	7	-21950.0542	-36495.7647	7687434.8505
4	2	46925.2743	8298.3528	4153357.0950
	3	-46925.2743	-8298.3528	4400389.1686
5	3	47200.9716	9401.1420	4400389.1686
	4	-47200.9716	-9401.1420	-5290086.1649
6	4	39783.8473	8659.4964	3787326.3657
	6	-39783.8473	-8659.4964	5138678.2195
7	6	40729.7624	-12443.1574	-5138678.2195
	7	-40729.7624	12443.1574	-7687434.8505

4.5 DYNAMIC PROBLEMS

To establish the dynamic version of the principle of virtual displacements we employ d'Alembert's principle, which can be written as

$$F = ma \tag{4.38}$$

or

$$F - ma = 0$$

where m is mass and a acceleration.

For a continuous body the inertia forces given by D'Alembert's principle can be written as

$$-\rho \, \frac{d^2 u}{dt^2}$$

and

$$-\rho \, \frac{d^2 v}{dt^2}$$

in the equilibrium equations, and are body forces similar to b_x and b_y. Hence the equilibrium equations can now be written

$$\frac{\partial \sigma_x}{\partial x} + \frac{\partial \tau}{\partial y} + b_x - \rho \, \ddot{u} = 0$$

$$\frac{\partial \tau}{\partial x} + \frac{\partial \sigma_y}{\partial y} + b_y - \rho \, \ddot{v} = 0 \tag{4.39}$$

where

$$\ddot{u} = \frac{d^2 u}{dt^2}, \text{ and } \ddot{v} = \frac{d^2 v}{dt^2}$$

The stress and displacement boundary conditions are assumed to be the same as seen previously, i.e. time independent.

As the inertia forces act as body forces we can simply extend the Principle of Virtual Displacements, Equation (4.20), for a given time t, as follows

$$\iint \{\sigma_x \, \delta\epsilon_x + \sigma_y \, \delta\epsilon_y + \tau \, \delta\gamma\} \, dA + \iint \{\rho \, \ddot{u} \, \delta u + \rho \, \ddot{v} \, \delta v\} \, dA$$

$$= \int_{S_\sigma} \{\bar{p}_x \, \delta u + \bar{p}_y \, \delta v\} \, dS + \iint \{b_x \, \delta u + b_y \, \delta v\} \, dA \qquad (4.40)$$

In many engineering problems we are interested in finding the natural frequencies of the structure. These are those frequencies for which once started, the system would in theory continue to vibrate indefinitely. In order to find the frequencies we assume that there are no external forces acting on the system (i.e. $p = b = 0$). This gives

$$\iint \{\sigma_x \, \delta\epsilon_x + \sigma_y \, \delta\epsilon_y + \tau \, \delta\gamma\} \, dA + \iint \{\rho \, \ddot{u} \, \delta u + \rho \, \ddot{v} \, \delta v\} \, dA = 0 \qquad (4.41)$$

For the case of linearly elastic constitutive relationships the first term in Equation (4.41) is proportional to the stiffness of the structure and the second term to the mass.

For a discrete system Equation (4.41) can be written,

$$\mathbf{KU} + \mathbf{M\ddot{U}} = 0 \qquad (4.42)$$

This formula represents a set of second order equations in time, whose solution can be written,

$$\mathbf{U} = \begin{Bmatrix} u_1 \\ u_2 \\ \cdot \\ \cdot \\ \cdot \end{Bmatrix} = \begin{Bmatrix} A_1 \cos \omega t + B_1 \sin \omega t \\ A_2 \cos \omega t + B_2 \sin \omega t \\ \cdot \\ \cdot \\ \cdot \end{Bmatrix} \qquad (4.43)$$

Hence the accelerations are

$$\mathbf{\ddot{U}} = -\omega^2 \, \mathbf{U} \qquad (4.44)$$

where ω is the circular frequency.

Equation (4.42) becomes

$$(\mathbf{K} - \omega^2 \, \mathbf{M})\mathbf{U} = \mathbf{0} \qquad (4.45)$$

which can be solved as an eigenvalue–eigenvector problem. Each eigenvalue

ω_i represents a natural frequency for the system. The corresponding eigen-vectors, U_i, are the 'shapes' associated with the frequencies.

Example 4.7
Consider the case of a simply supported 'prismatic beam' without any external loading. For this case the internal energy part of Equation (4.41) reduces to

$$EI \int_0^l \left(\frac{d^2 v}{dx^2}\right) \left(\frac{d^2 \delta v}{dx^2}\right) dx \tag{a}$$

The inertia terms are given by the integration of

$$A \int_0^l \rho \, \ddot{v} \, \delta v \, dx \tag{b}$$

where A = cross-sectional area of the beam (assumed to be constant). Note that only inertia due to vertical deflections has been considered.

The equilibrium relationships can now be written,

$$EI \int_0^l \left(\frac{d^2 v}{dx^2}\right) \left(\frac{d^2 \delta v}{dx^2}\right) dx + \rho A \int_0^l \ddot{v} \, \delta v \, dx = 0 \tag{c}$$

We now assume that the vertical deflections for the beam are given, as in Example 4.6, by

$$v = \Phi U^e$$

where

$$U^e = [v_1 \quad \theta_1 \quad v_2 \quad \theta_2]$$

and

$$\ddot{v} = \Phi \ddot{U}^e \tag{d}$$

Substituting formula (d) into (c) we obtain

$$\delta U^{e,T} \int_0^l B^T EI\, B\, dx\, U^e + \delta U^{e,T} \rho A \int_0^l \Phi^T \Phi\, dx\, \ddot{U}^e = 0 \qquad (e)$$

or

$$K^e U^e + M^e \ddot{U}^e = 0 \qquad (f)$$

The stiffness matrix, K^e, (for a detailed deduction of the stiffness matrix see Example 4.6), is given by

$$K^e = \frac{EI}{l^3}
\begin{bmatrix}
12 & 6l & -12 & 6l \\
 & 4l^2 & -6l & 2l^2 \\
 & & 12 & -6l \\
\text{symm} & & & 4l^2
\end{bmatrix} \qquad (g)$$

and the mass matrix can be written after multiplication and integration as

$$M^e = \rho A \int_0^l \Phi^T \Phi\, dx = \frac{\rho A l}{420}
\begin{bmatrix}
156 & 22l & 54 & -13l \\
 & 4l^2 & 13l & -3l^2 \\
 & & 156 & -22l \\
\text{symm} & & & 4l^2
\end{bmatrix} \qquad (h)$$

Equation (f) has a harmonic solution of the type shown in Equation (4.43), hence $\ddot{U}^e = -\omega^2\, U^e$, which gives

$$(K^e - \omega^2 M^e)U^e = 0 \qquad (i)$$

Consider now that the simply supported beam is composed by only one element. For this case the vertical displacements v_1 and v_2 (at 0 and l) are zero

Hence, taking Equations (g) and (h) into consideration, we can write Equation (i) as

$$\left\{ \frac{EI}{l} \begin{bmatrix} 4 & 2 \\ 2 & 4 \end{bmatrix} - \frac{\rho A \omega^2 l^3}{420} \begin{bmatrix} 4 & -3 \\ -3 & 4 \end{bmatrix} \right\} \begin{Bmatrix} \theta_0 \\ \theta_l \end{Bmatrix} = \begin{Bmatrix} 0 \\ 0 \end{Bmatrix} \tag{j}$$

We can define a λ parameter, $\lambda = \rho A l^4 \omega^2 / 420 EI$ and write (j) as

$$\begin{bmatrix} 4 - 4\lambda & 2 + 3\lambda \\ 2 + 3\lambda & 4 - 4\lambda \end{bmatrix} \begin{Bmatrix} \theta_0 \\ \theta_l \end{Bmatrix} = \begin{Bmatrix} 0 \\ 0 \end{Bmatrix} \tag{k}$$

The determinant of Equation (k) is

$$D = 12 - 44\lambda + 7\lambda^2 = 0 \tag{l}$$

$\therefore \lambda_1 = 2/7$, $\lambda_2 = 6$, which gives,

$$\frac{\rho A l^4 \omega^2}{EI} = 120 \text{ or } 2520$$

The exact solution for this case is actually 97.4 and 1559. Note that we have not found the *exact* solution because we used an approximation for v_2, i.e. 'cubic' variation, while the actual shape is more complex. We may improve on the results by taking a larger number of elements, as will be seen later on.

Example 4.8
Consider the case of a multi-storey building as shown in Figure 4.11. The building is assumed to vibrate in the y direction.

The in-plane stiffness of the floors is large by comparison with the stiffness of the columns. Hence, we can assume that the floors will only displace horizontally during motion, i.e. there will not be any end rotations on the columns. The stiffness of a beam element is given by

$$K^e = \frac{EI_c}{l^3} \begin{bmatrix} 12 & -12 \\ -12 & 12 \end{bmatrix} \tag{a}$$

The stiffness for the two-floor structures shown in Figure 4.11, is given by

$$KU = \frac{EI}{l^3} \begin{bmatrix} 12 & -12 \\ -12 & 24 \end{bmatrix} \begin{Bmatrix} v_1 \\ v_2 \end{Bmatrix} \tag{b}$$

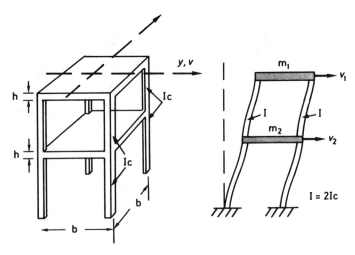

Figure 4.11   Multi-storey building

The mass of the columns can be neglected by comparison with that of the floors which give the following inertia forces

$$\mathbf{M\ddot{U}} = \rho V \begin{bmatrix} 1 & 0 \\ 0 & 1 \end{bmatrix} \begin{Bmatrix} \ddot{v}_1 \\ \ddot{v}_2 \end{Bmatrix} \tag{c}$$

$V = hb^2$ is the volume of the slabs; h = thickness and b = side length of the floors.

Equations (b) and (c) give

$$\frac{EI}{l^3} \begin{bmatrix} 12 & -12 \\ -12 & 24 \end{bmatrix} \begin{Bmatrix} v_1 \\ v_2 \end{Bmatrix} + \rho V \begin{bmatrix} 1 & 0 \\ 0 & 1 \end{bmatrix} \begin{Bmatrix} \ddot{v}_1 \\ \ddot{v}_2 \end{Bmatrix} = \begin{Bmatrix} 0 \\ 0 \end{Bmatrix} \tag{d}$$

This equation can be written as an eigenvalue–eigenvector problem

$$\begin{bmatrix} 1-\lambda & -1 \\ -1 & 2-\lambda \end{bmatrix} \begin{Bmatrix} v_1 \\ v_2 \end{Bmatrix} = \begin{Bmatrix} 0 \\ 0 \end{Bmatrix} \tag{e}$$

$$\lambda = \frac{\rho V l^3}{12EI} \omega^2$$

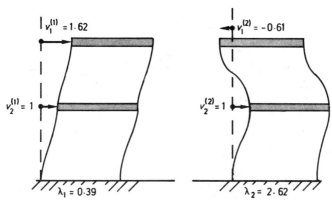

Figure 4.12   Natural frequencies and corresponding nodal frequencies

Hence,

$$1 - 3\lambda + \lambda^2 = 0$$

or

$$\lambda_1 = \frac{3 - \sqrt{5}}{2} \simeq 0.39$$

and

$$\lambda_2 = \frac{3 + \sqrt{5}}{2} \simeq 2.62 \tag{f}$$

The corresponding eigenvectors are

for $\lambda_1 \rightarrow v_1^{(1)} = +1.62, \quad v_2^{(1)} = +1.$

for $\lambda_2 \rightarrow v_1^{(2)} = -0.61, \quad v_2^{(2)} = +1.$ (g)

The forms of these eigenvectors are plotted in Figure 4.12.

It is interesting to compare the energies required to produce the shapes corresponding to λ_1, λ_2.

The internal or strain energy of the structure can be written as

$$V = \tfrac{1}{2} \{v_1 \quad v_2\} \frac{EI}{l^3} \begin{bmatrix} 12 & -12 \\ -12 & 24 \end{bmatrix} \begin{Bmatrix} v_1 \\ v_2 \end{Bmatrix}$$

so that

$$V = \frac{EI}{l^3} \{12v_1^2 + 24v_2^2 - 24v_1 v_2\} \tag{h}$$

For $\lambda_1 = 0.39$, $v_1^{(1)} = 1.62$ and $v_2^{(1)} = 1$ we have

$$V^{(1)} = \frac{EI}{2l^3} (16.8) \tag{i}$$

For $\lambda_2 = 2.62$, $v_1^{(2)} = -0.61$ and $v_2^{(2)} = 1$, we obtain

$$V^{(2)} = \frac{EI}{2l^3} (43.07) \tag{j}$$

We see that $V^{(1)} < V^{(2)}$, hence it will require less energy to excite the structure in the λ_1 frequency. Because of this it is usual to check that the frequencies of any exciting forces are far removed from the first few natural frequencies of the structure. However in many cases for which this condition is difficult to satisfy (for instance if the exciting force has many different frequency components) we only require that the exciting frequency components corresponding to these first few frequencies introduces a small amount of energy (i.e. their amplitudes are small).

4.6   PROGRAM FOR DYNAMIC ANALYSIS OF PLANE FRAME SYSTEMS

If we wish to develop a program for the computation of the natural frequencies and modes of vibration for plane frame structures, the same as in the case of the member stiffness matrix of Example 4.6, the member local mass matrix, Equation (h) of Example 4.7, will have to be expanded to include the axial deformation terms, and then rotated to the global reference system.

To compute the mass terms corresponding to the axial deformation, i.e. those of the truss member, we use the formula

$$\mathbf{M}_t = \rho A \int_0^l \mathbf{\Phi}^T \mathbf{\Phi} \, dx \tag{a}$$

For the u displacement we have that

$$u = u_1 \left(1 - \frac{x}{l}\right) + u_2 \frac{x}{l} \tag{b}$$

Hence,

$$\Phi = \left[\left(1 - \frac{x}{l}\right), \frac{x}{l} \right] \tag{c}$$

and the mass matrix may be written after integration as

$$M_t = \rho Al \begin{bmatrix} \dfrac{1}{3} & \dfrac{1}{6} \\[2mm] \dfrac{1}{6} & \dfrac{1}{3} \end{bmatrix} \tag{d}$$

The mass matrix for the beam element is

$$M = \frac{\rho Al}{420} \begin{bmatrix} 140 & 0 & 0 & 70 & 0 & 0 \\ & 156 & 22l & 0 & 54 & -13l \\ & & 4l^2 & 0 & 13l & -3l^2 \\ & \text{symm} & & 140 & 0 & 0 \\ & & & & 156 & -22l \\ & & & & & 4l^2 \end{bmatrix} \tag{e}$$

The rotation of the member local mass matrix to the global reference system, is again performed using the rotation matrix, Equation (o) of Example 4.6, according to

$$M^{g,e} = R^T M^e R \tag{f}$$

To implement a program for dynamic analysis of plane frame systems, we will take as a base the program for static analysis of plane frame systems, explained before. Because of simplicity, however, we will not use the symmetric banded storage scheme for the total system matrices. Rather, we will store the total stiffness and mass matrices as full matrices. In that way we will be able to use the programs for eigenvalue and eigenvector calculations given in Chapter 2.

The changes to be introduced are described below.

4.6.1 Data structure

The real variable G is now used to store the mass density ρ.

The array definitions, with regard to those of the program for static analysis of plane frame systems, are now as follows

CON: Unchanged.
X: Unchanged.
Y: Unchanged.
PROP: Unchanged.
TK: It now stores the total stiffness matrix as a symmetric but not banded matrix. Therefore it will have NRMX rows and NRMX columns.
ELST: Unchanged.
V: Unchanged.

The arrays IB, AL, FORC, and REAC, are not needed any more, and are therefore eliminated. On the other hand, the following new arrays are introduced.

IUNK: Array containing the row and column positions in the total matrices, for each nodal unknown. The element in IUNK for a prescribed displacement, will be equal to zero. Thus, if IUNK(NDF*(J − 1) + I) is different from zero gives the row and column positions of the Ith unknown of the Jth node. If zero, it indicates that unknown is prescribed and therefore is not taken into account in the governing system of equations.

TM: Array used to store the total mass matrix. Its structure is similar to that of array TK.

ELMA: Array used to store the element mass matrix. Its structure is similar to that of array ELST.

H: Auxiliary array used in program EIGG, of structure similar to that of arrays TK and TM.

All programs which have modifications, or are new, are given below. The program BTAB3, STIFF, DECOG, INVCH, MATMB, JACOB, and EIGG, were explained before and are therefore not included again.

Program 39   Main program

The dimensions of the arrays used is modified according to the data structure changes described above, and in such a way that the program can treat plane frames of a maximum of 50 nodes and 100 bars. The basic parameters are initialized accordingly. Note, however, that the integer variables NCMX, NLN, and MS, not needed any more, still remain in COMMON. This was done so to minimize the changes to be introduced.

The main program calls the subprograms INPUT for data input, ASSEMB to assemble the total stiffness and mass matrices, EIGG to compute the eigenvalues and eigenvectors, and OUTPT for the output of results. Its FORTRAN code is given below.

```
C
C                    PROGRAM 39 - MAIN PROGRAM
C
C     ****    PROGRAM FOR DYNAMIC ANALYSIS OF FRAME SYSTEMS *****
C
        COMMON NRMX,NCMX,NDFEL,NN,NE,NLN,NBN,NDF,NNE,N,MS,IN,IO,E,G
        DIMENSION X(50),Y(50),CON(200),PROP(200),IUNK(50),TK(150,150),
       *TM(150,150),H(150,150),ELST(6,6),ELMA(6,6),V(150)
C
C  INITIALIZATION OF PROGRAM PARAMETERS
C
C  NRMX   = ROW DIMENSION FOR THE STIFFNESS
C           MATRIX
C  NDF    = NUMBER OF DEGREES OF FREEDOM PER
C           NODE
C  NNE    = NUMBER OF NODES PER ELEMENT, EQUAL
C           TO 2 FOR BAR SYSTEMS
C  NDFEL  = TOTAL NUMBER OF DEGREES OF FREEDOM
C           FOR ONE ELEMENT
C
        NRMX=150
        NDF=3
        NNE=2
        NDFEL=NDF*NNE
C
C  ASSIGN DATA SET NUMBERS TO IN, FOR INPUT,
C  AND IO, FOR OUTPUT
C
        IN=5
        IO=6
C
C  APPLY THE ANALYSIS STEPS
C
C   INPUT
C
        CALL INPUT(X,Y,CON,PROP,IUNK)
C
C  ASSEMBLE TOTAL STIFFNESS MATRIX IN ARRAY TK,
C  AND TOTAL MASS MATRIX IN ARRAY TM
C
        CALL ASSEM(X,Y,CON,PROP,TK,TM,ELST,ELMA,IUNK)
C
C  COMPUTE NATURAL MODES AND FREQUENCIES
C
        CALL EIGG(TK,TM,H,V,0.000001,N,NRMX)
C
C   OUTPUT
C
        CALL OUTPT(TK,TM)
C
        CALL EXIT
        END
```

Program 40   Data input (INPUT)

The cards read by the subprogram for input of data have, with regard to the previous case, the following changes.

(1) *Basic Parameters Card.* It contains the number of nodes NN, the number of elements NE, the number of boundary nodes NBN, the modulus of elasticity E, and the mass density G, with format 3I10, 2F10.2. Note that since no loads are considered the variable NLN is not read.

(2) *Nodal Coordinates Cards.* Unchanged.

(3) *Element Connectivity and Properties Cards.* Unchanged.

(4) *Nodal Loads Cards.* Eliminated.

(5) *Support Data Cards.* Same as before but prescribed unknown values are not read. For each boundary node there will be a card containing the node number, and the indicators of the status of the unknowns displacements u, v, and θ, equal to 0 if prescribed, and to 1 if unknown, with format 4I10.

The FORTRAN code for the input program, considering those modifications, is given below. Note that the array IUNK is first initialized for all its elements to contain the number 1. Upon reading the boundary data, the elements corresponding to prescribed displacements are changed to 0.

Finally, after all boundary data are read, IUNK is modified so that elements corresponding to unknown displacements contain their row and column positions in the arrays TK and TM.

```
      SUBROUTINE INPUT(X,Y,CON,PROP,IUNK)
C
C         PROGRAM 40 - INPUT PROGRAM
C
      COMMON NRMX,NCMX,NDFEL,NN,NE,NLN,NBN,NDF,NNE,N,MS,IN,IO,E,G
      DIMENSION X(1),Y(1),CON(1),PROP(1),IUNK(1),IC(3),W(3)
C
C  W = AN AUXILIARY VECTOR TO TEMPORARELY STORE A SET OF
C         ELEMENT PROPERTIES
C  IC = AUXILIARY ARRAY TO STORE TEMPORARELY THE CONNECTIVITY
C         OF AN ELEMENT, AND THE BOUNDARY UNKNOWNS STATUS
C         INDICATORS
C
C  READ BASIC PARAMETERS
C
C NN  = NUMBER OF NODES
C NE  = NUMBER OF ELEMENTS
C NBN = NUMBER OF BOUNDARY NODES
C E   = MODULUS OF ELASTICITY
C G   = DENSITY
C
      WRITE(IO,20)
   20 FORMAT(' ',130('*'))
      READ(IN,1) NN,NE,NBN,E,G
      WRITE(IO,21) NN,NE,NBN,E,G
   21 FORMAT(//' INTERNAL DATA'//' NUMBER OF NODES        :',I5/' NUM
     *R OF ELEMENTS      :',I5/' NUMBER OF SUPPORT NODES :',I5/' MODUL
     * OF ELASTICITY :',F15.0/' DENSITY :',14X,F15.4//' NODAL COORDINA
     *S'/7X,'NODE',6X,'X',9X,'Y')
    1 FORMAT(3I10,2F10.2)
C
C  READ NODAL COORDINATES IN ARRAY X AND Y
C
```

```
    READ(IN,2) (I,X(I),Y(I),J=1,NN)
    WRITE(IO,2) (I,X(I),Y(I),I=1,NN)
  2 FORMAT(I10,2F10.2)

READ ELEMENT CONNECTIVITY IN ARRAY CON
AND ELEMENT PROPERTIES IN ARRAY PROP

    WRITE(IO,22)
 22 FORMAT(/' ELEMENT CONNECTIVITY AND PROPERTIES'/4X,'ELEMENT',3X,'ST
   *ART NODE   END NODE',5X,'AREA',  5X,'M. OF INERTIA')
    DO 3 J=1,NE
    READ(IN,4) I,IC(1),IC(2),W(1),W(2)
    WRITE(IO,34) I,IC(1),IC(2),W(1),W(2)
 34 FORMAT(3I10,2F15.5)
  4 FORMAT(3I10,2F10.2)
    N1=NNE*(I-1)
    PROP(N1+1)=W(1)
    PROP(N1+2)=W(2)
    CON(N1+1)=IC(1)
  3 CON(N1+2)=IC(2)

READ BOUNDARY CONDITIONS AND INITIALIZE IUNK TO CONTAIN 1 FOR
UNKNOWN DISPLACEMENTS AND 0 FOR PRESCRIBED DISPLACEMENTS

    N=NN*NDF
    DO 7 I=1,N
  7 IUNK(I)=1
    WRITE(IO,24)
 24 FORMAT(/' BOUNDARY CONDITION DATA'/27X,'STATUS'
   *      /19X,'(0:PRESCRIBED, 1:FREE)'/7X,'NODE',8X,'U',9X,'V',8X,'RZ'
   *)
    DO 6 I=1,NBN
    READ(IN,5) J,(IC(K),K=1,NDF)
    WRITE(IO,5) J,(IC(K),K=1,NDF)
  5 FORMAT(4I10)
    K1=NDF*(J-1)
    DO 6 K=1,NDF
    K2=K1+K
  6 IUNK(K2)=IC(K)

MODIFY IUNK PLACING ACTUAL ROW AND COLUMN NUMBER, INSTEAD OF 1, FOR
UNKNOWN DISPLACEMENTS. COMPUTE TOTAL NUMBER OF UNKNOWN DISPLACEMENTS
AND STORE IT IN INTEGER N

    K=0
    DO 10 I=1,N
    IF(IUNK(I))8,10,8
  8 K=K+1
    IUNK(I)=K
 10 CONTINUE
    N=K
    RETURN
    END
```

Program 41   Assembling of the total stiffness and mass matrices (ASSEM)

In addition to the total stiffness matrix, the subprogram ASSEM now has to assemble also the total mass matrix. It will first clear the arrays TK and TM, which will store those matrices. Then it will loop on the members of the structure, and for each of them will call the subprogram STIFF, to compute the element stiffness matrix, EMASS, to compute the element mass matrix, and ELASS, to add the element contributions to the arrays TK and TM.

The corresponding FORTRAN code is given below.

```
      SUBROUTINE ASSEM(X,Y,CON,PROP,TK,TM,ELST,ELMA,IUNK)
C
C             PROGRAM 41
C  ASSEMBLING OF THE TOTAL STIFFNESS AND MASS MATRICES
C
      COMMON NRMX,NCMX,NDFEL,NN,NE,NLN,NBN,NDF,NNE,N,MS,IN,IO,E,G
      DIMENSION X(1),Y(1),PROP(1),CON(1),TK(NRMX,NRMX),TM(NRMX,NRMX)
     *ELST(NDFEL,NDFEL),ELMA(NDFEL,NDFEL),IUNK(1),AL(1)
C
C CLEAR THE STIFFNESS AND MASS MATRICES
C
      DO 10 I=1,N
      DO 10 J=1,N
      TM(I,J)=0.
   10 TK(I,J)=0.
C
C LOOP ON THE ELEMENTS AND ASSEMBLE TOTAL STIFFNESS AND MASS MATRICES
C
      DO 20 NEL=1,NE
C
C COMPUTE THE ELEMENT STIFFNESS MATRIX
C
      CALL STIFF(NEL,X,Y,PROP,CON,ELST,AL)
C
C COMPUTE THE ELEMENT MASS MATRIX
C
      CALL EMASS(NEL,X,Y,PROP,CON,ELMA)
C
C ADD THE ELEMENT STIFFNESS AND MASS MATRICES
C TO THE TOTAL MATRICES
C
   20 CALL ELASS(NEL,CON,IUNK,ELST,ELMA,TK,TM)
      RETURN
      END
```

Program 42   Computation of the element mass matrix (EMASS)

The subprogram EMASS has the same structure as that of the subprogram STIFF, but rather than compute the element stiffness matrix it computes the element mass matrix, according to expression (h) of Example 4.7, storing it in array ELMA, as shown in the FORTRAN code given below. Note that due to symmetry only the upper triangular part needs to be stored.

```
      SUBROUTINE EMASS(NEL,X,Y,PROP,CON,ELMA)
C
C PROGRAM 42
C
      COMMON NRMX,NCMX,NDFEL,NN,NE,NLN,NBN,NDF,NNE,N,MS,IN,IO,E,G
      DIMENSION X(1),Y(1),PROP(1),CON(1),ELMA(NDFEL,NDFEL),ROT(6,6),V(
C
C  NEL  = CURRENT ELEMENT NUMBER
```

```
C   N1    = NUMBER OF START NODE
C   N2    = NUMBER OF END NODE
C   AX    = AREA OF ELEMENT CROSS-SECTION

      L=NNE*(NEL-1)
      N1=CON(L+1)
      N2=CON(L+2)
      AX=PROP(L+1)
C
C   COMPUTE LENGTH OF ELEMENT, AND SINE AND COSINE OF ITS LOCAL
C   X AXIS, AND STORE  IN D, SI, AND CO, RESPECTIVELY
C
      DX=X(N2)-X(N1)
      DY=Y(N2)-Y(N1)
      D=SQRT(DX**2+DY**2)
      CO=DX/D
      SI=DY/D
      DO 1 I=1,6
      DO 1 J=1,6
      ELMA(I,J)=0.
    1 ROT(I,J)=0.
C
C   FORM ELEMENT ROTATION MATRIX
C
      ROT(1,1)=CO
      ROT(1,2)=SI
      ROT(2,1)=-SI
      ROT(2,2)=CO
      ROT(3,3)=1.
      DO 2 I=1,3
      DO 2 J=1,3
    2 ROT(I+3,J+3)=ROT(I,J)
C
C   COMPUTE THE ELEMENT LOCAL MASS MATRIX
C
      COEF=G*AX*D/420.
      ELMA(1,1)=COEF*140.
      ELMA(1,4)=COEF*70.
      ELMA(2,2)=COEF*156.
      ELMA(2,3)=COEF*22*D
      ELMA(2,5)=COEF*54.
      ELMA(2,6)=  -13.*D*COEF
      ELMA(3,2)=ELMA(2,3)
      ELMA(3,3)=COEF*4.*D*D
      ELMA(3,5)=COEF*13.*D
      ELMA(3,6)=-COEF*3.*D*D
      ELMA(4,1)=ELMA(1,4)
      ELMA(4,4)=ELMA(1,1)
      ELMA(5,2)=ELMA(2,5)
      ELMA(5,3)=ELMA(3,5)
      ELMA(5,5)=ELMA(2,2)
      ELMA(5,6)=-COEF*22.*D
      ELMA(6,2)=ELMA(2,6)
      ELMA(6,3)=ELMA(3,6)
      ELMA(6,5)=ELMA(5,6)
      ELMA(6,6)=ELMA(3,3)
C
C   ROTATE THE ELEMENT MASS MATRIX TO GLOBAL COORDINATES
C
      CALL BTAB3(ELMA,ROT,V,NDFEL,NDFEL)
      RETURN
      END
```

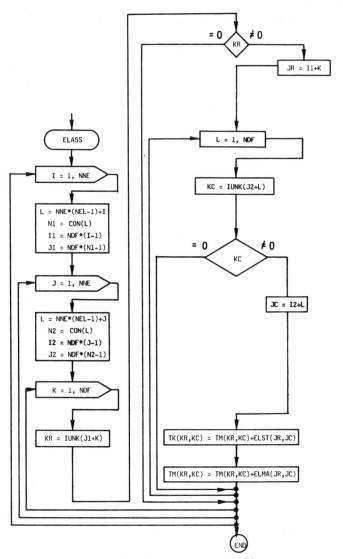

Figure 4.13   Flow-chart for Program ELASS

Program 43   Addition of the element matrices into the total matrices (ELASS

The new subprogram ELASS now has to add into the total matrices both the element stiffness and mass contributions. It will differ considerably from the previous subprogram of the same name (Program 30), since we now have to operate with a symmetric but not banded storage scheme. In addition, the assembling will be done in a compact way, such that the homogeneous boundary conditions are automatically taken into account in the total mass and stiffness matrices. This means that only the rows and columns corresponding to unknown displacement will be included, while the rows and columns corresponding to zero displacements are not. Otherwise, zeroing these rows and columns and placing a unit number in the pertinent diagonal coefficients, we will be introducing spurious, non-relevant modes of vibration.

The subprogram ELASS will be organized as shown in the flow-chart in Figure 4.13. The flow-chart for full matrix assembling, Figure 3.15, was taken as a base. The integers N1 and N2 contain the node numbers corresponding to the hyper-row and hyper-column in TK and TM to be operated upon. The integers J1 and J2 contain the number of rows and columns before those pertaining to the unknowns of nodes N1 and N2, respectively, in arrays TK and TM. I1 and I2 are similar, but for matrices ELST and ELMA. The value assigned to KR, obtained from array IUNK, is the row number, in TK and TM, of the Kth unknown for node N1. If zero, that displacement is prescribed equal to zero, and therefore it is not taken into account. Similar is the case of the integer KC, corresponding to the column number in TK and TM for the Lth unknown of node N2.

The FORTRAN code for subprogram ELASS, prepared according to the flow-chart just described, is given below.

```
      SUBROUTINE ELASS(NEL,CON,IUNK,ELST,ELMA,TK,TM)
C
C             PROGRAM 43
C ADDITION OF THE ELEMENT STIFFNESS AND MASS
C CONTRIBUTIONS TO THE TOTAL STIFFNESS AND
C MASS MATRICES
C
      COMMON NRMX,NCMX,NDFEL,NN,NE,NLN,NBN,NDF,NNE,N,MS,IN,IO,E,G
      DIMENSION CON(1),IUNK(1),ELST(NDFEL,NDFEL),ELMA(NDFEL,NDFEL),
     *TK(NRMX,NRMX),TM(NRMX,NRMX)
      DO 100 I=1,NNE
C
C NEL = NUMBER OF THE CURRENT ELEMENT
C N1  = NUMBER OF THE START NODE
C
      L=NNE*(NEL-1)+I
      N1=CON(L)
      I1=NDF*(I-1)
      J1=NDF*(N1-1)
      DO 100 J=1,NNE
      L=NNE*(NEL-1)+J
C
C N2   = NUMBER OF THE END NODE
```

```
C
      N2=CON(L)
      I2=NDF*(J-1)
      J2=NDF*(N2-1)
      DO 100 K=1,NDF
C
C KR    = ROW NUMBER IN TK AND TM FOR THE KTH UNKNOWN OF NODE N1
C
      KR=IUNK(J1+K)
      IF(KR)11,100,11
C
C UNKNOWN IS RELEVANT. PROCEED.
C
   11 JR=I1+K
      DO 100 L=1,NDF
C
C KC    = COLUMN NUMBER IN TK AND TM FOR LTH UNKNOWN OF NODE N2
C
      KC=IUNK(J2+L)
      IF(KC)14,100,14
C
C UNKNOWN IS RELEVANT. PROCEED.
C
   14 JC=I2+L
C
C ADD ELEMENT COEFFICIENTS TO TOTAL MATRICES
C
   16 TK(KR,KC)=TK(KR,KC)+ELST(JR,JC)
      TM(KR,KC)=TM(KR,KC)+ELMA(JR,JC)
  100 CONTINUE
      RETURN
      END
```

Program 44   Output program (OUTPT)

The new subprogram OUTPT prints the natural frequencies computed from
the eigenvalues and the natural modes of vibration. Its FORTRAN code is
the following:

```
      SUBROUTINE OUTPT(TK,TM)
C
C PROGRAM 44
C
      COMMON NRMX,NCMX,NDFEL,NN,NE,NLN,NBN,NDF,NNE,N,MS,IN,IO,E,G
      DIMENSION TK(NRMX,NRMX),TM(NRMX,NRMX)
      WRITE(IO,1)
    1 FORMAT(' ',130('*'))
C
C COMPUTE THE NATURAL FREQUENCIES, TAKING THE
C SQUARE ROOTS OF THE EIGENVALUES
C
      DO 10 I=1,N
   10 TK(I,I)=SQRT(TK(I,I))
C
C PRINT THE NATURAL FREQUENCIES AND PERIODS
C
```

```
   WRITE(IO,2)
 2 FORMAT(//' RESULTS'//' NATURAL FREQUENCIES'//6X,'NUMBER  FREQUENCI
  *ES        PERIODS')
   DO 3 I=1,N
   T=6.2831854/TK(I,I)
 3 WRITE(IO,4) I,TK(I,I),T
 4 FORMAT(I10,2E15.6)

PRINT NATURAL MODES

   WRITE(IO,5)
 5 FORMAT(//' NATURAL MODES')
   DO 6 I=1,N
 6 WRITE(IO,7) I,(TM(J,I),J=1,N)
 7 FORMAT(/' MODE NUMBER :',I4/(8E15.6))
   WRITE(IO,1)
   RETURN
   END
```

Example 4.9
As an illustration   of the application of the program for dynamic analysis we consider the case of the structure shown in Figure 4.14, which shows the over- all dimensions and the node and member numbering.

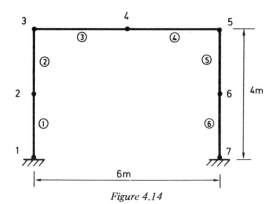

Figure 4.14

Note that two members are used to represent each of the columns and the beam. This is done to obtain a better approximation.

The Young's modulus is 2 500 000 N/cm^2, and the mass density is 0.0025 kg/cm^3. The cross-sectional area and the moment of inertia are $A = 100$ cm^2 and $I = 833.33$ cm^4 for the columns and $A = 150$ cm^2 and $I = 2812.50$ cm^4 for the beam.

```
2                    200.
3                    400.
4      300.          400.
5      600.          400.
6      600.          200.
7      600.
1            1            2    100.         833.33
2            2            3    100.         833.33
3            3            4    150.        2812.50
4            4            5    150.        2812.50
5            5            6    100.         833.33
6            6            7    100.         833.33
1
7
```

The output produced by the program is

```
**********************************************************************

INTERNAL DATA

NUMBER OF NODES                  :   7
NUMBER OF ELEMENTS               :   6
NUMBER OF SUPPORT NODES          :   2
MODULUS OF ELASTICITY    :            2500000.
DENSITY    :                              0.0025

NODAL COORDINATES
      NODE        X             Y
      1           0.00          0.00
      2           0.00        200.00
      3           0.00        400.00
      4         300.00        400.00
      5         600.00        400.00
      6         600.00        200.00
      7         600.00          0.00

ELEMENT CONNECTIVITY AND PROPERTIES
   ELEMENT      START NODE   END NODE   AREA       M. OF INERTIA
   1            1            2          100.00000    833.33000
   2            2            3          100.00000    833.33000
   3            3            4          150.00000   2812.50000
   4            4            5          150.00000   2812.50000
   5            5            6          100.00000    833.33000
   6            6            7          100.00000    833.33000

BOUNDARY CONDITION DATA
                          STATUS
                (0: PRESCRIBED, 1: FREE)
      NODE        U             V          RZ
      1           0             0          0
      7           0             0          0

**********************************************************************
```

RESULTS

NATURAL FREQUENCIES

NUMBER	FREQUENCIES	PERIODS
1	.383264E+03	.153939E-01
2	.317179E+03	.198096E-01
3	.374836E+03	.167625E-01
4	.159274E+03	.394489E-01
5	.118801E+03	.528882E-01
6	.115392E+03	.644507E-01
7	.757884E+02	.829043E-01
8	.452781E+02	.138769E+00
9	.552125E+02	.113800E+00
10	.370140E+02	.159742E+00
11	.197658E+02	.317882E+00
12	.123605E+02	.508326E+00
13	.117901E+02	.532920E+00
14	.464982E+01	.135128E+01
15	.149398E+01	.420568E+01

NATURAL MODES

MODE NUMBER : 1
```
-.110156E-01   .103113E+00   .489818E-03  -.414448E-01  -.998720E-01   .216922E-02   .389706E-02   .106133E-06
 .552499E-03  -.414448E-01   .998720E-01   .216924E-02  -.110186E-01  -.103113E+00   .489796E-03
```

MODE NUMBER : 2
```
 .123741E-01  -.390359E-01  -.600434E-03  -.841127E-01   .159527E-01   .270359E-03   .104578E+00   .283935E-06
 .381355E-03  -.841126E-01  -.159521E-01   .270379E-03   .123737E-01   .390360E-01  -.600513E-03
```

MODE NUMBER : 3
```
 .125984E-01  -.109827E+00  -.571874E-03   .143628E-01   .990553E-01  -.179616E-02  -.165749E-05   .107148E-01
 .679655E-07  -.143659E-01   .990541E-01   .179628E-02  -.126010E-01  -.109828E+00   .571973E-03
```

MODE NUMBER : 4
```
 .105963E-01  -.176428E-01  -.567588E-03  -.987254E-01  -.199729E-01   .795231E-03  -.956246E-05  -.133077E-01
 .119169E-06   .987064E-01  -.199709E-01  -.795257E-03  -.105927E-01  -.176416E-01   .567572E-03
```

MODE NUMBER : 5
```
 .215328E-01   .619736E-01  -.116178E-02   .363694E-02   .919726E-01  -.245048E-02   .764073E-02   .613634E-06
-.950610E-03   .363817E-02  -.919753E-01  -.245040E-02   .214831E-01  -.619755E-01  -.116136E-01
```

MODE NUMBER : 6

.188283E-01	.613148E-01	-.103841E-02	-.117862E-01	.926348E-01	-.183770E-02	-.16377E-05	.149909E-01
.122428E-06	.117836E-01	.926347E-01	.183780E-02	-.188287E-01	.613145E-01	.103767E-02	

MODE NUMBER : 7

-.224232E-01	.145353E-01	.168647E-02	-.193230E-02	.258462E-01	.226044E-02	-.254452E-02	-.177398E-04
.263475E-02	-.193172E-02	-.258510E-01	.226047E-02	-.224120E-01	-.145295E-01	.168630E-02	

MODE NUMBER : 8

-.465670E-02	-.954670E-03	-.324517E-02	-.124323E-01	-.183900E-02	.573371E-03	-.136509E-01	-.321094E-04
.103160E-02	-.124360E-01	.184146E-02	.570445E-03	-.462613E-02	.953846E-03	-.325095E-02	

MODE NUMBER : 9

-.178527E-01	.711702E-02	.301397E-02	-.199837E-02	.133849E-01	.156773E-02	.860484E-05	-.511639E-01
-.135561E-05	.201571E-02	.133704E-01	-.156808E-02	.180846E-01	.710693E-02	-.300877E-02	

MODE NUMBER : 10

.256082E-01	-.173527E-02	.205247E-02	-.165913E-02	-.337777E-02	-.156127E-02	-.103882E-02	.563801E-01
.535036E-06	.145549E-02	-.337845E-02	.159925E-02	-.262064E-01	-.173471E-02	-.205074E-02	

MODE NUMBER : 11

-.479570E-01	.108031E-02	-.383711E-03	.445503E-02	.214461E-02	.909205E-03	.453598E-02	.248806E-03
-.123916E-02	.446479E-02	-.214878E-02	.901538E-03	-.499577E-01	-.108340E-02	-.381279E-03	

MODE NUMBER : 12

.110748E+00	-.193467E-03	-.445868E-04	.437833E-03	-.383547E-03	.160506E-03	.890384E-04	-.330415E-01
.932810E-05	-.254562E-03	-.315291E-03	-.172072E-03	-.109461E+00	-.149103E-03	.495891E-04	

MODE NUMBER : 13

-.100915E+00	-.585791E-04	.547441E-04	.196582E-01	-.123605E-03	-.513392E-03	.197826E-01	.279933E-03
.365408E-02	.196585E-01	.166051E-03	-.528893E-03	-.101453E+00	.805301E-04	.592403E-04	

MODE NUMBER : 14

-.256367E-01	-.107797E-03	.105030E-03	.105196E-02	-.213203E-03	-.423395E-03	.104135E-02	-.932573E-01
-.479031E-05	.102774E-02	-.232965E-03	.424989E-03	.248286E-01	-.119043E-03	-.116361E-03	

MODE NUMBER : 15

-.273440E-01	.152762E-04	.204155E-03	-.583640E-01	.195496E-04	.519459E-04	-.583859E-01	-.169768E-02
-.282860E-04	-.584007E-01	-.293776E-05	.554938E-04	-.270313E-01	.393434E-05	.209451E-03	

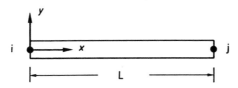

Figure 4.15

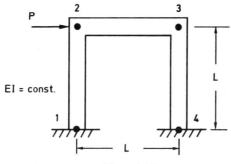

Figure 4.16

Exercises

(1) Derive the stiffness matrix for the axial deformation member of Figure 4.15, considering that the axial displacement has the following linear variation

$$u = \left(1 - \frac{x}{L}\right) u_i + \frac{x}{L} u_j$$

Compare with the stiffness matrix given in Chapter 3.

(2) Repeat the derivation of the previous problem for a bar having the following linear variation in cross-sectional area,

$$A(x) = \left(1 - \frac{x}{L}\right) A_i + \frac{x}{L} A_j$$

(3) Given the frame of Figure 4.16, write the governing system of equations in terms of stiffness coefficients, specialize it for the case in which $u_2 = u_3$, and find u_2.

(4) Derive the bending stiffness matrix for the beam shown in Figure 4.17.

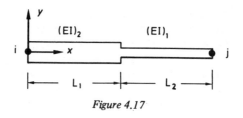

Figure 4.17

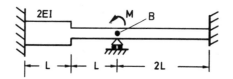

Figure 4.18

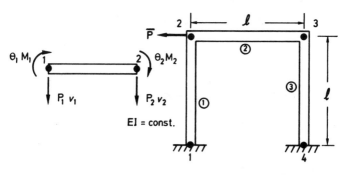

Figure 4.19

(5)   Compute the rotation in point B, of the continuous beam in Figure 4.18.

(6)   Solve the frame shown in Figure 4.19 by using the stiffness method, applying symmetry considerations. The element stiffness matrix is

$$
\begin{Bmatrix} P_1 \\ M_1 \\ P_2 \\ M_2 \end{Bmatrix} = \frac{EI}{l^3}
\begin{bmatrix}
12 & 6l & -12 & 6l \\
 & 4l^2 & -6l & 2l^2 \\
 & \text{symm} & 12 & -6l \\
 & & & 4l^2
\end{bmatrix}
\begin{Bmatrix} v_1 \\ \theta_1 \\ v_2 \\ \theta_2 \end{Bmatrix}
$$

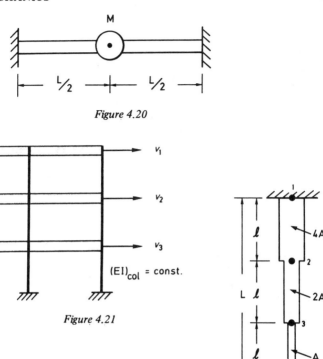

Figure 4.20

Figure 4.21

Figure 4.22

(7) Using the beam stiffness matrix shown in the above example, find the fundamental frequency of the beam shown in Figure 4.20. Assume that the mass M is concentrated at $L/2$.

(8) Write the governing equations for the frame shown in Figure 4.21, assuming that the rigidity of the floors is very much larger than that of the columns.

(9) Consider the hanging bar divided into 3 elements shown in Figure 4.22. Find the displacements at 4 by assuming a linear variation of displacement in each element and uniform distribution of body forces in each element plus the concentrated force $\bar{P}$. (Note that the body forces will be proportional to the area of each element.)

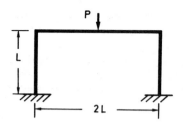

Figure 4.23

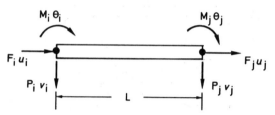

Figure 4.24

(10) Give a mathematical idealization of the frame shown in Figure 4.23 using symmetry to reduce the size of the problem. The stiffness matrix for the element shown in Figure 4.24 is given by

$$
\begin{bmatrix} F_i \\ P_i \\ M_i \\ F_j \\ P_j \\ M_j \end{bmatrix} = \frac{EI}{L} \begin{bmatrix} \dfrac{A}{I} & 0 & 0 & -\dfrac{A}{I} & 0 & 0 \\ & \dfrac{12}{L^2} & \dfrac{6}{L} & 0 & -\dfrac{12}{L^2} & \dfrac{6}{L} \\ & & 4 & 0 & -\dfrac{6}{L} & 2 \\ & \text{symmetric} & & \dfrac{A}{I} & 0 & 0 \\ & & & & \dfrac{12}{L^2} & -\dfrac{6}{L} \\ & & & & & 4 \end{bmatrix} \begin{bmatrix} u_i \\ v_i \\ \theta_i \\ u_j \\ v_j \\ \theta_j \end{bmatrix}
$$

Form the stiffness matrix for the frame and apply the boundary conditions. Give the matrix equation to be solved to find the deflected shape of the frame. Do not attempt a solution by hand.

(11) Find the global stiffness matrix of the structure shown in Figure 4.25 by using conditions of symmetry. Indicate the matrix equations to be solved to determine the displacements (or rotation) and reactions of the system,

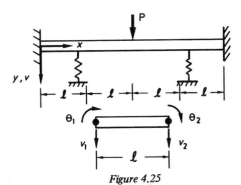

Figure 4.25

The stiffness matrix of a beam element is the same as in Problem 6 above.

Bibliography

McCallion, H., *Vibration of Linear Mechanical Systems*, Longman (1973)

Warburton, G. B., *The Dynamical Behaviour of Structures*, Pergamon Press (1964)

Hurty, W. C. and Rubinstein, M. F., *Dynamics of Structures*, Prentice Hall (1963)

Tauchert, T. R., *Energy Principles in Structural Mechanics*, McGraw-Hill (1974)

Rubinstein, M. F., *Structural Systems – Statics, Dynamics and Stability*, Prentice Hall (1970)

Chapter 5

Approximate methods of solution

5.1 INTRODUCTION

We have already mentioned in Chapter 4 the possibility of using approximate functions for problems in which the exact functions are difficult to obtain or cumbersome to use. For instance, in the case of a beam, the exact expression for terms in the mass matrix would involve hyperbolic functions but they were approximated by polynomials (Section 4.5).

Problems in continuum mechanics require the solution of a system of differential equations with certain boundary conditions. Usually it is impossible to find the exact solution of this system and we have to apply an approximate method.

There are many different types of approximate methods, some of which may be known to the reader from previous numerical methods courses (e.g. least squares, collocation). The two most important methods for engineering applications are the Rayleigh–Ritz and Galerkin method. The first of these was used by Rayleigh to solve vibration problems and was further developed by Ritz who gave to it a mathematical basis and applied it to a large range of problems. The method is based on the existence of a functional, such as the total potential energy, which can then be minimized. The second technique is due to Galerkin, a Russian mathematician, who presented it as an error minimization method. The main advantage of Galerkin's method is that it works with the governing equations of the problem, i.e. does not need a functional. We will see that under certain conditions both methods are equivalent although they are different in general.

5.2 THE RAYLEIGH–RITZ METHOD

We have tried in Chapter 4, to determine a function, out of the range of *all* admissible functions, which makes the potential energy functional a minimum. The Rayleigh–Ritz method consists instead of proposing a *limited* range of admissible functions. Thus we do not obtain the actual minimum but a solution as near to it as the approximate functions allow. The choice of these functions is all important in order to obtain a good approximation. In many engineering problems the shape of the function is reasonably well known and the procedure is very accurate. One also can go on increasing the number of terms in the approximating functions and this will generally improve the solution.

To illustrate the method let us consider the problem of determining the unknown function $u(x)$ which corresponds to a stationary value of a functional, such as the total potential energy

$$\Pi_p = \int_{x_1}^{x_2} I\left(u_e, \frac{du_e}{dx}, x\right) dx \tag{5.1}$$

with boundary conditions $u_e(x_1) = u_e(x_2) = 0$. u_e is the exact solution. I is the integrand, integrated over the one-dimensional domain x.

We now assume that the solution can be approximated by a series of functions satisfying the boundary conditions, but with certain undetermined α_i parameters

$$u_e \simeq u = \alpha_1 \phi_1 + \alpha_2 \phi_2 + \ldots + \alpha_n \phi_n \tag{5.2}$$

The ϕ_i are prescribed functions of x which are *linearly independent,* and each function individually satisfies the boundary conditions

$$\phi_i(x_1) = 0, \quad \phi_i(x_2) = 0, \quad i = 1, 2, \ldots, n \tag{5.3}$$

For the case of functional (5.1) the functions must be continuous but their first derivatives can be discontinuous. Substituting for u and requiring that $\delta \Pi_p = 0$ for equilibrium, leads to

$$\delta \Pi_p = \frac{\partial \Pi_p}{\partial \alpha_1} \delta \alpha_1 + \frac{\partial \Pi_p}{\partial \alpha_2} \delta \alpha_2 + \ldots + \frac{\partial \Pi_p}{\partial \alpha_n} \delta \alpha_n \tag{5.4}$$

as Π_p is a function of the undetermined α_i parameters. As $\delta \alpha_i$ are arbitrary, Equation (5.4) implies that

$$\frac{\partial \Pi_p}{\partial \alpha_1} = 0; \quad i = 1, 2, \ldots, n \tag{5.5}$$

These equations will be linear in α_i when Π_p is a quadratic function of u and du/dx.

The necessary conditions for convergence of the method are

(1)   The approximating functions must be continuous to one order less than the highest derivative in the integrand. This condition gives a defined integrand.

(2)   The functions must individually satisfy the essential boundary conditions on S_u.

Functions satisfying the above conditions are said to be *admissible*. Finally,

(3)   The sequence of functions must be *complete*. Let $u(x)$ be an admissible function. By completeness we mean that the 'mean square error' vanishes in the limit, i.e.

$$\lim_{n \to \infty} \int_{x_1}^{x_2} \left(u_e - \sum_{i=1}^{n} \alpha_i \phi_i \right)^2 dx = 0 \qquad (5.6)$$

Polynomials and trigonometric functions are suitable choices in the Rayleigh–Ritz method.

In order to assess the convergence, one has to take at least two trial solutions. When the method is applied to a functional such as Π_p, for which the stationary point is also a *relative minimum*, one can measure convergence by comparing successive values of the functional obtained with the following sequence:

$$u^{(1)} = \alpha_1^{(1)} \phi_1$$

$$u^{(2)} = \alpha_1^{(2)} \phi_1 + \alpha_2^{(2)} \phi_2$$

$$
\begin{array}{ccc}
. & . & . \\
. & . & . \\
. & . & .
\end{array}
\qquad (5.7)
$$

$$u^{(j)} = \alpha_1^{(j)} \phi_1 + \alpha_2^{(j)} \phi_2 \ldots + \alpha_j^{(j)} \phi_j$$

Since the jth expansion includes all the functions contained in the previous expansions and Π_p is minimised at each step, it follows that

$$\Pi_p^{(1)} \geqslant \Pi_p^{(2)} \geqslant \ldots \geqslant \Pi_p^{(j)} \qquad (5.8)$$

We call Equations (5.7) a minimizing sequence. Employing a minimizing sequence ensures monotonic convergence of the functional. Note, however, that the functions must be *admissible* and comprise a *complete* sequence in order that the solution converges to the exact solution.

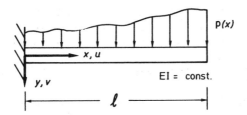

Figure 5.1   Cantilever beam

Example 5.1

Let us consider the case of the cantilever beam shown in Figure 5.1 as an illustration. The total length of the beam is L and a distributed load $p(x)$ is acting on it.

Taking only the bending energy into account, the total potential energy functional can be written as

$$\Pi_p \int_0^l \left\{ \frac{EI}{2} \left(\frac{d^2v}{dx^2} \right)^2 - pv \right\} dx \tag{a}$$

where E is the modulus of elasticity of the material and I the moment of inertia of the section. The principle of minimum potential energy requires that

$$\delta \Pi_p = 0 \tag{b}$$

which can also be written (see Example 4.5) as

$$\delta \Pi_p = \int_0^l \left\{ EI \frac{d^4v}{dx^4} - p \right\} \delta v \, dx + EI \left[\frac{d^2v}{dx^2} \frac{d\delta v}{dx} \right]_0^l - EI \left[\frac{d^3v}{dx^3} \delta v \right]_0^l = 0 \tag{c}$$

Hence the governing equation in the domain is the equilibrium equation for the beam, i.e.

$$EI \frac{d^4v}{dx^4} = p \tag{d}$$

The boundary conditions are

$$\left[EI\frac{d^2v}{dx^2}\frac{d\delta v}{dx}\right]_0^l = \left[M\frac{d\delta v}{dx}\right]_0^l = 0 \tag{e}$$

$$-\left[EI\frac{d^3v}{dx^3}\delta v\right]_0^l = [Q\,\delta v]_0^l = 0$$

where M is the bending moment and Q is the shear force. In particular, the boundary conditions for the cantilever beam are

at $x = 0$, $v = \dfrac{dv}{dx} = 0$

at $x = l$, $M = Q = 0$ \tag{f}

The variations have to be such that

at $x = 0 \rightarrow \delta v = 0$, $\dfrac{d\delta v}{dx} = 0$ \tag{g}

Let us now apply the Rayleigh–Ritz method with the functional (a). If we propose as a first approximation

$$v = \alpha_1 + \alpha_2 x \tag{h}$$

it is easy to see that in order to satisfy the displacement boundary conditions $v = dv/dx = 0$ in (f), we need $\alpha_1 = \alpha_2 \equiv 0$, which means that the function (h) is of no use. We need at least a quadratic function, i.e. the first approximation is

$$v = \alpha_1 x^2 \tag{i}$$

Note that Equation (i) satisfies the displacement boundary conditions and has a continuous first derivative (i.e. is *admissible*). In addition we will see later on that if more functions of the same set are taken for the approximation, the results will converge towards the correct solution (i.e. *completeness* is satisfied).

Substituting (i) into functional (a) we obtain

$$\Pi_p = \int_0^l \left\{ \frac{EI}{2} (2\alpha_1)^2 - p\,\alpha_1\,x^2 \right\}\,dx.$$

Minimizing for constant p we have

$$\frac{\partial \Pi_p}{\partial \alpha_1} = 4EI\alpha_1 \int_0^l dx - p \int_0^l x^2\,dx = 0$$

Therefore

$$\alpha_1 = \frac{pl^2}{12EI} \tag{j}$$

with displacements

$$v = \alpha_1 x^2 = \frac{pl^2}{12EI}\,x^2 \tag{k}$$

We can now calculate the moments and the value of Π_p functional, i.e.

$$M = EI\,\frac{d^2v}{dx^2} = \frac{pl^2}{6}$$

$$\Pi_p = \int_0^l \frac{EI}{2} \left(\frac{pl^2}{6EI} \right)^2\,dx - \int_0^l x^2\,\frac{p^2 l^2}{12EI}\,dx$$

$$= -\frac{4p^2 l^5}{288EI} \tag{l}$$

The deflection at the tip is

$$v_l = \frac{pl^4}{12EI}$$

which is 50% in error when compared against the exact value of $pl^4/8EI$. Note that we have obtained a constant value for the M moment, which is far from the exact solution.

We can now obtain a better estimate of the solution by adding one more term in the approximation, i.e.

$$v = \alpha_1 x^2 + \alpha_2 x^3 \tag{m}$$

Note that the new x^3 term satisfies the displacement boundary conditions (i.e. is part of a *minimizing* sequence, which ensures *monotonic* convergence of the energy functional). We have now

$$\Pi_p = \int\limits_0^l \left\{ \frac{EI}{2} (4\alpha_1^2 + 24\alpha_1 \alpha_2 x + 36\alpha_2^2 x^2) - p(\alpha_1 x^2 + \alpha_2 x^3) \right\} \, dx \tag{n}$$

The minimization gives

$$\frac{\partial \Pi_p}{\partial \alpha_1} = \int\limits_0^l \left\{ \frac{EI}{2} (8\alpha_1 + 24\alpha_2 x) - px^2 \right\} \, dx = 0$$

$$\frac{\partial \Pi_p}{\partial \alpha_2} = \int\limits_0^l \left\{ \frac{EI}{2} (24\alpha_1 x + 72\alpha_2 x^2) - px^3 \right\} \, dx = 0 \tag{o}$$

The solution of this system is

$$\alpha_1 = \frac{5pl^2}{24EI}, \quad \alpha_2 = -\frac{pl}{12EI} \tag{p}$$

The deflection and moment can now be calculated, i.e.

$$v = \frac{pl}{12EI} \left(\frac{5}{2} lx^2 - x^3 \right)$$

$$M = \frac{pl}{12} (5l - 6x)$$

The tip deflection is

$$v_l = \frac{pl^4}{8EI} \tag{q}$$

The moment at $x = 0$ and the potential energy functional are

$$M_0 = \frac{5pl^2}{12}, \quad \Pi_p = \frac{-7pl^5}{288EI} \tag{r}$$

Note that the deflection v_l is *exact* but the moment at $x = 0$ is still different from the exact value $pl^2/2$.

If we now take a fourth order function for v,

$$v = \alpha_1 x^2 + \alpha_2 x^3 + \alpha_3 x^4 \tag{s}$$

we obtain

$$\alpha_1 = \frac{pl^2}{4EI}, \quad \alpha_2 = -\frac{pl}{6EI}, \quad \alpha_3 = \frac{p}{24EI} \tag{t}$$

and $v = \dfrac{px^2}{EI}\left(\dfrac{l^2}{4} - \dfrac{lx}{6} + \dfrac{x^2}{24}\right)$ $\tag{u}$

This is now the *exact* solution of the problem. If we take one more term in the expression for v (i.e. $+\alpha_4 x^5$) and minimize, we will find that $\alpha_4 = 0$. That is, we obtain the exact solution again.

The results for the case $E = I = p = l = 1$ are presented in Table 5.1. We can see that the energy converges monotonically, i.e.

$$\Pi_p^{(1)} > \Pi_p^{(2)} > \Pi_{p_{\text{exact}}} \tag{v}$$

It is also evident that the approximate solutions obtained for v are always better than those for dv/dx and much better than the ones for M, which is proportional to d^2v/dx^2. This occurs because the derivations increase the errors in the solution.

We have also shown that the approximate solutions exactly fulfil the displacement but not the force boundary conditions (i.e. at $x = l$, $M \neq 0$ and $Q \neq 0$ for the approximations). The error in the force boundary conditions at the tip of the cantilever tends however, to decrease for the more refined functions.

Example 5.2

In the previous example we have considered a uniformly distributed load and that the approximating functions were polynomials. Let us now look at the

Table 5.1

| | | v | | | | dv/dx | | | | | M | | | |
x =	0	1/3	2/3	1	0	1/3	2/3	1	0	1/3	2/3	1	Πp
2nd order function	0	0.009	0.037	0.083	0	0.055	0.111	0.167	0.167	0.167	0.166	0.166	−0.014
3rd order function	0	0.020	0.068	0.125	0	0.111	0.166	0.167	0.417	0.250	0.083	−0.083	−0.024
Exact solution	0	0.022	0.069	0.125	0	0.117	0.160	0.167	0.5	0.222	0.055	0	−0.025

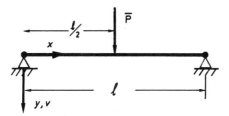

Figure 5.2   Simply supported beam under a concentrated load

case of a beam with a concentrated load at the centre and try to find the approximate solution using trigonometric series (Figure 5.2).

We take the following approximating function for the vertical deflections

$$v = \alpha_1 \sin \frac{x}{l}\pi + \alpha_2 \sin \frac{x}{l}3\pi = \alpha_1 \phi_2 + \alpha_2 \phi_2 \tag{a}$$

Note that this function satisfies the boundary conditions and is symmetric. If we compute the internal energy and the potential of the load we have

$$\Pi_p = V + \Omega = \tfrac{1}{2}EI \int_0^l \left(\frac{d^2v}{dx^2}\right)^2 dx - \bar{P}v_c \tag{b}$$

(where v_c = deflection at the centre)

$$\Pi_p = \frac{EI}{2} \int_0^l \left\{\alpha_1 \left(\frac{\pi}{l}\right)^2 \sin\frac{x}{l}\pi + \alpha_2 \left(\frac{3\pi}{l}\right)^2 \sin\left(\frac{x}{l}3\pi\right)\right\}^2 dx$$

$$- \bar{P}\left\{\alpha_1 \sin\left(\frac{\pi}{2}\right) + \alpha_2 \sin\left(\frac{3\pi}{2}\right)\right\}$$

$$= \frac{EI}{2} \int_0^l \left\{\alpha_1^2 \left(\frac{\pi}{l}\right)^4 \sin^2\left(\frac{x}{l}\pi\right) + \alpha_2^2 \left(\frac{3\pi}{l}\right)^4 \sin^2\left(\frac{x3\pi}{l}\right)\right.$$

$$\left. + 2\alpha_1 \alpha_2 \left(\frac{\pi}{l}\right)^2 \left(\frac{3\pi}{l}\right)^2 \sin\left(\frac{x\pi}{l}\right) \sin\left(\frac{x3\pi}{l}\right)\right\} dx - \bar{P}\{\alpha_1 - \alpha_2\} \tag{c}$$

This expression can be integrated taking into account the following orthogonality conditions

$$\int_0^l \sin\left(\frac{x}{l}\pi\right) \sin\left(\frac{3x}{l}\pi\right) dx = 0 \tag{d}$$

or in general

$$\int_0^l \sin\frac{n\pi x}{l} \sin\frac{m\pi x}{l} dx = 0, \quad (m \neq n)$$

and also that

$$\int_0^l \sin^2\left(\frac{x}{l}\pi\right) dx = l/2$$

$$\int_0^l \sin^2\left(\frac{3x}{l}\pi\right) dx = l/2 \tag{e}$$

or in general,

$$\int_0^l \sin^2\left(\frac{n\pi x}{l}\right) dx = l/2$$

Hence we can write

$$\Pi_p = V + \Omega = \frac{EI}{2} \left[\alpha_1^2 \left(\frac{\pi}{l}\right)^4 \frac{l}{2} + \alpha_2^2 \left(\frac{3\pi}{l}\right)^4 \frac{l}{2} \right] - \bar{P} \left(\alpha_1 - \alpha_2\right) \qquad \text{(f)}$$

Note that α_1, α_2 are *independent* of each other. Then

$$\frac{\partial \Pi_p}{\partial \alpha_1} = \frac{EI}{2} \left\{ 2\alpha_1 \left(\frac{\pi}{l}\right)^4 \frac{l}{2} \right\} - \bar{P} = 0,$$

is a function of α_1 only, and

$$\frac{\partial \Pi_p}{\partial \alpha_2} = \frac{EI}{2} \left\{ 2\alpha_2 \left(\frac{3\pi}{l}\right)^4 \frac{l}{2} \right\} + \bar{P} = 0, \qquad \text{(g)}$$

is a function of α_2 only.
    We find that,

$$\alpha_1 = 2 \frac{\bar{P}}{EI} \frac{l^3}{\pi^4}, \quad \alpha_2 = -2 \frac{\bar{P}}{EI} \frac{l^3}{\pi^4} \left(\frac{1}{3}\right)^4 \qquad \text{(h)}$$

The solution for the centre deflection is

$$v_c = 2 \frac{\bar{P}}{EI} \frac{l^3}{\pi^4} \left(1 + \frac{1}{3^4}\right) \qquad \text{(i)}$$

We can generalise this solution for 'n' terms, obtaining

$$v_c = 2 \frac{\bar{P}l^3}{\pi^4 EI} \left(1 + \frac{1}{3^4} + \frac{1}{5^4} + \ldots\right) \qquad \text{(j)}$$

This is a convergent series such that when $n \to \infty$ we get the exact solution

$$(v_c)_{\text{exact}} = \frac{\bar{P}l^3}{EI} \cdot \frac{1}{48} \qquad \text{(k)}$$

Example 5.3. Two-dimensional continuum
The approximate methods are widely used in two and three dimensional continuum as in these cases it is practically impossible to find exact solutions.

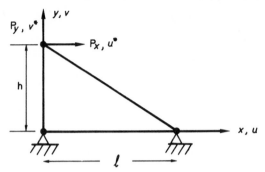

Figure 5.3   Triangular element

The steps to follow in two- or three-dimensional cases are the same as those we have already seen but the problems are now more complicated because of the increasing number of independent parameters. To illustrate the application of approximate solutions for the continuum we will consider the simple case of a triangular element in plane strain (Figure 5.3).

Calculate the u^*, v^* deflections of the tip of the plate *assuming* linear displacements for u and v.

$$u = a_1 + a_2 x + a_3 y$$

$$v = b_1 + b_2 x + b_3 y$$

The boundary conditions are

$$x = 0, \quad y = 0 \rightarrow \quad \begin{cases} u = 0, & \therefore a_1 = 0 \\ v = 0, & \therefore b_1 = 0 \end{cases}$$

$$y = 0 \rightarrow \quad \begin{cases} u = 0, & \therefore a_2 x = 0 \rightarrow a_2 = 0 \\ v = 0, & \therefore b_2 x = 0 \rightarrow b_2 = 0 \end{cases}$$

Thus,

$$y = h \rightarrow u = a_3 h = u^* \rightarrow a_3 = \frac{u^*}{h}$$

$$v = b_3 h = v^* \rightarrow b_3 = \frac{v^*}{h}$$

The displacement functions are

$$u = u^* \left(\frac{y}{h} \right), \quad v = v^* \left(\frac{y}{h} \right)$$

The strains are

$$\epsilon_x = \frac{\partial u}{\partial x} = 0, \quad \epsilon_y = \frac{\partial v}{\partial y} = \left(\frac{v^*}{h} \right)$$

$$\gamma = \frac{\partial u}{\partial y} + \frac{\partial v}{\partial x} = \left(\frac{u^*}{h} \right)$$

The internal energy (considering unit thickness) is

$$V = \tfrac{1}{2} \iint (\sigma_x \epsilon_x + \sigma_y \epsilon_y + \tau \gamma) \, dx \, dy$$

The stress–strain relationships are

$$\sigma_x = \frac{E}{(1+v)(1-2v)} \{(1-v)\epsilon_x + v\epsilon_y\}$$

$$\sigma_y = \frac{E}{(1+v)(1-2v)} \{(1-v)\epsilon_y + v\epsilon_x\}$$

$$\tau = \frac{E}{2(1+v)} \gamma$$

For $v = 0$, $\sigma_x = E \epsilon_x$, $\sigma_y = E \epsilon_y$, $\tau = (E/2)\gamma$.
For $v = 0$, we have

$$V = \tfrac{1}{2} E \iint \{ \epsilon_y^2 + \tfrac{1}{2}\gamma^2 \} \, dx \, dy$$

$$= \tfrac{1}{2} E \iint \left\{ \left(\frac{v^*}{h} \right)^2 + \tfrac{1}{2} \left(\frac{u^*}{h} \right)^2 \right\} \, dx \, dy$$

$$= \tfrac{1}{2} E \frac{l}{2h} (v^{*2} + \tfrac{1}{2}u^{*2})$$

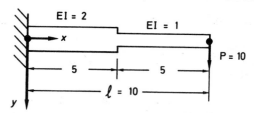

Figure 5.4   Cantilever beam with two different rigidities

The potential of loads can be written

$$\Omega = -(P_x u^* + P_y v^*)$$

Thus we have after minimizing,

$$\frac{\partial \Pi_p}{\partial v^*} = \frac{\partial V}{\partial v^*} + \frac{\partial \Omega}{\partial v^*} = 0 \rightarrow P_y = \frac{E}{2} \frac{l}{h} v^*, \quad \therefore v^* = P_y \left(\frac{2h}{EI} \right)$$

$$\frac{\partial \Pi_p}{\partial u^*} = \frac{\partial V}{\partial u^*} + \frac{\partial \Omega}{\partial u^*} = 0 \rightarrow P_x = \frac{E}{2} \frac{l}{h} \frac{u^*}{2}, \quad \therefore u^* = P_x \left(\frac{4h}{EI} \right)$$

Example 5.4

Let us now consider the case of a cantilever beam with two different rigidities, shown in Figure 5.4. The exact solution for the tip deflection of the beam is,

$$v_l = 1875 \tag{a}$$

If 2nd to 4th order polynomials such as those of Example 5.1 are used as approximating functions over *all the beam*, applying Rayleigh–Ritz, we obtain the following results:

Second order polynomial

$$v^{(1)} = \frac{100}{6} x^2, \qquad \therefore v_l^{(1)} = 1666 \tag{b}$$

Third order polynomial

$$v^{(2)} = \frac{800}{3} x^2 - \frac{20}{33} x^3, \qquad \therefore v_l^{(2)} = 1818 \tag{c}$$

Fourth order polynomial

$$v^{(3)} = \frac{11550}{513} x^2 - \frac{120}{513} x^3 - \frac{10}{513} x^4, \quad \therefore v_l^{(3)} = 1822 \tag{d}$$

Comparing the approximate results it is evident that the improvement of (c) over (b) or (d) over (c) is very small and still far from the exact solution. This suggests that there is something inherently wrong with the way in which we have attacked the problem. What has happened is that at $x = l/2$ we should have had a *discontinuity* in the curvature (or second derivative) to allow for the *continuity* of the moment, i.e.

$$M = EI \frac{d^2 v}{dx^2} \tag{e}$$

as EI changes abruptly from 2 to 1. This discontinuity cannot be represented with *one* polynomial for the whole beam.

To improve our results we need to use two different functions; the first valid for $0 \leqslant x \leqslant l/2$, which will be called v_1 and the second valid for $l/2 \leqslant x \leqslant l$, called v_2. At $x = l/2$, the following displacement conditions are imposed

$$\left. \begin{array}{l} v_1 = v_2 \\[2mm] \dfrac{dv_1}{dx} = \dfrac{dv_2}{dx} \end{array} \right\} \qquad \text{at } x = l/2 \tag{f}$$

Let us now continue with the solution, using 2nd and 3rd order functions for the parts 1 and 2.

Second order functions

$$v_1 = \alpha_1 x^2$$

$$v_2 = \alpha_1 \left(lx - \frac{l^2}{4} \right) + \alpha_2 \left(\frac{l^2}{4} - lx + x^2 \right)$$

$$\Pi_p = \int_0^{l/2} 4\alpha_1^2 \, dx + \int_{l/2}^{l} 2\alpha_2^2 \, dx - \frac{Pl^2}{4} (3\alpha_1 + \alpha_2) \tag{g}$$

Table 5.2

Solution	$v_{(x=l)}$	$v_{(x=l/2)}$	$M_{(x=l)}$	$M_{2(x=l/2)}$	$M_{1(x=l/2)}$	$M_{(x=0)}$
1	1666	416	33.5	33.5	67	67
2	1818	530	12	30.5	61	97
3	1822	520	7	32	64	92
4	1718	469	25	25	75	75
Exact solution	1875	521	0	50	50	100

Minimizing Π_p with respect to α_1 and α_2 we have that

$$v_1 = \frac{3Pl}{16}\, x$$

$$v_2 = \frac{3Pl}{16}\left(lx - \frac{l^2}{4}\right) + \frac{Pl}{8}\left(\frac{l^2}{8} - lx + x^2\right) \tag{h}$$

Third order functions (Solution 4 in Table 5.2)

$$v_1 = \alpha_1 x^2 + \alpha_2 x^3$$

$$v_2 = \alpha_1\left(lx - \frac{l^2}{4}\right) + \alpha_2\left(\frac{3}{4}l^2 x - \frac{l^3}{4}\right) + \alpha_3\left(\frac{l^2}{4} - lx + x^2\right)$$

$$+ \alpha_4\left(\frac{l^3}{4} - \frac{3l^2}{4} + x^3\right) \tag{i}$$

After carrying out all the algebraic operations we find that

$$v_1 = \frac{1}{4}Pl\,x^2 - \frac{P}{12}x^3$$

$$v_2 = \frac{1}{4}Pl\left(lx - \frac{l^2}{4}\right) - \frac{P}{12}\left(\frac{3}{4}l^2 x - \frac{l^3}{4}\right)$$

$$+ \frac{8}{16}Pl\left(\frac{l^2}{4} - lx - x^2\right) - \frac{P}{6}\left(\frac{l^3}{4} - \frac{3}{4}l^2 x + x^3\right) \tag{j}$$

Equations (j) give the exact solution for the problem. The results obtained are summarized in Table 5.2.

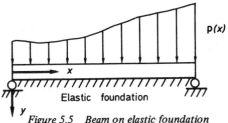

Figure 5.5   Beam on elastic foundation

This type of solution for which the domain integration is divided in different regions, with a different approximate function for each of them, is a localized Rayleigh–Ritz technique. The scheme, with some variations, is the one used in finite elements.

5.3   THE GALERKIN METHOD

Although Rayleigh–Ritz is a powerful method, it requires the existence of a functional (generally an energy functional), which sometimes may be difficult or impossible to obtain. Most engineering problems are expressed, not in terms of a functional, but in terms of a set of governing equations plus boundary conditions. The approximate solution of this set may be attempted by any of the weighted residual methods, of which the more interesting for our applications is Galerkin's method.

Galerkin's method approximates the solution of a given differential equation (or set of them) by substituting in it a trial function (which can be made to satisfy all the boundary conditions). The equation is then not equal to zero but produces a residual. This residual is weighted by the modes of the approximate solution and the product assumed to vanish over a given interval.

In order to understand the differences between the Rayleigh–Ritz and Galerkin's method we can consider the case of a prismatic beam on an elastic foundation as shown in Figure 5.5.

The total potential energy of the beam is

$$\Pi_p = \frac{EI}{2} \int_0^l \left(\frac{d^2 v}{dx^2}\right)^2 dx + \tfrac{1}{2} \int_0^l k v^2 \, dx - \int_0^l \overline{p} v \, dx \tag{5.9}$$

The first variation of this functional gives

$$\delta \Pi_p = EI \int_0^l \frac{d^2 v}{dx^2} \frac{d^2 \delta v}{dx^2} \, dx + \int_0^l k v \, \delta v \, dx - \int_0^l \overline{p} \, \delta v \, dx = 0 \tag{5.10}$$

Integrating by parts we have

$$\int_0^l \left\{ EI \frac{d^4v}{dx^4} + kv - \overline{p} \right\} \delta v\,dx + \left[EI \frac{d^2v}{dx^2} \frac{d\delta v}{dx} - EI \frac{d^3v}{dx^3} \delta v \right]_{x=0}^{x=l} = 0 \quad (5.11)$$

The term under the integral is the differential equation of an element dx of the beam under the load $\overline{p}dx$. The second term is the work done by the boundary forces. In the Galerkin method it is sometimes usual (but *not necessary*) to satisfy the boundary conditions in such a way that the second term disappears. That is, at $x = 0, l$, consider that we have the following homogeneous conditions

$$\frac{d^2v}{dx^2} \text{ is known or } \frac{d\delta v}{dx} = 0 \text{ and } \frac{d^3v}{dx^3} \text{ is known or } \delta v = 0 \qquad (5.12)$$

Thus, more conditions than in the Rayleigh–Ritz method are for the moment satisfied. Note that the Rayleigh–Ritz method only requires satisfaction of the *displacement* or *essential* boundary conditions. We will later on see that it is also possible to satisfy only the essential boundary conditions in Galerkin's method. When all the boundary conditions are satisfied, Equation (5.11) can be written

$$\int_0^l \underbrace{\left[EI \frac{d^4v}{dx^4} + kv - \overline{p} \right]}_{(=\epsilon)} \delta v\,dx = 0$$

$$\qquad (5.13)$$

or
$$\int_0^l \epsilon\, \delta v\,dx = 0$$

Note that the term ϵ represents the 'error' or residual due to having substituted an approximate value for v into the governing differential equation.

If the function v is substituted by a trial function of the type

$$v = \sum_{i=1}^n \alpha_i \phi_i(x) \qquad (5.14)$$

we obtain a system of equations of the form

$$\int \epsilon \, \phi_i \, dx = 0 \quad i = 1, 2, \ldots, n \tag{5.15}$$

Equation (5.15) represents the weighting of the residual by the modes or shapes of the approximate solution.

In a more general case we can have moments and shear forces at the nodes equal to

$$\bar{M} = EI \frac{d^2 v}{dx^2} = M, \quad \bar{Q} = -EI \frac{d^3 v}{dx^3} = Q$$

Hence Equation (5.13) can be written as

$$\int_0^l \left[EI \frac{d^4 v}{dx^4} + kv - \bar{p} \right] \delta v dx$$

$$= \left[\left(\bar{M} - EI \frac{d^2 v}{dx^2} \right) \frac{d \delta v}{dx} + \left(\bar{Q} + EI \frac{d^3 v}{dx^3} \right) \delta v \right]_{x=0}^{x=l} \tag{5.16}$$

Note that the term on the left hand side in (5.16) involves the governing equation inside the domain of the beam, the right hand side represents the stress boundary conditions. Equation (5.16) represents another way of writing the Principle of Virtual Displacements (see Chapter 4, Example 4.5).

Equation (5.16) has the advantage over Equation (5.13) that we can now use approximate functions which satisfy the displacement boundary conditions only, i.e. the stress boundary conditions do not need to be identically satisfied.

In more general terms the displacement boundary conditions are called *essential* and the force boundary conditions are called *natural*. The natural boundary conditions are found by integrating the governing equations by parts. This gives a term which is the specialization of the governing equations on the boundary, that is the *natural* boundary conditions.

We can generalize this result for a two dimensional continuum for which the starting formula for Galerkin's method is the second expression of the

Principle of Virtual Displacements, i.e.

$$\iint \left[\left\{ \frac{\partial \sigma_x}{\partial x} + \frac{\partial \tau}{\partial y} + b_x \right\} \delta u + \left\{ \frac{\partial \tau}{\partial x} + \frac{\partial \sigma_y}{\partial y} + b_y \right\} \delta v \right] dx \, dy$$

$$= \int_{S_\sigma} \left[\{ p_x - \bar{p}_x \} \delta u + \{ p_y - \bar{p}_y \} \delta v \right] dS \tag{5.17}$$

Although in the above Galerkin's method has been deduced from a Rayleigh–Ritz type functional, the former is more general than the latter as it does not require the existence of a functional. In fact Rayleigh–Ritz is a particular case of Galerkin's method.

Example 5.5
Using Galerkin's method, let us try to solve the equation

$$\frac{d^2 u}{dx^2} + u + x = 0 \tag{a}$$

with homogeneous boundary conditions $u(0) = u(1) = 0$.
   The approximate solution can be taken as

$$u \simeq \phi_1 \alpha_1 + \phi_2 \alpha_2 + \ldots \phi_n \alpha_n$$

or

$$u \simeq x(1 - x)(\alpha_1 + \alpha_2 x + \ldots \alpha_n x^{n-1}) \tag{b}$$

We will consider the first and second approximations, i.e.

$$u^{(1)} = x(1 - x)\alpha_1^{(1)}, \quad u^{(2)} = x(1 - x)(\alpha_1^{(2)} + \alpha_2^{(2)} x)$$

For the first approximating function the Galerkin's type statement can be written as

$$\int_0^1 \left\{ \frac{d^2 u^{(1)}}{dx^2} + u^{(1)} + x \right\} \phi_1 \, dx = 0 \tag{c}$$

and for the second,

$$\int_0^1 \left\{ \frac{d^2u^{(2)}}{dx^2} + u^{(2)} + x \right\} \phi_1 \, dx = 0$$

$$\int_0^1 \left\{ \frac{d^2u^{(2)}}{dx^2} + u^{(2)} + x \right\} \phi_2 \, dx = 0 \qquad (d)$$

For the first approximation we obtain

$$\int_0^1 \left[-2\alpha_1^{(1)} + x(1-x)\alpha_1^{(1)} + x \right] x(1-x)dx = 0 \qquad (e)$$

which gives after integration

$$\frac{6}{10}\alpha_1^{(1)} - \frac{2}{12} = 0$$

$$\alpha_1^{(1)} = \frac{5}{18} \qquad (f)$$

and

$$u^{(1)} = \frac{5}{18} x(1-x)$$

After substituting $u^{(2)}$ in Equation (d) and integrating with respect to $\alpha_1^{(2)}$ and $\alpha_2^{(2)}$ we obtain the following system for the second approximation:

$$\frac{3}{10}\alpha_1^{(2)} + \frac{3}{20}\alpha_2^{(2)} = \frac{1}{12}$$

$$\frac{3}{20}\alpha_1^{(2)} + \frac{13}{105}\alpha_2^{(2)} = \frac{1}{20} \qquad (g)$$

Table 5.3

x	u_{exact}	$u^{(1)}$	$u^{(2)}$
0.25	0.044014	0.0521	0.0440
0.50	0.069747	0.0694	0.0698
0.75	0.060056	0.0521	0.0600

Thus

$$\alpha_1^{(2)} = \frac{71}{369} \qquad \alpha_2^{(2)} = \frac{7}{41} \tag{h}$$

The approximate solution is

$$u^{(2)} = x(1-x)\left[\frac{71}{369} + \frac{7}{41}x\right] \tag{i}$$

Comparing the value of the two first approximations with those of the exact solution at a series of points we can draw Table 5.3. The exact solution of (a) with boundary conditions $u(0) = u(1) = 0$ is

$$u = \frac{\sin x}{\sin 1} - x$$

Example 5.6
Consider Poisson's equation

$$\frac{\partial^2 u}{\partial x^2} + \frac{\partial^2 u}{\partial y^2} = c \tag{a}$$

with homogeneous boundary conditions $u = 0$ at $x = 0, a$ and $y = 0, b$ (Figure 5.6). We can take as a first approximation

$$u = \alpha x(x-a)y(y-b) \tag{b}$$

The Galerkin type statement can be written

$$\int_0^a \int_0^b \left(\frac{\partial^2 u}{\partial x^2} + \frac{\partial^2 u}{\partial y^2} - c\right) \delta u \, dx \, dy = 0 \tag{c}$$

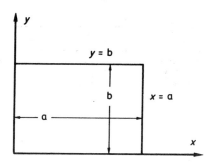

Figure 5.6   Two-dimensional domain for Poisson's equation

where $\delta u = \delta \alpha x(x - a)y(y - b)$. After substituting Equation (b) we find

$$-\frac{\alpha}{90}[a^3b^3(a^2 + b^2)] - \frac{ca^3b^3}{36} = 0 \tag{d}$$

$$\alpha = \frac{-5}{2}\frac{c}{a^2 + b^2} \tag{e}$$

Hence

$$u = \frac{-5c}{2(a^2 + b^2)}(x^2 - ax)(y^2 - by) \tag{f}$$

For the centre point

$$\left(x = \frac{a}{2}, \quad y = \frac{b}{2}\right)$$

we have

$$u_c = \frac{-5}{32}c\frac{a^2b^2}{a^2 + b^2} \tag{g}$$

Let us now solve the same example but using trigonometric series to approximate u. We assume, taking into consideration symmetry,

$$u = \sum_k \sum_l \alpha_{kl} \sin\frac{k\pi x}{a} \sin\frac{l\pi y}{b} \tag{h}$$

This expression satisfies the boundary conditions and the sinusoidal functions are orthogonal, that is

$$\int\limits_0^a \sin \frac{m\pi x}{a} \sin \frac{n\pi x}{a} \, dx = 0, \qquad \text{for } m \neq n$$

$$= \frac{a}{2}, \qquad \text{for } m = n \tag{i}$$

After substituting Equation (h) into (c) and taking into account orthogonality we have the following expression for each α_{kl}:

$$-\alpha_{kl} \int\limits_0^a \int\limits_0^b \left(\frac{\pi^2 k^2}{a^2} + \frac{\pi^2 l^2}{b^2} \right) \sin^2 \left(\frac{k\pi}{a} x \right) \sin^2 \left(\frac{l\pi}{b} y \right) \, dx \, dy$$

$$-\int\limits_0^a \int\limits_0^b c \sin \left(\frac{k\pi}{a} x \right) \sin \left(\frac{l\pi}{b} y \right) \, dx \, dy = 0 \tag{j}$$

This equation gives after integration

$$\alpha_{kl} = \frac{-16a^2 b^2}{\pi^4 kl(b^2 k^2 + a^2 l^2)} c \tag{k}$$

Note that each new α_{kl} coefficient does not involve the others. Hence the final result can be written

$$u = -\sum_k \sum_l c \frac{16a^2 b^2}{\pi^4 kl(b^2 k^2 + a^2 l^2)} \sin \frac{k\pi x}{a} \sin \frac{l\pi y}{b} \tag{l}$$

When the number of terms tends to infinity we obtain the exact solution. For the case $a = b$, we have at the centre point ($x = a/2$, $y = b/2$)

$$u_c = -\left(8 + \frac{16}{15} + \frac{8}{81} + \ldots \right) \frac{a^2}{\pi^4} c = \frac{-36.64}{\pi^4} \left(\frac{a}{2} \right)^2 c \tag{k}$$

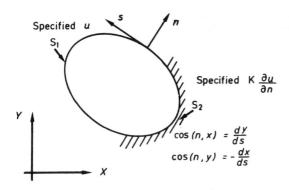

Figure 5.7 Two-dimensional body

Reduction of the order of differentiation

We will show here how Galerkin's expressions can be transformed to obtain formulae with a lower order of differentiation than in the original equilibrium equations (i.e. Equation (5.16) for a beam or Equation (5.17) for the two dimensional continuum).

Let us consider again the case of Poisson's equation with a given coefficient K. (We shall see the physical meaning of this equation in Chapters 6 and 7.)

$$K \left(\frac{\partial^2 u}{\partial x^2} + \frac{\partial^2 u}{\partial y^2} \right) = c \tag{5.18}$$

The boundary conditions for the problem are of two types:
(1) Essential boundary conditions, $u = \overline{u}$ where $\overline{u}$ is a specified value of the potential on S_1 (Figure 5.7).
(2) Natural boundary conditions of the type, flux equal to a given value, $q = K(\partial u/\partial n) = \overline{q}$ on S_2, where n is the normal to the boundary.

As shown in the previous example the trial function for Galerkin's expression can be assumed to satisfy both types of boundary conditions. In this case we have to satisfy only the equilibrium Equation (5.18) which can be written

$$\iint \left\{ K \left(\frac{\partial^2 u}{\partial x^2} + \frac{\partial^2 u}{\partial y^2} \right) - c \right\} \delta u \, dx \, dy = 0 \tag{5.19}$$

Using Green's theorem on (5.19) we obtain

$$\iint \left\{ K \left(\frac{\partial u}{\partial x} \frac{\partial \delta u}{\partial x} + \frac{\partial u}{\partial y} \frac{\partial \delta u}{\partial y} \right) + c \, \delta u \right\} \, dx \, dy$$

$$= K \int_S \frac{\partial u}{\partial n} \delta u \, dS \tag{5.20}$$

If the flux is known at the boundary,

$$\bar{q} = K \frac{\partial u}{\partial n} \tag{5.21}$$

we can satisfy it in an approximate way by writing

$$\iint \left\{ K \left(\frac{\partial^2 u}{\partial x^2} + \frac{\partial^2 u}{\partial y^2} \right) - c \right\} \delta u \, dx \, dy = \int_{S_2} \left(K \frac{\partial u}{\partial n} - \bar{q} \right) \delta u \, dS \tag{5.22}$$

Equation (5.22) is Galerkin's expression for the case in which we approximate the differential equation *and* the boundary conditions on S_2.

We could write Equation (5.22) in a different way by integrating it by parts.

$$\iint \left[K \left\{ \frac{\partial u}{\partial x} \frac{\partial \delta u}{\partial x} + \frac{\partial u}{\partial y} \frac{\partial \delta u}{\partial y} \right\} + c \delta u \right] dx \, dy = \int \bar{q} \, \delta u \, dS \tag{5.23}$$

which could also be written as the variation of the following functional

$$\Pi(u) = \frac{1}{2} \iint \left[K \left\{ \left(\frac{\partial u}{\partial x} \right)^2 + \left(\frac{\partial u}{\partial y} \right)^2 \right\} + cu \right] dx \, dy = \int_{S_2} \bar{q} \, u \, dS \tag{5.24}$$

We could apply the Rayleigh–Ritz technique directly on the Π functional (5.24) or alternatively Galerkin's method could be applied in either Equation (5.22) or (5.23). Note however, that if we use Equation (5.23) the trial function can be of lower order than in Equation (5.22), as the latter contains first derivatives and the former second derivatives. It can be easily proved that for this case formulae (5.22) and (5.24) are equivalent to minimizing the functional (5.24).

Example 5.7
Consider a simply supported beam in free vibrations for which case we can write the following equilibrium equation for a dx,

$$EI \frac{d^4 v}{dx^4} + A \rho \ddot{v} = 0 \tag{a}$$

with the boundary conditions $\bar{M} = M = 0$, $v = 0$ at $x = 0, l$.
Galerkin's statement for this case can be written as (see Equation (5.16))

$$\int_0^l \left\{ EI \frac{d^4 v}{dx^4} + A \rho \ddot{v} \right\} \delta v \, dx = \left[-EI \frac{d^2 v}{dx^2} \frac{d\delta v}{dx} \right]_{x=0}^{x=l} \tag{b}$$

Integrating Equation (b) twice by parts and taking into account the boundary conditions (we are assuming the functions δv satisfy the displacement boundary condition $v = 0$, hence $\delta v = 0$ at x and l) we have

$$\int_0^l EI \frac{d^2 v}{dx^2} \frac{d^2 \delta v}{dx^2} \, dx + \int_0^l A \rho \ddot{v} \delta v \, dx = 0 \tag{c}$$

Note that the order of the derivatives in Equation (c) has been reduced when compared against those of Equation (b). Equation (c) can be seen as the Principle of Virtual Displacements for the dynamic case.
We consider the following approximate solution which satisfies the displacement or essential boundary conditions

$$v = \alpha_1 x(x - l) + \alpha_2 x^2 (x - l) \tag{d}$$

Let us assume all the constants are equal to 1 for simplicity, i.e. $E = I = \rho = A = l = 1$. Thus substituting Equation (d) into (c) we obtain after integration

$$[4\alpha_1 \delta\alpha_1 + 4\alpha_2 \delta\alpha_2 + 2\alpha_1 \delta\alpha_2 + 2\alpha_2 \delta\alpha_1]$$

$$+ \left[\left(\frac{1}{30}\right) \ddot{\alpha}_1 \delta\alpha_1 + \left(\frac{1}{105}\right) \ddot{\alpha}_2 \delta\alpha_2 + \left(\frac{1}{60}\right) \ddot{\alpha}_1 \delta\alpha_2 \right.$$

$$+ \left(\frac{1}{60} \right) \ddot{\alpha}_2 \, \delta\alpha_1 \Bigg] \tag{e}$$

The solution for this system of equations is such that

$$\frac{d^2}{dt^2} (\alpha_i) = -\omega^2 \alpha_i$$

Hence Equation (e) gives the following system of equations

$$\{\delta\alpha_1 \quad \delta\alpha_2\} \begin{bmatrix} 4 - \dfrac{\omega^2}{30} & 2 - \dfrac{\omega^2}{60} \\[3mm] 2 - \dfrac{\omega^2}{60} & 4 - \dfrac{\omega^2}{105} \end{bmatrix} \begin{Bmatrix} \alpha_1 \\ \alpha_2 \end{Bmatrix} = \begin{Bmatrix} 0 \\ 0 \end{Bmatrix} \tag{f}$$

We have to solve the following eigenvalue problem

$$\begin{vmatrix} 4 - \dfrac{\omega^2}{30} & 2 - \dfrac{\omega^2}{60} \\[3mm] 2 - \dfrac{\omega^2}{60} & 4 - \dfrac{\omega^2}{105} \end{vmatrix} = 0 \tag{g}$$

This gives,

$$\omega_1^2 = 120, \quad \omega_2^2 = 2520 \tag{h}$$

The exact solution is $\omega_n^2 = (n\pi)^4$, for $n = 1, 2, \ldots$

Exercises

(1) Consider the beam-spring system shown in Figure 5.8. Solve for the displacement of point A in the direction of the force P with the Rayleigh–Ritz method. Consider two trial functions

$$v_1 = \alpha_1 \sin \frac{\pi x}{l}$$

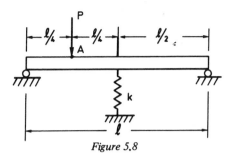

Figure 5.8

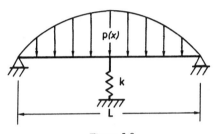

Figure 5.9

$$v_2 = \alpha_1 \sin \frac{\pi x}{l} + \alpha_2 \sin \frac{2\pi x}{l}$$

Discuss convergence for the displacement under the load.

(2) Consider a simple supported prismatic beam under free vibrations, with essential boundary conditions

$$v(0) = v(l) = 0$$

Apply the Rayleigh–Ritz method to find the frequency of vibration, first for the nodal shape $x(x - l)$ and then for $x^2(x - l)$. Compare against the exact solution

$$\omega^2 = n^4 \pi^4 \frac{EI}{\pi \rho l^4}$$

(3) Apply the Rayleigh–Ritz method to the beam shown in Figure 5.9 to calculate the deflection at the centre. The beam is simply supported at each end ($v(0) = v(L) = 0$) and the energy functional is

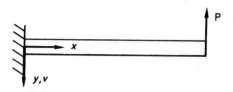

Figure 5.10

$$\pi_p = \int_0^L \left\{ \frac{1}{2} EI \left(\frac{d^2v}{dx^2} \right)^2 - pv \right\} dx + \frac{1}{2} kv_{L/2}^2$$

where $p(x) = x(L - x)$. Use two trial solutions and comment on the convergence of the results.

(4) Find the solution for the two dimensional Poisson's equation

$$\frac{\partial^2 u}{\partial x^2} + \frac{\partial^2 u}{\partial y^2} = c \qquad\qquad\qquad (a)$$

with boundary conditions $u = 0$ at $x = \pm a$, $y = \pm b$ using Galerkin's method. Take as a first approximation

$$u^{(1)} = \alpha_1(a^2 - x^2)(b^2 - y^2) \qquad\qquad\qquad (b)$$

and as the second approximation

$$u^{(2)} = \alpha_1(a^2 - x^2)(b^2 - y^2) + \alpha_2 x^2(a^2 - b^2)(b^2 - y^2) \qquad (c)$$

Comment on the choice of functions (b) and (c) and the convergence of the results.

(5) Consider the fixed–free beam shown in Figure 5.10. Describe how you would include the natural boundary conditions at the free end in order to solve using the Galerkin method.

(6) Solve the differential equation

$$\frac{d^2u}{dx^2} + \frac{du}{dx} + x = 0 \qquad\qquad\qquad (a)$$

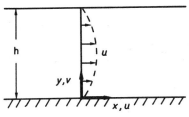

Figure 5.11

with boundary conditions $u(0) = u(1) = 0$, using Galerkin.

(7) Consider the case of flow in a channel of unit depth for the case vertical velocity $v = 0$ (Figure 5.11). The momentum equation governing the problem is

$$-\frac{\partial p}{\partial x} + \mu \frac{\partial^2 u}{\partial y^2} = 0$$

where μ is the viscosity of the fluid and u the velocity in the x direction. Find this velocity using Galerkin's method.

Bibliography

Finlayson, B. A., *The Method of Weighted Residuals and Variational Principles*, Academic Press (1972)

Rayleigh, J. W., *The Theory of Sound*, Dover (1945)

Brebbia, C. A. and Tottenham, H. (Editors), *Variational Methods in Engineering*, Vols. I and II, Southampton University Press (1973)

Brebbia, C. A. and Ferrante, A. J. *The Finite Element Technique*, U.R.G.S. Press, Porto Alegre, Brazil (1975)

Chapter 6

The finite element technique

6.1 INTRODUCTION

One of the major problems for the direct application of both the Rayleigh–Ritz and Galerkin methods is the adequate selection of the global approximate solutions. They must not only satisfy the essential boundary conditions, but must also conveniently represent other characteristics of the problem. On the other hand, we have seen that to improve the approximate numerical solutions it is necessary to work with trial functions of ever increasing order. The labour involved in handling those functions may become considerable, which in many cases renders these methods unfeasible. These difficulties prevent their generalized use for practical problems.

The concept of the finite element method consists of subdividing the domain of integration in a discrete number of small regions, of finite dimensions, and to locally apply, under certain conditions, one of the approximate methods described above. Thus, over each region, called finite element, it will be possible to adopt simple functions to represent the local behaviour of that region. Then, to obtain better approximate solutions it will not be necessary to increase the order of the trial functions, but to work with a finer subdivision, of smaller dimensions.

The first step consists of subdividing the domain of integration of the problem in a number of regions, forming a mesh of finite elements. The finite element mesh is defined by the type and number of finite element used. A finite element type is characterized by its geometric shape and the local trial functions used to express its local behaviour. For plane problems the geometric shapes used are mainly triangles, rectangles and quadrilaterals, which may have straight or curved sides. For three-dimensional problems tetrahedra, hexahedra, and parallelepiped elements are used. Some problems require linear elements, which may be straight or curved. For problems with axial symmetry toroidal elements of various cross sectional shapes are used.

Over each element there will be several nodal points, differing in number and location according to the element type. These may be internal or external

250

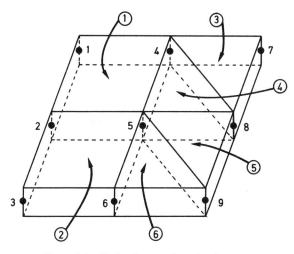

Figure 6.1   Finite element discretization

nodal points, the latter materializing the 'connection' of the element to other surrounding elements, in the sense that on a nodal point the values of the problem variables must be the same, no matter which of the trial functions are used to compute them, provided they correspond to elements incident on that nodal point.

There is no fixed rule which can be generally used to determine the number of elements, and thus of nodes, which should be used in a finite element mesh. Clearly, the finer the mesh the better will be the numerical solution obtained. However, the computer time needed will also increase, eventually to undesirable proportions. The key question is to find the coarsest, or 'cheapest', possible mesh, which will produce a numerical solution within the desired precision. This question has no easy answer, since it depends both on the type of problem, and on the type of element used. The design of a finite element mesh, which will be done using as regular elements as possible, will have to be based on previous experience in the use of the finite element method. Convergence tests for different finite element types, available in the technical literature, are always helpful. When in doubt, a problem can be solved with several different meshes, with increasing number of elements and nodal points, and the comparison of the numerical solution obtained can help in evaluating approximately the precision achieved.

Once a finite element mesh is selected, elements and nodal points will be numbered, for reference, in the remaining steps. The element and nodal numbering are included into an element-node connectivity table, by listing the nodes associated with each of the elements in the finite element mesh. As an illustration, Figure 6.1 shows a typical two-dimensional problem. In this

Table 6.1

Element	n_1	n_2	n_3	n_4	Type
1	1	2	5	4	Rectangular
2	2	3	6	5	Rectangular
3	4	8	7	–	Triangular
4	4	5	8	–	Triangular
5	5	9	8	–	Triangular
6	5	6	9	–	Triangular

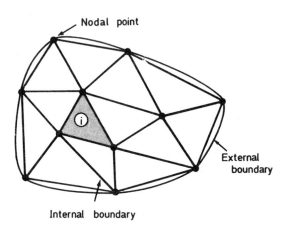

Figure 6.2

case the nodes are placed at the mid-surface of the plate, and correspond to
the element corners. In general, we could also have nodes along the element
boundaries and eventually inside the elements, as it will be discussed later.
Table 6.1 shows the element-node connectivity table for the mesh. Note that
the nodes have to be listed in the same direction (clockwise or anticlockwise)
for all elements, but it does not matter which is the starting node. In the
example, the anticlockwise direction is used.

Once the element mesh has been selected, it is required to adopt the
approximate expansion for the problem variables over each element. To
clarify the basic scheme of the finite element method, let us assume that for a
given problem there exists a functional Π, in terms of just one problem variable
u, such that the solution of the problem is given by the variational statement
$\delta\Pi = 0$, subject to some boundary conditions on the problem variables. This
defines the continuous model. The domain of integration is then divided into
m regions, or finite elements, of certain geometric shape. Figure 6.2, for
instance, suggests the subdivision of a plane domain by a mesh of triangular
elements. Over each finite element we can write the partial expression of the

problem functional, which will be generally called Π^e for element e, such that:

$$\Pi = \sum_{e=1}^{m} \Pi^e(u) \tag{6.1}$$

Assuming an approximate solution, over element e, of the form:

$$u^e = \sum_{k=1}^{r} \alpha_k^e \, \psi_k^e = A\,\alpha \tag{6.2}$$

where

$$A = [\psi_1^e \quad \psi_2^e \ldots \psi_r^e]$$
$$\alpha = \{\alpha_1^e \quad \alpha_2^e \ldots \alpha_r^e\} \tag{6.3}$$

we can replace Equation (6.1) by the approximate expression

$$\Pi_a = \sum_{e=1}^{m} \Pi_a^e(\alpha_k^e), \quad k = 1, 2, \ldots, r \tag{6.4}$$

It must be noticed that Equation (6.2) is only valid over element e, such that over each element a trial function of the same form as (6.2) will be adopted, but the unknown parameters α_k^e will differ from element to element. Eventually the functions ψ_k^e may also be different from element to element, although in practice they are normally the same.

In the Rayleigh-Ritz, Galerkin, and other similar trial function methods, one would proceed to set up the problem governing system of equations in terms of the unknown parameters α_k^e. In the finite element method, however, a variable transformation is performed, such that the discrete model unknowns are the values of the problem variables at the nodal points. To see this, let us assume that n^e is the number of nodes per element. Then evaluating Equation (6.2) for each of the nodes pertaining to element e one obtains

$$u_1^e = A_1 \alpha$$

$$u_2^e = A_2\alpha$$

$$\vdots \qquad \vdots$$

$$u_{ne}^e = A_n\alpha \tag{6.5}$$

or, in matrix form

$$U^e = C\,\alpha \tag{6.6}$$

where

$$U^e = \left\{ u_1^e \quad u_2^e \ldots u_n^e \right\}$$

is the element unknown vector, and

$$C = \left\{ \begin{array}{c} A_1 \\ A_2 \\ \cdot \\ \cdot \\ \cdot \\ A_n \end{array} \right\} \tag{6.7}$$

where A_j^1 is the value of A at node j.

If the functions ψ_k^e are properly selected, and the number of element unknowns equals the number of unknown parameters α_k^e (i.e. $n = r$ in this case), C is a square and regular matrix. Then Equation (6.6) can be inverted so that

$$\alpha = C^{-1}\,U^e \tag{6.8}$$

which replaced in (6.2) gives

$$u^e = (A\,C^{-1})\,U^e = \Phi^{e,T}\,U^e \tag{6.9}$$

where

$$\Phi^{e,T} = A\,C^{-1} \tag{6.10}$$

Equation (6.9) defines the value of the problem variable u^e at any point over element e, in terms of the values of the same variable at the element nodal points. In extended form, it can be written as

$$u^e = \Phi^{e,T} U^e = \sum_{k=1}^{r} \phi_k^e u_k^e \qquad (6.11)$$

where the functions ϕ_k^e are called 'shape functions'. It should be noticed that in the finite element method, whenever possible, rather than using Equation (6.2) one starts directly with expressions of the type (6.11).

Based on Equation (6.11), rather than (6.2) we write

$$\Pi_a = \sum_{e=1}^{m} \Pi_a^e \left(\sum_{k=1}^{r^e} \phi_k^e u_k^e \right) = \sum \Pi_a^e (u_k^e) \qquad (6.12)$$

where r is the number of interpolation functions.

Finally, we apply the stationary conditions for the approximate functional Π_a,

$$\delta \Pi_a = \sum_{e=1}^{m} \delta \Pi_a^e(u_k^e) = \sum_{e=1}^{m} \sum_{k=1}^{r} \frac{\partial \Pi_a^e}{\partial u_k^e} \delta u_k^e = 0 \qquad (6.13)$$

Since the variations of u_k^e are arbitrary

$$\sum_{e=1}^{m} \left(\sum_{k=1}^{r} \frac{\partial \Pi_a^e}{\partial u_k^e} \right) = 0 \qquad (6.14)$$

which is a system of equations whose solution permits the values of the unknowns nodal variables u_k^e to be found. The solution over each element is then completely known.

The development above was based on a Rayleigh–Ritz scheme, but the finite element method can be based as well on schemes following the Galerkin, or even other approximate methods.

The practical application of the finite element method is feasible only if electronic computers are used and all the computations are expressed in matrix form. Starting at the element level the behaviour of each element is defined in terms of element matrices. The global behaviour is then defined in terms

of global matrices, representing the governing system of equations, obtained
by superposition of the element matrices. The matrix operations involved
in a finite element solution are ideally suited for computer automation. They
can be subdivided according to the following basic steps:

(1) Discretization of the continua, or selection of the finite element
    mesh (normally done outside of the computer).
(2) Evaluation of the element matrices.
(3) Assembling of the governing system of equations.
(4) Introduction of the boundary conditions.
(5) Solution of the system of equations.
(6) Computation of secondary results, from the problem variables of the
    approximate solution.

These are similar to the analysis steps for member systems, discussed in
Chapters 3 and 4.

The critical step in the finite element method is the selection of the approxi
mate behaviour defined by Equation (6.2) or (6.11). The approximate functions on each element have to satisfy the admissibility and completeness conditions for the problem, similar to those discussed in Rayleigh–Ritz.

Admissibility now implies that the problem variables be continuous across
the element boundary (i.e. essential boundary conditions). In addition the
functions have to be continuous up to one order less than the higher derivative in functional.

Completeness implies that when the elements tend to become infinitesimal,
and the derivatives existing in the functional tend to become constant, the
approximating functions must be capable of representing those constant
derivatives exactly.

It is generally recognized that the first application of the finite element
method, as described above, is due to Courant who in 1943 developed a formulation based on the principle of minimum potential energy for the solution of the Saint Venant torsion problem, adopting linear approximate expansions for the warping function over triangular regions or finite elements. Some
years later Synge and also Praguer, generalized Courant's formulation giving
it the name of the hypercircle method. Most of the applications of these
developments were for the solution of mathematical problems, without becoming known in general in engineering.

The first engineering application of the finite element method was due
to Turner, Martin, Clough and Topp[1], who in 1956 presented triangular
and rectangular finite elements for plane stress, in connection with stress
analysis problems related to the aerospace industry. Their formulation was
made possible thanks to the matrix analysis theory previously presented by
Argyris[2].

In this case, the finite element method was introduced not according to
the 'variational approach', but as an extension to continuous problems of the
matrix structural analysis schemes, already known for member systems.

Accordingly, a continuous problem would be considered by subdividing it into a collection of finite regions called finite elements. Each element would be studied individually, isolated from the others, with the object of establishing a matrix relationship between displacements and stresses or forces at the element nodal points, in terms of stiffness coefficients. Then, the same as in member systems, the problem system of equations would be assembled from the element contributions, according to the element connectivity. Introducing the displacement boundary conditions and solving the governing system of equations, the nodal displacements become known. Returning again to the element level, from the element nodal displacements, element stresses or forces could be computed. The computation of the element stiffness matrix was based on the principle of virtual displacements.

The introduction of the finite element method generated a great deal of interest and soon many new applications to solid mechanics problems were presented. Thus, Melosh[3], Adini[4], Clough and Tocher[5], and others, applied the method to plate bending; Argyris[6] to three-dimensional solids of arbitrary geometry; Grafton and Strome[7] to axisymmetric shells under axisymmetric loadings; Percy, et al[8], to axisymmetric shells under arbitrary loadings; Connor and Brebbia[9] to shallow shells, etc.

These first developments were all based on the 'physical approach' described above. It was only after 1962, that a more exhaustive study of the theoretical bases of the finite element method and variational principles, particularly by Melosh[10], and Pian and Tong[11], allowed its generalization according to the variational approach. This also permitted a better analysis of the conditions for convergence, which was studied, among others, by Oden[12], Arantes de Oliveira[13], and Strang and Fix[14].

In the last decade the finite element method became generally accepted as one of the most powerful computational tools available for the solution of solid mechanics problems. In addition, its applications were extended to many different engineering fields, including hydraulics, heat transfer, soil mechanics, lubrication problems, etc.

In the remainder of this chapter, and in the next chapter, several different applications of the finite element method will be presented.

6.2   EXTENDED LAPLACE EQUATION

We will now take the case of a field problem, governed by an equation of the type

$$\frac{\partial}{\partial x}\left(h_x \frac{\partial u}{\partial x}\right) + \frac{\partial}{\partial y}\left(h_y \frac{\partial u}{\partial y}\right) + \lambda u = c, \text{ in a domain } A. \tag{6.15}$$

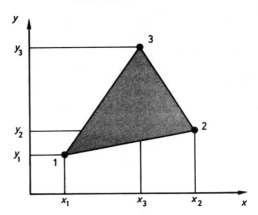

Figure 6.3
Triangular element

with boundary conditions

$$u = \overline{u}, \text{ on } S_1 \tag{6.16}$$

where S_1 defines the part of the boundary where the function u is prescribed and

$$q = h_x \frac{\partial u}{\partial x} \alpha_{nx} + h_y \frac{\partial u}{\partial y} \alpha_{ny} = \overline{q}, \text{ on } S_2,$$

where S_2 denotes the part of the boundary where the flux q is prescribed. α_{nx} and α_{ny} are the direction cosines of the normal to S with respect to x and y, λ is a constant and h_x, h_y, c and $\overline{q}$ are functions of position. The total boundary is $S = S_1 + S_2$.

In particular when $\lambda = 0$, Equation (6.15) is Poisson's equation. If $\lambda = c = 0$ then we have Laplace's equation. Equation (6.15) is the mathematical representation of a great number of problems in solids and fluids.

Equations (6.15) and (6.16) can be written in Galerkin's form by generalizing Example 5.7.

$$\iint \left\{ h_x \frac{\partial u}{\partial x} \frac{\partial \delta u}{\partial x} + h_y \frac{\partial u}{\partial y} \frac{\partial \delta u}{\partial y} - \lambda u \, \delta u + c \delta u \right\} dx \, dy = \int_{S_2} \overline{q} \, \delta u \, ds \tag{6.17}$$

Let us first consider Laplace's equation, i.e.

$$\iint \left\{ h_x \frac{\partial u}{\partial x} \frac{\partial \delta u}{\partial x} + h_y \frac{\partial u}{\partial y} \frac{\partial \delta u}{\partial y} \right\} dx \, dy = \int_{S_2} \overline{q} \, \delta u \, ds \tag{6.18}$$

We will apply Equation (6.17) on the triangular domain shown in Figure 6.3.

We can propose a function

$$u = \alpha_1 + \alpha_2 x + \alpha_3 y \tag{6.19}$$

which varies linearly in the domain and produces 3 unknowns per element. These unknowns can be related to nodal values at (1, 2, 3) nodes.

The derivatives of Equation (6.19) are

$$\frac{\partial u}{\partial x} = \alpha_2, \quad \frac{\partial u}{\partial y} = \alpha_3 \tag{6.20}$$

and

$$\frac{\partial \delta u}{\partial x} = \delta\alpha_2, \quad \frac{\partial \delta u}{\partial y} = \delta\alpha_3$$

Now the left-hand side of Equation (6.18) can be written

$$\iint \{h_x \alpha_2 \delta\alpha_2 + h_y \alpha_3 \delta\alpha_3\} \, dx \, dy = A \ \{\delta\alpha_1 \ \delta\alpha_2 \ \delta\alpha_3\} \begin{bmatrix} 0 & 0 & 0 \\ 0 & h_x & 0 \\ 0 & 0 & h_y \end{bmatrix} \begin{Bmatrix} \alpha_1 \\ \alpha_2 \\ \alpha_3 \end{Bmatrix}$$

$$= \delta\alpha^T H \alpha \tag{6.21}$$

Here the results of the integration only produces a term A = area of the element, because the integrand is constant with respect to x, y. We want now to express (6.21) as a function of nodal unknowns u_1, u_2, u_3. Specializing (6.19) for the corner nodes we have

$$\begin{Bmatrix} u_1 \\ u_2 \\ u_3 \end{Bmatrix} = \begin{bmatrix} 1 & x_1 & y_1 \\ 1 & x_2 & y_2 \\ 1 & x_3 & y_3 \end{bmatrix} \begin{Bmatrix} \alpha_1 \\ \alpha_2 \\ \alpha_3 \end{Bmatrix} \tag{6.22}$$

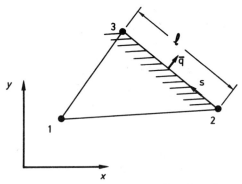

Figure 6.4   Prescribed $\bar{q}$ on boundary

The inverse of Equation (6.22) gives the relationship between α_i and $\mathbf{U}^e$ and is

$$
\begin{Bmatrix} \alpha_1 \\ \alpha_2 \\ \alpha_3 \end{Bmatrix} = \begin{bmatrix} c_{11} & c_{12} & c_{13} \\ c_{21} & c_{22} & c_{23} \\ c_{31} & c_{32} & c_{33} \end{bmatrix} \mathbf{U}^e = \mathbf{C}\, \mathbf{U}^e \tag{6.23}
$$

where

$$
\begin{aligned}
&c_{11} = (x_2 y_3 - x_3 y_2)/\Delta, \quad c_{12} = (x_3 y_1 - x_1 y_3)/\Delta, \quad c_{13} = (x_1 y_2 - x_2 y_1)/\Delta \\
&c_{21} = (y_2 - y_3)/\Delta, \quad\quad\quad c_{22} = (y_3 - y_1)/\Delta, \quad\quad\quad c_{23} = (y_1 - y_2)/\Delta \\
&c_{31} = (x_3 - x_2)/\Delta, \quad\quad\quad c_{32} = (x_1 - x_3)/\Delta, \quad\quad\quad c_{33} = (x_2 - x_1)/\Delta
\end{aligned} \tag{6.24}
$$

and

$$
\Delta = (x_2 y_3 + x_1 y_2 + x_3 y_1) - (x_2 y_1 + x_1 y_3 + y_2 x_3) = 2A \text{ (twice area)}
$$

Note that the $\mathbf{C}$ in Equation (6.23) is equivalent to $\mathbf{C}^{-1}$ of Equation (6.8) but from here on the superscript is not written for simplicity.

Finally, we have for Equation (6.21)

$$
\delta\alpha^T\, \mathbf{H}\, \alpha \Rightarrow \delta\mathbf{U}^{e,T}(\mathbf{C}^T\, \mathbf{H}\, \mathbf{C})\mathbf{U}^e = \delta\mathbf{U}^{e,T}\, \mathbf{K}^e\, \mathbf{U}^e \tag{6.25}
$$

The right-hand side of Equation (6.18) only exists for the boundary S_2 of the body. If we have a constant value of q on side 2–3, for instance (Figure 6.4), this integral becomes

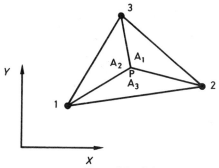

Figure 6.5   Sub-triangles

$$\int_{node\ 2}^{node\ 3} \overline{q}\ \delta u\ dS = \overline{q}\left(\frac{l}{2}\delta u_2 + \frac{l}{2}\delta u_3\right) \tag{6.26}$$

That is, the values of u can be averaged on nodes 2 and 3. This is not always valid but can be done when $\overline{q}$ is constant and u is linear.

Equation (6.18) can now be written

$$\delta U^{e,T} K^e U^e = \delta U^{e,T} P^e \tag{6.27}$$

where

$$P^e = \begin{Bmatrix} 0 \\ 1 \\ 1 \end{Bmatrix} \frac{l}{2}\ \overline{q}, \quad K^e = C^T H C$$

Therefore

$$K^e U^e = P^e \tag{6.28}$$

Equation (6.28) represents the equilibrium equations for each element.

6.2.1 Triangular coordinates

It is not always convenient to work with x,y coordinates as was done in the above example. The homogeneous or triangular coordinates allow for easier representation of the interpolation functions and simpler integrations, when working with triangular finite elements.

Let us consider the triangle shown in Figure 6.5. A point P, of coordinates

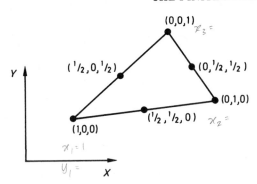

Figure 6.6   Homogeneous coordinates

(x,y) defines, together with the corners of the triangle, three subtriangles of areas A_1, A_2 and A_3, such that:

$$A = A_1 + A_2 + A_3 \qquad (6.29)$$

where A is total area of the triangle $(1, 2, 3)$. Since the values of A_1, A_2 and A_3 are unique for each point in the triangle, we conclude that these values can be used to locate the position of points on the triangle (i.e. as coordinates). In practice, however, it is better to work with normalized values. The homogeneous or triangular coordinates are then defined as:

$$\xi_1 = \frac{A_1}{A}; \quad \xi_2 = \frac{A_2}{A}; \quad \xi_3 = \frac{A_3}{A} \qquad (6.30)$$

Note that from (6.29) and (6.30) one finds

$$\xi_1 + \xi_2 + \xi_3 = 1 \qquad (6.31)$$

which means that not all three triangular coordinates are independent. In fact, being a reference system for a 2-dimensional space, we could work with just two of the triangular coordinates computing the third one, if needed, from Equation (6.31). It is natural, however, and also more convenient, to work with all three triangular coordinates, when dealing with points defined on a triangle. Figure 6.6 shows the value of coordinates (ξ_1, ξ_2, ξ_3) for some relevant points.

To establish the relationship between the triangular and the Cartesian coordinates, one first remembers that the area of a triangle of corners $(1, 2, 3)$

is given by

Example 1:

$$\frac{1}{2}\left[(x_2 y_3 - x_3 y_2) - x_1(y_3 - y_2) + (x_3 - x_2)\right]$$

$$A = \frac{1}{2} \begin{vmatrix} 1 & x_1 & y_1 \\ 1 & x_2 & y_2 \\ 1 & x_3 & y_3 \end{vmatrix} \qquad = -\frac{1}{2}\left[(3\times2 - 3\times0) - 0 + 0\right]$$

(6.32)

$$= 3$$

Then, expressions (6.30) become: *example 2 :*

$$= \frac{1}{2}\left[(5\times5 - 5\times0) - 0 + 0\right] = 12.5$$

$$\xi_1 = \frac{1}{2A} \begin{vmatrix} 1 & x & y \\ 1 & x_2 & y_2 \\ 1 & x_3 & y_3 \end{vmatrix} ; \qquad \xi_2 = \frac{1}{2A} \begin{vmatrix} 1 & x_1 & y_1 \\ 1 & x & y \\ 1 & x_3 & y_3 \end{vmatrix} ;$$

$$\xi_3 = \frac{1}{2A} \begin{vmatrix} 1 & x_1 & y_1 \\ 1 & x_2 & y_2 \\ 1 & x & y \end{vmatrix}$$

(6.33)

or

$$\xi_1 = \frac{1}{2A}\left[(x_2 y_3 - y_2 x_3) + x(y_2 - y_3) + y(x_3 - x_2)\right]$$

$$\xi_2 = \frac{1}{2A}\left[(x_3 y_1 - x_1 y_3) + x(y_3 - y_1) + y(x_1 - x_3)\right]$$

$$\xi_3 = \frac{1}{2A}\left[(x_1 y_2 - x_2 y_1) + x(y_1 - y_2) + y(x_2 - x_1)\right]$$

(6.34)

Comparing with Equation (6.24), it can be seen that

$$\xi_1 = c_{11} + x\,c_{21} + y\,c_{31}$$

$$\xi_2 = c_{12} + x\,c_{22} + y\,c_{32}$$

$$\xi_3 = c_{13} + x\,c_{23} + y\,c_{33}$$

(6.35)

or

$$
\begin{Bmatrix} \xi_1 \\ \xi_2 \\ \xi_3 \end{Bmatrix} = \begin{bmatrix} c_{11} & c_{21} & c_{31} \\ c_{12} & c_{22} & c_{32} \\ c_{13} & c_{23} & c_{33} \end{bmatrix} \begin{Bmatrix} 1 \\ x \\ y \end{Bmatrix}
\tag{6.36}
$$

Inverting this last expression one obtains the relationship between the Cartesian and the triangular coordinates, as

$$
\begin{Bmatrix} 1 \\ x \\ y \end{Bmatrix} = \begin{bmatrix} 1 & 1 & 1 \\ x_1 & x_2 & x_3 \\ y_1 & y_2 & y_3 \end{bmatrix} \begin{Bmatrix} \xi_1 \\ \xi_2 \\ \xi_3 \end{Bmatrix}
\tag{6.37}
$$

or

$$
x = \sum_{i=1}^{3} x_i \xi_i
$$

$$
y = \sum_{i=1}^{3} y_i \xi_i
\tag{6.38}
$$

Note that the first equation from (6.37) is the same as (6.31).

If it is desired to use triangular coordinates to develop the formulation of a triangular finite element, derivatives and integrals of functions defined in terms of triangular coordinates will have to be computed. For derivatives, using the chain rule, one obtains

$$
\frac{\partial f(\xi_1, \xi_2, \xi_3)}{\partial x} = \frac{\partial f}{\partial \xi_1} \frac{\partial \xi_1}{\partial x} + \frac{\partial f}{\partial \xi_2} \frac{\partial \xi_2}{\partial x} + \frac{\partial f}{\partial \xi_3} \frac{\partial \xi_3}{\partial x}
$$

$$
= \frac{\partial f}{\partial \xi_1} c_{21} + \frac{\partial f}{\partial \xi_2} c_{22} + \frac{\partial f}{\partial \xi_3} c_{23} = \sum_{i=1}^{3} \frac{\partial f}{\partial \xi_i} c_{2i}
$$

$$
\frac{\partial f(\xi_1, \xi_2, \xi_3)}{\partial y} = \frac{\partial f}{\partial \xi_1} \frac{\partial \xi_1}{\partial y} + \frac{\partial f}{\partial \xi_2} \frac{\partial \xi_2}{\partial y} + \frac{\partial f}{\partial \xi_3} \frac{\partial \xi_3}{\partial y} =
$$

$$= \frac{\partial f}{\partial \xi_1} c_{31} + \frac{\partial f}{\partial \xi_2} c_{32} + \frac{\partial f}{\partial \xi_3} c_{33} = \sum_{i=1}^{3} \frac{\partial f}{\partial \xi_i} c_{3i} \qquad (6.39)$$

Higher derivatives can be computed by repeated application of Equations (6.39).

The integration of polynomial functions defined in terms of the triangular coordinates is very simple, and can be made using the expression

$$\int\int \xi_1^i \, \xi_2^j \, \xi_3^k \, \mathrm{d}A = \frac{i! \, j! \, k!}{(i+j+k+2)!} \, 2A \qquad (6.40)$$

For instance

$$\int\int (2\xi_1^3 + 4\xi_2 \, \xi_3 + 1)\mathrm{d}A = 2 \frac{3! \, 0! \, 0!}{5!} \, 2A + 4 \frac{0! \, 1! \, 1!}{4!} \, 2A + \frac{0! \, 0! \, 0!}{2!} \, 2A$$

$$= \frac{23}{15} A$$

Similar simple expressions are available for integration in one dimension. For instance, an integral along the side 1−2 of the triangle can be computed using the formula

$$\int \xi_1^i \, \xi_2^j \, \mathrm{d}s = \frac{i! \, j!}{(i+j+1)!} \, l_{1-2} \qquad (6.41)$$

where l_{1-2} is the length of side 1−2. Note that on side 1−2, $\xi_3 = 0$

Let us now go back to the element of Figure 6.3, and Equation (6.17). The approximate linear expansion for u can be written in terms of the triangular coordinates, as

$$u = \xi_1 u_1 + \xi_2 u_2 + \xi_3 u_3 = \sum_{k=1}^{3} \xi_k u_k \qquad (6.42)$$

where the ξ_k are the interpolation or shape functions for the linear case, and u_k are the nodal values of u. By comparison with (6.11) one sees that

$$\phi_k = \xi_k \qquad (6.43)$$

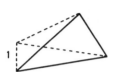 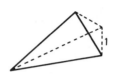 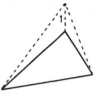

Figure 6.7   Shape functions

where the super-subscript e from ϕ_k^e has been dropped. Figure 6.7 shows graphically the shape or interpolation functions (6.43).

The derivatives of u, according to Equation (6.39), will be given by

$$\frac{\partial u}{\partial x} = u_1 c_{21} + u_2 c_{22} + u_3 c_{23}$$

$$\frac{\partial u}{\partial y} = u_1 c_{31} + u_2 c_{32} + u_3 c_{33}$$

Substituting these values in the variational statement the first two terms on the right hand side become

$$\delta \mathbf{U}^{e,T} A \left\{ h_x \begin{bmatrix} c_{21}^2 & c_{22}c_{21} & c_{23}c_{21} \\ \text{symm.} & c_{22}^2 & c_{23}c_{22} \\ & & c_{23}^2 \end{bmatrix} \right.$$

$$\left. + h_y \begin{bmatrix} c_{31}^2 & c_{32}c_{31} & c_{33}c_{31} \\ \text{symm.} & c_{32}^2 & c_{33}c_{32} \\ & & c_{33}^2 \end{bmatrix} \right\} \begin{Bmatrix} u_1 \\ u_2 \\ u_3 \end{Bmatrix}$$

$$= \delta \mathbf{U}^{e,T} \mathbf{K}^e \mathbf{U}^e \tag{6.44}$$

where $\mathbf{K}^e$ is the same matrix as in the previous case.

Taking now the term

$$\lambda \iint u \, \delta u \, dA$$

and considering (6.42), one obtains

$$\lambda \iint (\delta u_1 \, \xi_1 + \delta u_2 \, \xi_2 + \delta u_3 \, \xi_3)(u_1 \, \xi_1 + u_2 \, \xi_2 + u_3 \, \xi_3) dA$$

which can be written in matrix form as

$$\iint \lambda u \, \delta u \, dA = \delta \mathbf{U}^{e,T} \lambda \iint \begin{bmatrix} \xi_1^2 & \xi_1 \xi_2 & \xi_1 \xi_3 \\ & \xi_2^2 & \xi_2 \xi_3 \\ \text{symm.} & & \xi_3^2 \end{bmatrix} dA \; \mathbf{U}^e$$

$$= \delta \mathbf{U}^{e,T} \frac{\lambda A}{12} \begin{bmatrix} 2 & 1 & 1 \\ & 2 & 1 \\ \text{symm.} & & 2 \end{bmatrix} \mathbf{U}^e$$

$$= \delta \mathbf{U}^{e,T} \lambda \, \mathbf{M}^e \, \mathbf{U}^e \tag{6.45}$$

The last term on the right hand side is

$$\iint c \, \delta u \, dA = \delta \mathbf{U}^{e,T} \iint c \begin{Bmatrix} \xi_1 \\ \xi_2 \\ \xi_3 \end{Bmatrix} dA \tag{6.46}$$

For a constant c one obtains

$$\delta \mathbf{U}^{e,T} c \frac{A}{3} \begin{Bmatrix} 1 \\ 1 \\ 1 \end{Bmatrix} = \delta \mathbf{U}^{e,T} \mathbf{P}_c \tag{6.47}$$

Finally the boundary flux vector can be computed. Assuming a constant $\bar{q}$ acting only on the boundary 2–3, one obtains

$$\int_{3-2} \bar{q} \, \delta u \, ds = \int_{3-2} \bar{q} (\delta u_1 \, \xi_1 + \delta u_2 \, \xi_2 + \delta u_3 \, \xi_3) ds \tag{6.48}$$

so that

$$\delta \mathbf{U}^{e,T} \, \overline{q} \, \frac{l}{2} \, \begin{Bmatrix} 0 \\ 1 \\ 1 \end{Bmatrix} = \delta \mathbf{U}^{e,T} \, \mathbf{P}_q \tag{6.49}$$

In summary, the element characteristic matrix equation is

$$\mathbf{K}^e \, \mathbf{U}^e - \lambda \, \mathbf{M}^e \, \mathbf{U}^e = \mathbf{P}^e \tag{6.50}$$

where

$$\mathbf{P}^e = \mathbf{P}_q - \mathbf{P}_c \tag{6.51}$$

6.2.2 Assembling the governing system of equations

Since the element matrices can be written as

$$(\mathbf{K}^e - \lambda \, \mathbf{M}^e) \, \mathbf{U}^e = \mathbf{P}^e \tag{6.52}$$

the total system of equations will adopt the form

$$(K - \lambda M)U = P \tag{6.53}$$

or, in extended form:

$$\left\{ \begin{bmatrix} k_{11} & k_{12} & \cdots & k_{1n} \\ k_{21} & k_{22} & \cdots & k_{2n} \\ \cdot & \cdot & & \cdot \\ \cdot & \cdot & & \cdot \\ \cdot & \cdot & & \cdot \\ k_{n1} & k_{n2} & \cdots & k_{nn} \end{bmatrix} - \lambda \begin{bmatrix} m_{11} & m_{12} & \cdots & m_{1n} \\ m_{21} & m_{22} & \cdots & m_{2n} \\ \cdot & \cdot & & \cdot \\ \cdot & \cdot & & \cdot \\ \cdot & \cdot & & \cdot \\ m_{n1} & m_{n2} & \cdot\cdot & m_{nn} \end{bmatrix} \right\} \begin{Bmatrix} u_1 \\ u_2 \\ \cdot \\ \cdot \\ \cdot \\ u_n \end{Bmatrix} = \begin{Bmatrix} P_1 \\ P_2 \\ \cdot \\ \cdot \\ \cdot \\ P_n \end{Bmatrix} \tag{6.54}$$

where n is the number of nodal points of the finite element mesh.

To illustrate the assembling of the system of Equation (6.54) let us consider the simple mesh indicated by Figure 6.8, having 4 triangular elements,

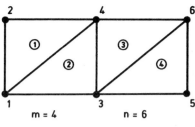

Figure 6.8

and 6 nodal points. The element matrix equations will be of the form:

Element 1 (nodal points 1, 4, 2)

$$\left\{\begin{bmatrix} k_{11}^1 & k_{14}^1 & k_{12}^1 \\ k_{41}^1 & k_{44}^1 & k_{42}^1 \\ k_{21}^1 & k_{24}^1 & k_{22}^1 \end{bmatrix} - \lambda \begin{bmatrix} m_{11}^1 & m_{14}^1 & m_{12}^1 \\ m_{41}^1 & m_{44}^1 & m_{42}^1 \\ m_{21}^1 & m_{24}^1 & m_{22}^1 \end{bmatrix}\right\} \begin{Bmatrix} u_1 \\ u_4 \\ u_2 \end{Bmatrix} = \begin{Bmatrix} p_1^1 \\ p_4^1 \\ p_2^1 \end{Bmatrix} \quad (6.55)$$

Element 2 (nodal points 3, 4, 1)

$$\left\{\begin{bmatrix} k_{33}^2 & k_{34}^2 & k_{31}^2 \\ k_{43}^2 & k_{44}^2 & k_{41}^2 \\ k_{13}^2 & k_{14}^2 & k_{11}^2 \end{bmatrix} - \lambda \begin{bmatrix} m_{33}^2 & m_{34}^2 & m_{31}^2 \\ m_{43}^2 & m_{44}^2 & m_{41}^2 \\ m_{13}^2 & m_{14}^2 & m_{11}^2 \end{bmatrix}\right\} \begin{Bmatrix} u_3 \\ u_4 \\ u_1 \end{Bmatrix} = \begin{Bmatrix} p_3^2 \\ p_4^2 \\ p_1^2 \end{Bmatrix} \quad (6.56)$$

Element 3 (nodal points 3, 6, 4)

$$\left\{\begin{bmatrix} k_{33}^3 & k_{36}^3 & k_{34}^3 \\ k_{63}^3 & k_{66}^3 & k_{64}^3 \\ k_{43}^3 & k_{46}^3 & k_{44}^3 \end{bmatrix} - \lambda \begin{bmatrix} m_{33}^3 & m_{36}^3 & m_{34}^3 \\ m_{63}^3 & m_{66}^3 & m_{64}^3 \\ m_{43}^3 & m_{46}^3 & m_{44}^3 \end{bmatrix}\right\} \begin{Bmatrix} u_3 \\ u_6 \\ u_4 \end{Bmatrix} = \begin{Bmatrix} p_3^3 \\ p_6^3 \\ p_4^3 \end{Bmatrix} \quad (6.57)$$

Element 4 (nodal points 5, 6, 3)

$$\left\{\begin{bmatrix} k_{55}^4 & k_{56}^4 & k_{53}^4 \\ k_{65}^4 & k_{66}^4 & k_{63}^4 \\ k_{35}^4 & k_{36}^4 & k_{33}^4 \end{bmatrix} - \lambda \begin{bmatrix} m_{55}^4 & m_{56}^4 & m_{53}^4 \\ m_{65}^4 & m_{66}^4 & m_{63}^4 \\ m_{35}^4 & m_{36}^4 & m_{33}^4 \end{bmatrix}\right\} \begin{Bmatrix} u_5 \\ u_6 \\ u_3 \end{Bmatrix} = \begin{Bmatrix} p_5^4 \\ p_6^4 \\ p_3^4 \end{Bmatrix} \quad (6.58)$$

Finally, adding up all the element contributions, by nodal points, we obtain

$$
\left\{
\begin{bmatrix}
k_{11}^1 + k_{11}^2 & k_{12}^1 & k_{13}^2 & & k_{14}^1 + k_{14}^2 & & \mathbf{0} \\
k_{21}^1 & k_{22}^2 & \cdot & & k_{24}^1 & & \mathbf{0} \\
k_{31}^2 & \cdot & k_{33}^2 + k_{33}^3 + k_{33}^4 & k_{34}^2 + k_{34}^3 & & k_{35}^4 & k_{36}^3 + k_{36}^4 \\
k_{41}^1 + k_{41}^2 & k_{42}^1 & k_{43}^2 + k_{43}^3 & k_{44}^1 + k_{44}^2 + k_{44}^3 & \cdot & k_{46}^3 \\
 & & k_{53}^4 & & \cdot & k_{55}^4 & k_{56}^4 \\
 \mathbf{0} & & k_{63}^3 + k_{63}^4 & k_{64}^3 & & k_{65}^4 & k_{66}^3 + k_{66}^4
\end{bmatrix}
\right.
$$

$$
\left. - \lambda
\begin{bmatrix}
 & \mathbf{0} & \\
 & \mathbf{M} & \\
 \mathbf{0} & &
\end{bmatrix}
\right\}
\begin{Bmatrix}
u_1 \\ u_2 \\ u_3 \\ u_4 \\ u_5 \\ u_6
\end{Bmatrix}
=
\begin{Bmatrix}
p_1^1 + p_1^2 \\
p_2^1 \\
p_3^2 + p_3^3 + p_3^4 \\
p_4^1 + p_4^2 + p_4^3 \\
p_5^4 \\
p_6^3 + p_6^4
\end{Bmatrix}
\tag{6.59}
$$

We can now see that the system (6.59) can be systematically formed, by adding:

(1)   In $\mathbf{K}$: k_{rs}^i in row r and column s
(2)   In $\mathbf{M}$: m_{rs}^i in row r and column s
(3)   In $\mathbf{P}$: p_r^i in row r

a procedure which can be very efficiently carried out by a computer. Note that this is the same scheme explained in Chapter 3 when assembling the governing equations for trusses.

6.2.3 Introduction of the boundary conditions

Once the system of equations has been assembled, we must proceed to introduce the essential boundary conditions, of the form $u = \bar{u}$. In the continuous model, these are conditions valid for the boundary surface, or line, called S_1, and their nature is also continuous. In the discrete model, however, we will have to apply them at the nodal points lying on the boundary S_1, and thus they become discrete boundary conditions.

The boundary conditions can be introduced in the system (6.54) by eliminating in the rows corresponding to the prescribed unknowns, and putting in the right-hand side the contributions of those prescribed unknowns. For instance, let us assume that for the problem of the Section 6.2.2, the boundary conditions are

$$u_1 = \overline{u}_1$$

$$u_2 = \overline{u}_2 \tag{6.60}$$

Then the system of Equations (6.54) becomes

$$\left\{ \begin{bmatrix} k_{33} & k_{34} & k_{35} & k_{36} \\ k_{43} & k_{44} & k_{45} & k_{46} \\ k_{53} & k_{54} & k_{55} & k_{56} \\ k_{63} & k_{64} & k_{65} & k_{66} \end{bmatrix} - \lambda \begin{bmatrix} m_{33} & m_{34} & m_{35} & m_{36} \\ m_{43} & m_{44} & m_{45} & m_{46} \\ m_{53} & m_{54} & m_{55} & m_{56} \\ m_{63} & m_{64} & m_{65} & m_{66} \end{bmatrix} \right\} \begin{Bmatrix} u_3 \\ u_4 \\ u_5 \\ u_6 \end{Bmatrix}$$

$$= \begin{Bmatrix} p_3 - (k_{31} - \lambda m_{31})\overline{u}_1 - (k_{32} - \lambda m_{32})\overline{u}_2 \\ p_4 - (k_{41} - \lambda m_{41})\overline{u}_1 - (k_{42} - \lambda m_{42})\overline{u}_2 \\ p_5 - (k_{51} - \lambda m_{51})\overline{u}_1 - (k_{52} - \lambda m_{52})\overline{u}_2 \\ p_6 - (k_{61} - \lambda m_{61})\overline{u}_1 - (k_{62} - \lambda m_{62})\overline{u}_2 \end{Bmatrix} \tag{6.61}$$

In practice, the scheme described above may be inconvenient, since it requires a reorganization of rows and columns, which would be particularly inefficient for large systems, or when working with more than one problem variable. Several other schemes are used, which can be more efficient, but are in general strongly dependent upon the program organization, and data structure selected. For example we can adopt the scheme discussed in Chapter 3.

6.2.4 Solution of the system of equations

Depending upon the type of system of equations to be solved, a different mathematical procedure will be required. Thus, for $\lambda = 0$, we have a system of the type

$$K\,U = P \tag{6.62}$$

which, for linear coefficients will require the solution of a linear system of simultaneous equations. For this there are many known efficient procedures which can be applied solving for U. If, however, $c = \overline{q} = 0$ and hence $P = 0$, we have

$$(K - \lambda M)U = 0 \tag{6.63}$$

which is an eigenvalue problem. There are also several known procedures to solve for the λ's and U's which satisfy (6.63). In other cases the system can be non-linear, time dependent, etc.

It is important to notice that in general the system matrices have special characteristics. Normally they are sparse banded matrices, with a small amount of elements different from zero. In many cases they are also symmetric matrices. All these characteristics must be taken into account when implementing a finite element program, to minimize the amount of computer memory required, and to improve the efficiency of the solution.

6.2.5 Computation of secondary results

The formulation of a finite element problem is made in terms of some problem variables, which are taken as the basic unknowns. Once the system of equations is solved, we obtain the values of those basic unknowns. However, the knowledge of the values of other variables may be even more important than those of the basic unknowns. For instance, in a fluid flow problem the solution can be formulated in terms of the stream lines, but we might be interested in knowing the flow velocities, or in a solid mechanics problem we can take the displacements as basic unknowns, but we will also be interested in knowing the stresses. Thus, the last step, once the system of equations is solved, requires a return to the element level, and to compute from the basic unknowns all other results which may be of interest. The way to proceed, in this case, depends on the type of problem and on the type of results sought.

6.3 A PROGRAM FOR THE SOLUTION OF THE EXTENDED LAPLACE EQUATION

In what follows a computer program for the solution of the extended Laplace equation is presented. The case considered is that given by Equation (6.53), to be solved using the simple triangular elements of Section 6.2.

The program for static analysis of truss systems, given in Chapter 3, will be taken as a base introducing the proper modifications, as indicated below.

6.3.1 Data structure

The meaning of the integer variables is unchanged, with the exception of NLN which is not needed any more. The real variables E and G will be used to store the constants h_x and h_y, respectively. Two new real variables C and ALA are introduced, to store the constants C and λ respectively, and will be placed in COMMON.

In comparison with the program of Chapter 3 the meaning of the arrays is now as follows.

CON: Unchanged.

IB: Unchanged.

X: Unchanged.

Y: Unchanged.

PROP: It will now store the boundary flux vectors $\bar{q}$ for each side of each element. Thus PROP($3*(L - 1) + I$) contains the value of $\bar{q}$ for the Ith side of element L. The order of the element sides is defined in accordance to the element connectivity. The first side goes from the first to the second node, the second side from the second to the third node, and the third side from the third to the first node.

AL: Same as before, containing the system vector of independent coefficients, one for each node.

TK: Unchanged.

ELST: Unchanged.

V: Unchanged.

REAC: It is first used, the same as before, to store the values of prescribed boundary unknowns, although in this case there will be just one value for each boundary node. Later it will contain the x and y derivatives of the basic unknown for each node. Thus, REAC($2*J - 1$) and REAC($2*J$) will contain the x and y derivatives, respectively, of the unknown for node J.

Note that the array FORC is not needed any more, and is therefore eliminated.

In what follows there is the description of all programs which have been modified, or are new. The programs ASSEM, ELAS, BOUND, and SLBSI, are the same as before, and are therefore not included.

Program 45   Main program

The array dimensions are altered to allow for a maximum of 100 nodes, 100 elements, and 30 boundary nodes. The integer NLN, although not needed, remains in COMMON to minimize the changes to be introduced.

The basic parameters are initialized in accordance to the characteristics of this type of problem. The sequence of subroutines called to apply the analysis steps is the same as before, although the array FORC is eliminated from the parameters list of program FORCE and OUTPT.

The corresponding FORTRAN code is given below.

```
C
C
C          PROGRAM 45 - MAIN PROGRAM
C   SOLUTION OF THE EXTENDED LAPLACE EQUATION
C
```

```
      COMMON NRMX,NCMX,NDFEL,NN,NE,NLN,NBN,NDF,NNE,N,MS,IN,IO,E,G,C,
      DIMENSION X(100),Y(100),CON(200),PROP(300),IB(50),TK(100,10),
     *AL(100),REAC(200),ELST(3,3),V(10)
C
C   INITIALIZATION OF PROGRAM PARAMETERS
C
C   NRMX  = ROW DIMENSION FOR THE TOTAL MATRIX OF THE PROBLEM
C   NCMX  = COLUMN DIMENSION FOR THE TOTAL MATRIX
C           OR MAXIMUN BAND-WIDTH ALLOWED
C   NDF   = NUMBER OF DEGREES OF FREEDOM PER
C           NODE
C   NNE   NUMBER OF NODES PER ELEMENT
C   NDFEL = TOTAL NUMBER OF DEGREES OF FREEDOM
C           FOR ONE ELEMENT
C
      NRMX=100
      NCMX=10
      NDF=1
      NNE=3
      NDFEL=NDF*NNE
C
C   ASSIGN DATA SET NUMBERS TO IN, FOR INPUT,
C   AND IO, FOR OUTPUT
C
      IN=5
      IO=6
C
C APPLY THE ANALYSIS STEPS
C
C   INPUT
C
      CALL INPUT(X,Y,CON,PROP,AL,IB,REAC)
C
C   ASSEMBLING OF THE TOTAL MATRIX FOR THE PROBLEM
C
      CALL ASSEM(X,Y,CON,PROP,TK,ELST,AL)
C
C   INTRODUCTION OF BOUNDARY CONDITIONS
C
      CALL BOUND(TK,AL,REAC,IB)
C
C   SOLUTION OF THE SYSTEM OF EQUATIONS
C
      CALL SLBSI(TK,AL,V,N,MS,NRMX,NCMX)
C
C   COMPUTATION OF SECONDARY RESULTS
C
      CALL FORCE(CON,PROP,REAC,X,Y,AL)
C
C   OUTPUT
C
      CALL OUTPT(AL,REAC)
C
      CALL EXIT
      END
```

Program 46   Input program (INPUT)

Considering the case of the program of Chapter 3, the input cards are now as follows.

(1) *Basic Parameters Card.* It includes the number of nodes NN, the number of elements NE, the number of boundary nodes NBN, and the constants h_x, h_y, c, and λ, with format 3I10, 4F10.4.

(2) *Nodal Coordinates Cards.* Unchanged.

(3) *Element Connectivity and Properties Cards.* There will be a card for each element containing the element number, the numbers of the three element nodes, and the values of the normal flux vector $\bar{q}$ for each of the three element sides, with format 4I10, 3F10.4.

(4) *Boundary Data Cards.* Same as before, but considering only one basic unknown per node. There will be one card for each boundary node, containing the node number, the status indicator for the nodal variable, 0 if prescribed and 1 if unknown and, if it is the case, the value of the prescribed unknown.

Note that, since they have no meaning in this case, the nodal loads cards were eliminated.

The FORTRAN code for the subprogram INPUT is given below.

```
      SUBROUTINE INPUT(X,Y,CON,PROP,AL,IB,REAC)

         PROGRAM 46 - INPUT PROGRAM

      COMMON NRMX,NCMX,NDFEL,NN,NE,NLN,NBN,NDF,NNE,N,MS,IN,IO,E,G,C,ALA
      DIMENSION X(1),Y(1),CON(1),PROP(1),AL(1),IB(1),REAC(1),W(3),IC(3)

  W = AN AUXILIARY VECTOR TO TEMPORARELY STORE A SET OF
         PRESCRIBED UNKNOWN VALUES AND FLUX ON ELEMENT SIDES
 IC = AUXILIARY ARRAY TO STORE TEMPORARELY THE CONNECTIVITY
         OF AN ELEMENT, AND THE BOUNDARY UNKNOWNS STATUS
         INDICATORS

READ BASIC PARAMETERS

 NN  = NUMBER OF NODES
 NE  = NUMBER OF ELEMENTS
 NBN = NUMBER OF BOUNDARY NODES
 E   = CONSTANT HX
 G   = CONSTANT HY
 C   = CONSTANT C
 ALA = CONSTANT LAMBDA

      WRITE(IO,20)
 20 FORMAT(' ',130('*'))
      READ(IN,1) NN,NE,NBN,E,G,C,ALA
      WRITE(IO,21) NN,NE,NBN,E,G,C,ALA
 21 FORMAT(//' INTERNAL DATA'//' NUMBER OF NODES         :',I5/' NUMBE
     *R OF ELEMENTS       :',I5/' NUMBER OF SUPPORT NODES :',I5/' CONSTAN
     *T HX :',F19.4/' CONSTANT HY :',F19.4/' CONSTANT C  :',F19.4/' CONS
     *TANT LAMBDA :',F15.4/ ' NODAL COORDINATES'/7X,'NODE',6X,'X',9X,'Y'
     *)
  1 FORMAT(3I10,4F10.4)
```

```
C
C   READ NODAL COORDINATES IN ARRAY X AND Y
C
        READ(IN,2) (I,X(I),Y(I),J=1,NN)
        WRITE(IO,2) (I,X(I),Y(I),I=1,NN)
      2 FORMAT(I10,2F10.2)
C
C   READ ELEMENT CONNECTIVITY IN ARRAY CON
C   AND THE BOUNDARY FLUX ON EACH OF THE ELEMENT SIDES
C
        WRITE(IO,22)
     22 FORMAT(/' ELEMENT CONNECTIVITY AND PROPERTIES'/4X,'ELEMENT',16X,
       *'NODES',18X,'QN1',12X,'QN2',12X,'QN3')
        DO 3 J=1,NE
        READ(IN,4) I,IC(1),IC(2),IC(3),W(1),W(2),W(3)
        WRITE(IO,34) I,IC(1),IC(2),IC(3),W(1),W(2),W(3)
        N1=NNE*(I-1)
        PROP(N1+1)=W(1)
        PROP(N1+2)=W(2)
        PROP(N1+3)=W(3)
        CON(N1+1)=IC(1)
        CON(N1+2)=IC(2)
      3 CON(N1+3)=IC(3)
      4 FORMAT(4I10,3F10.4)
     34 FORMAT(4I10,3F15.5)
C
C   COMPUTE N, ACTUAL NUMBER OF UNKNOWNS, AND CLEAR THE LOAD
C   VECTOR
C
        N=NN*NDF
        DO 5 I=1,N
      5 AL(I)=0.
C
C   READ BOUNDARY NODES DATA. STORE UNKNOWNS STATUS
C   INDICATORS IN ARRAY IB, AND PRESCRIBED UNKNOWNS
C   VALUES IN ARRAY REAC
C
        WRITE(IO,24)
     24 FORMAT(/' BOUNDARY CONDITION DATA'/23X,'STATUS',14X,'PRESCRIBED
       *LUE '/15X,'(0:PRESCRIBED, 1:FREE)'/7X,'NODE',14X,'U',28X,'U')
        DO 7 I=1,NBN
        READ(IN,8) J,(IC(K),K=1,NDF),(W(K),K=1,NDF)
        WRITE(IO,9) J,(IC(K),K=1,NDF),(W(K),K=1,NDF)
        L1=(NDF+1)*(I-1)+1
        L2=NDF*(J-1)
        IB(L1)=J
        DO 7 K=1,NDF
        N1=L1+K
        N2=L2+K
        IB(N1)=IC(K)
      7 REAC(N2)=W(K)
      8 FORMAT(2I10,2F10.4)
      9 FORMAT(I10,I16,22X,F10.4)
        RETURN
        END
```

Program 47 Computation of element matrices (STIFF)

The subprogram STIFF will now compute the element matrices K^e, M^e, P_c, an P_q, according to Equations (6.44), (6.45), (6.47), and (6.49), respectively.

The corresponding FORTRAN code is the following:

```
      SUBROUTINE STIFF(NEL,X,Y,PROP,CON,ELST,AL)
C
C            PROGRAM 47
C   COMPUTATION OF THE ELEMENT CHARACTERISTIC MATRIX AND VECTOR
C
      COMMON NRMX,NCMX,NDFEL,NN,NE,NLN,NBN,NDF,NNE,N,MS,IN,IO,E,G,C,ALA
      DIMENSION X(1),Y(1),CON(1),PROP(1),ELST(NDFEL,NDFEL),B(3),D(3),
     *AL(1)
C
C   NEL   = NUMBER OF CURRENT ELEMENT
C   N1,N2,N3 = NUMBERS OF FIRST, SECOND, AND THIRD ELEMENT NODE
C   D1,D2,D3 = LENGTH OF FIRST, SECOND, AND THIRD ELEMENT SIDES
C
      L=NNE*(NEL-1)+1
      N1=CON(L)
      N2=CON(L+1)
      N3=CON(L+2)
      D1=SQRT((X(N2)-X(N1))**2+(Y(N2)-Y(N1))**2)
      D2=SQRT((X(N3)-X(N2))**2+(Y(N3)-Y(N2))**2)
      D3=SQRT((X(N1)-X(N3))**2+(Y(N1)-Y(N3))**2)
C
C   DEFINE TRIANGULAR COORDINATES (SEE EQUATION 6.23)
C   ARRAYS B AND D CONTAIN SECOND AND THIRD COLUMN OF MATRIX C
C   A     = AREA OF ELEMENT
C
      B(1)=Y(N2)-Y(N3)
      B(2)=Y(N3)-Y(N1)
      B(3)=Y(N1)-Y(N2)
      D(1)=X(N3)-X(N2)
      D(2)=X(N1)-X(N3)
      D(3)=X(N2)-X(N1)
      A=(B(1)*D(2)-B(2)*D(1))/2.
      DO 1 I=1,3
      B(I)=B(I)/(2.*A)
    1 D(I)=D(I)/(2.*A)
      CCC=ALA*A/12.
C
C   COMPUTE ELEMENT MATRIX
C
      DO 10 I=1,3
      DO 10 J=1,3
   10 ELST(I,J)=A*(E*B(I)*B(J)+G*D(I)*D(J))-CCC
      DO 11 I=1,3
   11 ELST(I,I)=ELST(I,I)-CCC
C
C   COMPUTE ELEMENT VECTOR
C
      K=NNE*(NEL-1)
      CC=-C*A/3.
      D1=D1*PROP(K+1)/2
      D2=D2*PROP(K+2)/2
      D3=D3*PROP(K+3)/2
      AL(N1)=AL(N1)+CC+D1+D3
      AL(N2)=AL(N2)+CC+D1+D2
      AL(N3)=AL(N3)+D2+D3+CC
      RETURN
      END
```

The integers N1, N2, and N3, are the numbers of the three nodes of elemen' NEL. The real variables D1, D2, and D3, contain the length of the three element sides, while A contains the element area. The arrays **B** and **D** contain the second and third column of the matrix **C**, computed according to Equation (6.24).

Based on the variables described above, the subprogram STIFF compute the matrix expression $\mathbf{K}^e - \lambda \mathbf{M}^e$ storing the corresponding coefficients in array ELST. Then it computes the vectors $\mathbf{P}_c$ and $\mathbf{P}_q$, adding them into array AL.

Program 48   Routine for calculating the nodal variables derivatives (FORCE)

The subprogram FORCE now computes the x and y derivatives of the nodal variables, according to the following FORTRAN code.

```
      SUBROUTINE FORCE(CON,PROP,REAC,X,Y,AL)
C
C               PROGRAM 48
C  COMPUTATION OF SECUNDARY RESULTS
C
      COMMON NRMX,NCMX,NDFEL,NN,NE,NLN,NBN,NDF,NNE,N,MS,IN,IO,E,G,C,AL
      DIMENSION CON(1),PROP(1),FORC(1),REAC(1),X(1),Y(1),AL(1),B(3),D(
      E=0.
C
C  NEL  = NUMBER OF CURRENT ELEMENT
C  N1,N2,N3 = NUMBERS OF FIRST, SECOND, AND THIRD ELEMENT NODE
C
      DO 100 NEL=1,NE
      L=NNE*(NEL-1)+1
      N1=CON(L)
      N2=CON(L+1)
      N3=CON(L+2)
C
C  DEFINE TRIANGULAR COORDINATES (SEE EQUATION 6.23)
C  ARRAYS B AND D CONTAIN SECOND AND THIRD COLUMN OF MATRIX C
C  A    = AREA OF ELEMENT TIMES 2
C
      B(1)=Y(N2)-Y(N3)
      B(2)=Y(N3)-Y(N1)
      B(3)=Y(N1)-Y(N2)
      D(1)=X(N3)-X(N2)
      D(2)=X(N1)-X(N3)
      D(3)=X(N2)-X(N1)
      A=(B(1)*D(2)-B(2)*D(1))
      DO 1 I=1,3
      B(I)=B(I)/A
    1 D(I)=D(I)/A
C
C  COMPUTE DERIVATIVES OF PROBLEM VARIABLE AT NODAL POINTS
C
      L=2*(NEL-1)
      REAC(L+1)=AL(N1)*B(1)+AL(N2)*B(2)+AL(N3)*B(3)
      REAC(L+2)=AL(N1)*D(1)+AL(N2)*D(2)+AL(N3)*D(3)
C
C  COMPUTE ELEMENT CONTRIBUTION TO INTEGRAL OF PROBLEM VARIABLE
C
  100 E=E+A*(AL(N1)+AL(N2)+AL(N3))/6.
      RETURN
      END
```

The variables N1, N2, and N3, and the arrays **B** and **D** have the same meaning as in the previous program. The variable A now contains twice the element area. The derivatives of the nodal unknowns are computed according to the expressions

$$\frac{\partial u}{\partial x} = \sum_i u_i\, c_{2i}$$

$$\frac{\partial u}{\partial y} = \sum_i u_i\, c_{3i}$$

where the u_i are the values of the nodal variables for the three element nodes.
Since in some cases the value of the integral

$$\int_A u\,\mathrm{d}A$$

might be useful, this program also computes such a value. Notice that over an element we have

$$u = \xi_1 u_1 + \xi_2 u_2 + \xi_3 u_3$$

Then we find that over an element the value of the integral is

$$\int_A u\,\mathrm{d}A = \int_A (\xi_1 u_1 + \xi_2 u_2 + \xi_3 u_3)\mathrm{d}A = \frac{A}{3}(u_1 + u_2 + u_3)$$

Finally adding up the contributions from all elements we obtain the value of the integral over the complete domain of integration. The subprogram FORCE computes such value, storing it in the variable E.

Program 49   Output program (OUTPT)

The subprogram OUTPT prints the values of the nodal variables, of its derivatives, and of the integral

$$\int_A u\,\mathrm{d}A$$

according to the following FORTRAN code:

```
        SUBROUTINE OUTPT(AL,REAC)
C
C            PROGRAM 49 - OUTPUT PROGRAM
C
        COMMON NRMX,NCMX,NDFEL,NN,NE,NLN,NBN,NDF,NNE,N,MS,IN,IO,E,G,C,AL
        DIMENSION AL(1),REAC(1)
C
C   WRITE VALUES OF PROBLEM VARIABLE AT NODAL POINTS
C
        WRITE(IO,1)
      1 FORMAT(///1X,130('*')//' RESULTS'//' NODAL VARIABLES      '/7X,'NOD
       *,11X,'U')
        WRITE(IO,2) (I,AL(I),I=1,NN)
      2 FORMAT(I10,F15.4)
C
C   WRITE DERIVATIVES OF PROBLEM VARIABLE
C
        WRITE(IO,3)
      3 FORMAT(///'DERIVATIVES OF THE PROBLEM VARIABLE OVER EACH ELEMENT'
       *4X,'ELEMENT',9X,'X',14X,'Y')
        DO 4 I=1,NE
        K=2*(I-1)
      4 WRITE(IO,5) I,REAC(K+1),REAC(K+2)
      5 FORMAT(I10,2F15.5)
C
C   WRITE THE VALUE OF THE INTEGRAL OF THE PROBLEM VARIABLE
C
        WRITE(IO,7) E
      7 FORMAT(//' VALUE OF THE INTEGRAL :',F15.4)
        WRITE(IO,6)
      6 FORMAT(//1X,130('*'))
        RETURN
        END
```

Example 6.1
The Saint Venant torsion of prismatic beam is governed by a differential equation of the type

$$\frac{\partial}{\partial x}\left(\frac{1}{G}\frac{\partial \phi}{\partial x}\right) + \frac{\partial}{\partial y}\left(\frac{1}{G}\frac{\partial \phi}{\partial y}\right) = -2\theta \tag{a}$$

where G is the shear modulus and θ is the rate of twist. This expression is a special case of Equation (6.15), where

$$h_x = h_y = 1/G; \quad \lambda = 0; \quad c = -2\theta \tag{b}$$

The problem variable ϕ is a stress function, such that

$$\tau_{xz} = \frac{\partial \phi}{\partial y}; \quad \tau_{yz} = -\frac{\partial \phi}{\partial x} \tag{c}$$

ϕ must have a constant value on the boundary. For convenience it is normally assumed that such constant is equal to zero. Considering a homogeneous material, Equation (a) can be written as

$$\frac{\partial^2 \phi^*}{\partial x^2} + \frac{\partial^2 \phi^*}{\partial y^2} = -2 \qquad\qquad\qquad (d)$$

where

$$\phi^* = \phi/G\theta \qquad\qquad\qquad (e)$$

This problem can be solved for ϕ^*, but noticing that the rate of twist θ is so far unknown. We know that the torque M_t is

$$M_t = JG\theta = 2 \int_A \phi \, dx \, dy \qquad\qquad\qquad (f)$$

where J is the torsional rigidity. We can, therefore, compute J by evaluating the following integral

$$J = 2 \int_A \phi^* \, dx \, dy \qquad\qquad\qquad (g)$$

The rate of twist is then given by

$$\theta = \frac{M_t}{GJ} \qquad\qquad\qquad (h)$$

Finally one could compute the shear stresses as

$$\tau_{xz} = G\theta \, \frac{\partial \phi^*}{\partial y} ; \quad \tau_{yz} = -G\theta \, \frac{\partial \phi^*}{\partial x} \qquad\qquad\qquad (i)$$

Example 6.2
To illustrate the application of Program 49, we will study the problem of the torsion of a prismatic bar of elliptical cross-section, defined by the equation

$$\frac{x^2}{a^2} + \frac{y^2}{b^2} = 1$$

For this example we shall take $a = 2$ and $b = 1$. The finite element mesh selected, consisting of 33 nodes and 48 elements, is shown in Figure 6.9, which

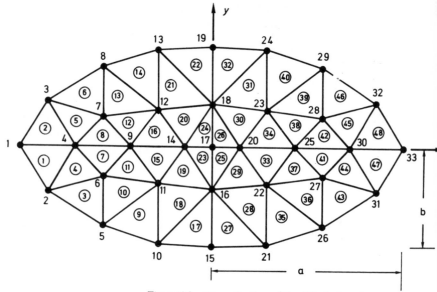

Figure 6.9 Discretization of the elliptical section

indicates also the nodal and element numbering. The values given to the program constants will be $\lambda = 0$, $h_x = h_y = 1$, and $c = -2$. The problem variable will be prescribed equal to 0 on all boundary nodes.

The input for the program, and the output produced, are given below. From the output of the program the rate of twist θ can be computed which, in turn, can be used to compute the shear stresses.

***** INPUT FOR LAPLACE EQUATION PROGRAM

```
33          48          16    1.        1.        -2.
 1    -2.
 2    -1.8          -0.4358
 3    -1.8           0.4358
 4    -1.5
 5    -1.2          -0.8
 6    -1.2          -0.35
 7    -1.2           0.35
 8    -1.2           0.8
 9    -0.9
10    -0.6          -0.954
11    -0.6          -0.45
12    -0.6           0.45
13    -0.6           0.954
14    -0.3
15                  -1.
16                  -0.45
17
18                   0.45
```

19		1.
20	0.3	
21	0.6	-0.954
22	0.6	-0.45
23	0.6	0.45
24	0.6	0.954
25	0.9	
26	1.2	-0.8
27	1.2	-0.35
28	1.2	0.35
29	1.2	0.8
30	1.5	
31	1.8	-0.4358
32	1.8	0.4358
33	2.	

1	4	1	2
2	1	4	3
3	2	5	6
4	2	6	4
5	7	3	4
6	8	3	7
7	9	4	6
8	4	9	7
9	5	10	11
10	5	11	6
11	6	11	9
12	12	7	9
13	7	12	8
14	12	13	8
15	14	9	11
16	9	14	12
17	10	15	16
18	10	16	11
19	11	16	14
20	18	12	14
21	18	13	12
22	13	18	19
23	14	16	17
24	18	14	17
25	16	20	17
26	20	18	17
27	21	16	15
28	16	21	22
29	16	22	20
30	23	18	20
31	24	18	23
32	18	24	19
33	25	20	22
34	20	25	23
35	26	22	21
36	22	26	27
37	22	27	25
38	28	23	25
39	29	23	28
40	29	24	23
41	30	25	27
42	25	30	28
43	26	31	27
44	27	31	30
45	32	28	30
46	32	29	28
47	33	30	31

```
      48           30           33           32
       1
       2
       3
       5
       8
      10
      13
      15
      19
      21
      24
      26
      29
      31
      32
      33
```

* *✷* * * * * * * *✷* *

INTERNAL DATA

```
NUMBER OF NODES          :      33
NUMBER OF ELEMENTS       :      48
NUMBER OF SUPPORT NODES  :      16
CONSTANT HX :                 1.0000
CONSTANT HY :                 1.0000
CONSTANT Q  :                -2.0000
CONSTANT LAMBDA :             0.0000
NODAL COORDINATES
     NODE        X            Y
       1       -2.00        0.00
       2       -1.80       -0.44
       3       -1.80        0.44
       4       -1.50        0.00
       5       -1.20       -0.80
       6       -1.20       -0.35
       7       -1.20        0.35
       8       -1.20        0.80
       9       -0.90        0.00
      10       -0.60       -0.95
      11       -0.60       -0.45
      12       -0.60        0.45
      13       -0.60        0.95
      14       -0.30        0.00
      15        0.00       -1.00
      16        0.00       -0.45
      17        0.00        0.00
      18        0.00        0.45
      19        0.00        1.00
      20        0.30        0.00
      21        0.60       -0.95
      22        0.60       -0.45
      23        0.60        0.45
      24        0.60        0.95
      25        0.90        0.00
      26        1.20       -0.80
      27        1.20       -0.35
      28        1.20        0.35
      29        1.20        0.80
      30        1.50        0.00
      31        1.80       -0.44
      32        1.80        0.44
      33        2.00        0.00
```

ELEMENT CONNECTIVITY AND PROPERTIES

ELEMENT	NODES			QN1	QN2	QN3
1	4	1	2	0.00000	0.00000	0.00000
2	1	4	3	0.00000	0.00000	0.00000
3	2	5	6	0.00000	0.00000	0.00000
4	2	6	4	0.00000	0.00000	0.00000
5	7	3	4	0.00000	0.00000	0.00000
6	8	3	7	0.00000	0.00000	0.00000
7	9	4	6	0.00000	0.00000	0.00000
8	4	9	7	0.00000	0.00000	0.00000
9	5	10	11	0.00000	0.00000	0.00000
10	5	11	6	0.00000	0.00000	0.00000
11	6	11	9	0.00000	0.00000	0.00000
12	12	7	9	0.00000	0.00000	0.00000
13	7	12	8	0.00000	0.00000	0.00000
14	12	13	8	0.00000	0.00000	0.00000
15	14	9	11	0.00000	0.00000	0.00000
16	9	14	12	0.00000	0.00000	0.00000
17	10	15	16	0.00000	0.00000	0.00000
18	10	16	11	0.00000	0.00000	0.00000
19	11	16	14	0.00000	0.00000	0.00000
20	18	12	14	0.00000	0.00000	0.00000
21	18	13	12	0.00000	0.00000	0.00000
22	13	18	19	0.00000	0.00000	0.00000
23	14	16	17	0.00000	0.00000	0.00000
24	18	14	17	0.00000	0.00000	0.00000
25	16	20	17	0.00000	0.00000	0.00000
26	20	18	17	0.00000	0.00000	0.00000
27	21	16	15	0.00000	0.00000	0.00000
28	16	21	22	0.00000	0.00000	0.00000
29	16	22	20	0.00000	0.00000	0.00000
30	23	18	20	0.00000	0.00000	0.00000
31	24	18	23	0.00000	0.00000	0.00000
32	18	24	19	0.00000	0.00000	0.00000
33	25	20	22	0.00000	0.00000	0.00000
34	20	25	23	0.00000	0.00000	0.00000
35	26	22	21	0.00000	0.00000	0.00000
36	22	26	27	0.00000	0.00000	0.00000
37	22	27	25	0.00000	0.00000	0.00000
38	28	23	25	0.00000	0.00000	0.00000
39	29	23	28	0.00000	0.00000	0.00000
40	29	24	23	0.00000	0.00000	0.00000
41	30	25	27	0.00000	0.00000	0.00000
42	25	30	28	0.00000	0.00000	0.00000
43	26	31	27	0.00000	0.00000	0.00000
44	27	31	30	0.00000	0.00000	0.00000
45	32	28	30	0.00000	0.00000	0.00000
46	32	29	28	0.00000	0.00000	0.00000
47	33	30	31	0.00000	0.00000	0.00000
48	30	33	32	0.00000	0.00000	0.00000

BOUNDARY CONDITION DATA

NODE	STATUS (0:PRESCRIBED, 1:FREE)	PRESCRIBED VALUE U
1	0	0.0000
2	0	0.0000
3	0	0.0000
5	0	0.0000
8	0	0.0000
10	0	0.0000
13	0	0.0000
15	0	0.0000
19	0	0.0000
21	0	0.0000
24	0	0.0000
26	0	0.0000

29	0	0.0000
31	0	0.0000
32	0	0.0000
33	0	0.0000

****

RESULTS

NODAL VARIABLES

NODE	U
1	0.0000
2	0.0000
3	0.0000
4	0.3414
5	0.0000
6	0.3925
7	0.3925
8	0.0000
9	0.6268
10	0.0000
11	0.5615
12	0.5615
13	0.0000
14	0.7895
15	0.0000
16	0.6650
17	0.7927
18	0.6650
19	0.0000
20	0.7895
21	0.0000
22	0.5615
23	0.5615
24	0.0000
25	0.6268
26	0.0000
27	0.3925
28	0.3925
29	0.0000
30	0.3414
31	0.0000
32	0.0000
33	0.0000

DERIVATIVES OF THE PROBLEM VARIABLE OVER EACH ELEMENT

ELEMENT	X	Y
1	0.68288	0.31339
2	0.68288	-0.31339
3	0.52948	0.87229
4	0.60138	0.36950
5	0.60138	-0.36950
6	0.52948	-0.87229
7	0.47566	0.26175
8	0.47566	-0.26175
9	0.28592	1.11399
10	0.42692	0.87229
11	0.34397	0.37462
12	0.34397	-0.37462
13	0.42692	-0.87229
14	0.28592	-1.11399
15	0.27108	0.32603
16	0.27108	-0.32603
17	0.09269	1.20904

18	0.17253	1.11399
19	0.17253	0.39173
20	0.17253	-0.39173
21	0.17253	-1.11399
22	0.09269	-1.20904
23	0.01075	0.28388
24	0.01075	-0.28388
25	-0.01075	0.28388
26	-0.01075	-0.28388
27	-0.09269	1.20904
28	-0.17253	1.11399
29	-0.17253	0.39173
30	-0.17253	-0.39173
31	-0.17253	-1.11399
32	-0.09269	-1.20904
33	-0.27108	0.32603
34	-0.27108	-0.32603
35	-0.28592	1.11399
36	-0.42692	0.87229
37	-0.34397	0.37462
38	-0.34397	-0.37462
39	-0.42692	-0.87229
40	-0.28592	-1.11399
41	-0.47566	0.26175
42	-0.47566	-0.26175
43	-0.52948	0.87229
44	-0.60138	0.36950
45	-0.60138	-0.36950
46	-0.52948	-0.87229
47	-0.68288	0.31339
48	-0.68288	-0.31339

VALUE OF THE INTEGRAL :   2.2786

The analytical solution for this problem gives $J = 5.026$ for the torsional rigidity.

Taking $G = \theta = 1$, one can compute

$$M_t = G\theta J = 5.026$$

Table 6.2 includes a comparison between the analytical and computed values for the problem variable, showing a very reasonable agreement.

Having the value of the integral of the problem variable over the area of the ellipse, one can compute the approximate value of the torsional rigidity J, obtaining

$$J = 2 \int_A \phi^* \, dA = 4.56$$

which is about 10% off the analytical value. It is clear that a better agreement is obtained for the problem variable than for the torsional rigidity, which is computed approximately from the approximate values of the problem variables.

Table 6.2

Node	x	y	Exact	Computed
2	−1.8	−0.4358		
4	−1.5	0	0.35	0.3414
6	−1.2	−0.35	0.414	0.3925
9	−0.9	0	0.638	0.6268
11	−0.6	−0.45	0.566	0.5615
14	−0.3	0	0.782	0.7895
16	0	−0.45	0.638	0.6650
17	0	0	0.800	0.7927

From the derivatives printed one can compute the values of the shear stresses, according to Equation (i) of Example 6.2. Notice that, according to formulation used, the shear stresses are constant over each element. It can be assumed that these values correspond to the baricenter of each triangle. Alternatively, the nodal shear stress average can be computed, from the shear stresses for the elements connected to each node, giving good approximations for the shear stresses at the nodal points.

6.4   TWO-DIMENSIONAL ELASTICITY PROBLEMS

There are two types of two-dimensional elasticity problems, called plane stress and plane strain, which can be solved using almost identical equations. The plane stress problems are encountered in the case of thin plates under in-plane forces, as suggested by Figure 6.10.

The basic assumptions are
(1)   The faces normal to the z axis are free of forces.
(2)   The surface forces acting on the cylindrical faces are functions of x and y only, and their component in the z direction is null (i.e. $p_z = 0$).
(3)   The body forces are functions of x and y only and their component in the z direction is null (i.e. $b_z = 0$).

In such a case it is assumed that the stress components σ_x, σ_y and τ are functions of x and y only, and that the remaining stress components are null (i.e. $\sigma_{xz} = \sigma_{yz} = \sigma_z = 0$). Then ϵ_x, ϵ_y and γ will also be functions of x and y only, and can be computed from the in-plane stress components, using the constitutive equations. Since those strain components depend only on u and v, one concludes that these are the relevant displacement components for plane stress problems.

Plane strain problems are encountered in the case of long prismatic bodies, as the one shown in Figure 6.11. Considering a slice of unit thickness, the basic assumptions are again the conditions (1), (2) and (3) discussed for plane

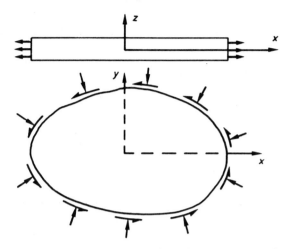

Figure 6.10   Thin plate 1

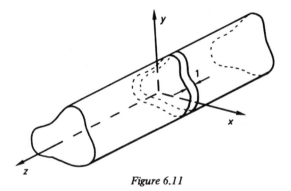

Figure 6.11

stress problems. Each slice, possibly excluding the ends, has similar geometry, loadings, and boundary conditions.

In plane strain problems, it is assumed that the displacement components u and v are functions of x and y only, and that the component w, in the z direction, is null. Then the only non zero strain components, ϵ_x, ϵ_y and γ will also be functions of x and y only, the same applying for σ_x, σ_y and τ.

Note that while for plane stress problems σ_z is null, for plane strain problems it is not, but can be computed from the in-plane stress components. Conversely, while w and ϵ_z are zero for plane strain problems, they are not zero for plane stress problems. Only in the case in which the Poisson coefficient ν is null, are plane stress and plane strain problems exactly identical.

According to the discussion above, the relevant vectors for plane stress/strain problems will be:

$$\mathbf{u} = \begin{Bmatrix} u \\ v \end{Bmatrix}; \qquad \mathbf{p} = \begin{Bmatrix} \bar{p}_x \\ \bar{p}_y \end{Bmatrix}; \qquad \mathbf{b} = \begin{Bmatrix} b_x \\ b_y \end{Bmatrix}$$

$$\boldsymbol{\epsilon} = \begin{Bmatrix} \epsilon_x \\ \epsilon_y \\ \gamma \end{Bmatrix}; \qquad \boldsymbol{\sigma} = \begin{Bmatrix} \sigma_x \\ \sigma_y \\ \tau \end{Bmatrix} \tag{6.64}$$

The total potential energy for a two dimensional elastic body, if no body forces exist, can be written as

$$\Pi_p = \frac{t}{2} \iint (\sigma_x \epsilon_x + \sigma_y \epsilon_y + \tau \gamma) \mathrm{d}x \, \mathrm{d}y - \int_{S_\sigma} (\bar{p}_x u + \bar{p}_y v) \mathrm{d}s \tag{6.65}$$

where t is the thickness. The strains are

$$\epsilon_x = \frac{\partial u}{\partial x}; \quad \epsilon_y = \frac{\partial v}{\partial y}; \quad \gamma = \frac{\partial u}{\partial y} + \frac{\partial v}{\partial x} \tag{6.66}$$

The strain—stress relationships are given by

$$\boldsymbol{\sigma} = \mathbf{D} \, \boldsymbol{\epsilon}$$

where for plane stress

$$\mathbf{D} = \frac{E}{1 - v^2} \begin{bmatrix} 1 & v & 0 \\ v & 1 & 0 \\ 0 & 0 & (1-v)/2 \end{bmatrix} \tag{6.67}$$

and for plane strain

$$\mathbf{D} = \frac{E}{(1+v)(1-2v)} \begin{bmatrix} (1-v) & v & 0 \\ v & (1-v) & 0 \\ 0 & 0 & (1-2v)/2 \end{bmatrix} \tag{6.68}$$

In what follows we will work with the plane stress relationships only. How ever, note that making the transformation

$$E' = \frac{E}{(1-v^2)}, \quad v' = \frac{v}{1-v} \tag{6.69}$$

from Equation (6.67) we obtain (6.68). This means that one can implement a program for plane stress problems, and by transforming the elastic constant data according to (6.69), can also use it to solve plane strain problems.

Introducing (6.67) into (6.65), and expressing the result in matrix notation

$$\Pi_p = \frac{t}{2} \iint \epsilon^T D \epsilon \, dxdy - \int_{S_\sigma} U^T \, Pds \tag{6.70}$$

Let us first work only with the term in the strain energy V. Thus

$$V = \frac{t}{2} \iint \epsilon^T D \epsilon \, dxdy \tag{6.71}$$

For the triangular element shown in Figure 6.12, we can assume simple linear expansions for u and v, so that

$$
\left.
\begin{aligned}
u &= u_1 \xi_1 + u_2 \xi_2 + u_3 \xi_3 \\
v &= v_1 \xi_1 + v_2 \xi_2 + v_3 \xi_3
\end{aligned}
\right\} \quad \text{or,}
$$

$$
\mathbf{u} = \left\{ \begin{matrix} u \\ v \end{matrix} \right\} = \begin{bmatrix} \xi_1 & 0 & \xi_2 & 0 & \xi_3 & 0 \\ 0 & \xi_1 & 0 & \xi_2 & 0 & \xi_3 \end{bmatrix} \left\{ \begin{matrix} u_1 \\ v_1 \\ u_2 \\ v_2 \\ u_3 \\ v_3 \end{matrix} \right\} \tag{6.72}
$$

$$= \Phi \, u^e$$

Then one can obtain the strains using the chain rule, as in Equation (6.39).

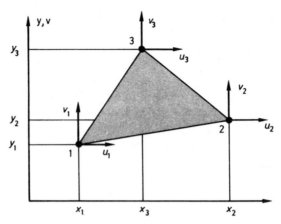

Figure 6.12   Triangular element

Rearranging, we have

$$
\epsilon = \left\{ \begin{array}{c} \dfrac{\partial u}{\partial x} \\[2mm] \dfrac{\partial v}{\partial y} \\[2mm] \dfrac{\partial u}{\partial y} + \dfrac{\partial v}{\partial x} \end{array} \right\} = \left[\begin{array}{cccccc} c_{21} & 0 & c_{22} & 0 & c_{23} & 0 \\[2mm] 0 & c_{31} & 0 & c_{32} & 0 & c_{33} \\[2mm] c_{31} & c_{21} & c_{32} & c_{22} & c_{33} & c_{23} \end{array} \right] \left\{ \begin{array}{c} u_1 \\ v_1 \\ u_2 \\ v_2 \\ u_3 \\ v_3 \end{array} \right\} \tag{6.73}
$$

or

$$
\epsilon = \mathbf{B}\,\mathbf{U}^e \tag{6.74}
$$

The next step is to substitute Equation (6.74) into (6.70), minimize it, and integrate. As all coefficients are constant the integration only involves multiplying the integrand by the area of the triangle. Thus, the stiffness matrix $\mathbf{K}^e$ can be written as

$$
t\,\delta\mathbf{U}^{e,T}\!\!\left(\iint \mathbf{B}^T\,\mathbf{D}\,\mathbf{B}\,\mathrm{d}A\right)\mathbf{U}^e = t\,\delta\mathbf{U}^{e,T}(A\,\mathbf{B}^T\,\mathbf{D}\,\mathbf{B})\mathbf{U}^e = \delta\mathbf{U}^{e,T}\,\mathbf{K}^e\,\mathbf{U}^e \tag{6.75}
$$

where

$$
\mathbf{K}^e = tA\,\mathbf{B}^T\,\mathbf{D}\,\mathbf{B} \tag{6.76}
$$

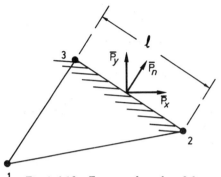

Figure 6.13   Forces on boundary 2-3

and

$$U^e = \{u_1 \quad v_1 \quad u_2 \quad v_2 \quad u_3 \quad v_3\} \tag{6.77}$$

The elements of the stiffness matrix K^e are given in Table 6.3.

To determine the load terms, consider a uniformly distributed load $\bar{P}_n$ acting on side 2–3, as shown in Figure 6.13. The right hand side of expression (6.70), after variation, can be written

$$\int_s \delta u^T \mathbf{p} \, ds = \int_s \delta U^{e,T} \Phi^T \mathbf{p} \, ds = \delta U^{e,T} \int_s \Phi^T \mathbf{p} \, ds = \delta U^{e,T} \mathbf{P}^e \tag{6.78}$$

where the equivalent nodal load vector is

$$\mathbf{P}^e = \int_s \Phi^T \mathbf{p} \, ds \tag{6.79}$$

Considering the load components $\bar{p}_x$ and $\bar{p}_y$ are constant, and that ξ_1 is zero over the side 2–3, we obtain

$$\mathbf{P}^e = \int_{l_{2-3}} \begin{bmatrix} 0 & 0 \\ 0 & 0 \\ \xi_2 & 0 \\ 0 & \xi_2 \\ \xi_3 & 0 \\ 0 & \xi_3 \end{bmatrix} \begin{Bmatrix} \bar{p}_x \\ \bar{p}_y \end{Bmatrix} ds \tag{6.80}$$

Table 6.3 ELEMENTS OF STIFFNESS MATRIX

	u_1	v_1	u_2	v_2	u_3	v_3
u_1	$c_{21}^2 + \left(\frac{1-v}{2}\right)c_{31}^2$	$vc_{21}c_{31} + c_{31}c_{21}\left(\frac{1-v}{2}\right)$	$c_{22}c_{21} + c_{31}c_{32}\left(\frac{1-v}{2}\right)$	$vc_{21}c_{32} + c_{22}c_{31}\left(\frac{1-v}{2}\right)$	$c_{23}c_{21} + c_{31}c_{33}\left(\frac{1-v}{2}\right)$	$vc_{21}c_{33} + c_{31}c_{23}\left(\frac{1-v}{2}\right)$
v_1		$c_{31}^2 + c_{21}^2\left(\frac{1-v}{2}\right)$	$c_{31}c_{22}v + c_{21}c_{32}\left(\frac{1-v}{2}\right)$	$c_{31}c_{32} + c_{21}c_{22}\left(\frac{1-v}{2}\right)$	$c_{31}c_{23}v + c_{21}c_{33}\left(\frac{1-v}{2}\right)$	$c_{31}c_{33} + c_{21}c_{23}\left(\frac{1-v}{2}\right)$
u_2			$c_{22}^2 + c_{32}^2\left(\frac{1-v}{2}\right)$	$vc_{22}c_{32} + c_{32}c_{22}\left(\frac{1-v}{2}\right)$	$c_{22}c_{23} + c_{32}c_{33}\left(\frac{1-v}{2}\right)$	$vc_{22}c_{33} + c_{32}c_{33}\left(\frac{1-v}{2}\right)$
v_2		Symm.		$c_{32}^2 + c_{22}^2\left(\frac{1-v}{2}\right)$	$c_{32}c_{23}v + c_{22}c_{33}\left(\frac{1-v}{2}\right)$	$c_{32}c_{33} + c_{22}c_{23}\left(\frac{1-v}{2}\right)$
u_3					$c_{23}^2 + c_{33}^2\left(\frac{1-v}{2}\right)$	$vc_{23}c_{33} + c_{33}c_{23}\left(\frac{1-v}{2}\right)$
v_3						$c_{33}^2 + \left(\frac{1-v}{2}\right)c_{23}^2$

$K^e = D_t \cdot$

$D = EtA/(1-v^2)$

which can be integrated using the formula

$$\int_0^l \xi_2^i \, \xi_3^j \, ds = \frac{i! \, j!}{(i+j+1)!} \, l \tag{6.81}$$

After integration we obtain

$$P^e = \left\{ 0 \quad 0 \quad \frac{\bar{p}_x l}{2} \quad \frac{\bar{p}_y l}{2} \quad \frac{\bar{p}_x l}{2} \quad \frac{\bar{p}_y l}{2} \right\} \tag{6.82}$$

The final relationships can be written,

$$K^e \, U^e = P^e \tag{6.83}$$

From the element matrices of type (6.83), the governing system of equations for the problem

$$K \, U = P \tag{6.84}$$

is assembled in a similar fashion, as in the case of member systems. The boundary conditions are introduced, and the system of equations is solved. Then, knowing the nodal displacements, and using

$$\epsilon = B \, U^e \tag{6.85}$$

the strains are computed. Finally, with

$$\sigma = D \, \epsilon \tag{6.86}$$

the stresses are also computed, and the solution is completed.

It is important to note that for this linear displacement triangular element, the strains and stresses will be constant over all the element. Obviously, to study regions with rapidly varying stresses a finite element mesh with a large number of these elements will be required, or higher order elements should be used (Section 6.6).

6.5   COMPUTER PROGRAM FOR FINITE ELEMENT ANALYSIS OF PLANE STRESS PROBLEMS

A program for finite element analysis of plane stress problems using simple triangular elements will be implemented. Again this program is based on the program for static analysis of truss systems included in Chapter 3. The modifications required to transform the truss program into this finite element program are given below.

Data structure

The meaning of the integer and real variables is the same as in the program of Chapter 3, except that the variable G is now used to store the Poisson's coefficient.

With regard to the arrays, their meaning is now as follows

CON: Same as before, except that it now allows for three element nodes instead of two.

IB: Unchanged.

X: Unchanged.

Y: Unchanged.

PROP: Same as before, except that PROP(L) now contains the element thickness for element L.

AL: Unchanged.

TK: Unchanged.

ELST: Unchanged.

V: Unchanged.

FORC: It will now contain the stresses at the baricenter of each element. $FORC(3*(L-1)+I)$, with $I = 1, 2, 3$ will contain the stresses σ_x, σ_y, and τ, respectively, for element L.

REAC: It will be first used to store, the same as before, the values of prescribed displacements for boundary nodes. Later it will contain the average stresses at the nodal points. Thus, $REAC(3*(J-1)+I)$, for $I = 1, 2, 3$ will contain the average stresses σ_x, σ_y, and τ, respectively, for node J.

The following new array is introduced

NODES: The position NODES(J) will contain the number of elements incident on node J. This rray is only used in the subprogram FORCE.

In what follows all new programs, and those which require modifications are described. The programs ASSEM, ELASS, BOUND, and SLBSI, are the same as before, and therefore will not be included.

Program 50   Main program

The array dimensions are adjusted to allow for the treatment of problems with a maximum of 100 nodes, 100 elements, and 30 boundary nodes. The maximum bandwidth allowed for the stiffness matrix is equal to 40.

The main program initializes the basic parameters for plane stress problems, and then performs the same sequence of subroutine calls as before, to apply the analysis steps. The FORTRAN code is the following:

```
C
C
C
C
C               PROGRAM 50 - MAIN PROGRAM

      COMMON NRMX,NCMX,NDFEL,NN,NE,NLN,NBN,NDF,NNE,N,MS,IN,IO,E,G
      DIMENSION X(100),Y(100),CON(300),PROP(100),IB(60),TK(200,40),
     *AL(200),FORC(300),REAC(300),ELST(6,6),V(40),NODES(100)
C
C INITIALIZATION OF PROGRAM PARAMETERS
C
C NRMX  = ROW DIMENSION FOR THE STIFFNESS
C           MATRIX
C NCMX  = COLUMN DIMENSION FOR THE STIFFNESS
C           MATRIX, OR MAXIMUM HALF BAND WIDTH
C           ALLOWED
C NDF   = NUMBER OF DEGREES OF FREEDOM PER
C           NODE
C NNE     NUMBER OF NODES PER ELEMENT
C NDFEL = TOTAL NUMBER OF DEGREES OF FREEDOM
C           FOR ONE ELEMENT
C
      NRMX=200
      NCMX=40
      NDF=2
      NNE=3
      NDFEL=NDF*NNE
C
C ASSIGN DATA SET NUMBERS TO IN, FOR INPUT,
C AND IO, FOR OUTPUT
C
      IN=5
      IO=6
C
C APPLY THE ANALYSIS STEPS
C
C INPUT
      CALL INPUT(X,Y,CON,PROP,AL,IB,REAC)
C
C ASSEMBLING OF TOTAL STIFFNESS MATRIX
C
      CALL ASSEM(X,Y,CON,PROP,TK,ELST,AL)
C
C INTRODUCTION OF BOUNDARY CONDITIONS
C
      CALL BOUND(TK,AL,REAC,IB)
C
C SOLUTION OF THE SYSTEM OF EQUATIONS
C
      CALL SLBSI(TK,AL,V,N,MS,NRMX,NCMX)
C
C COMPUTATION OF ELEMENT STRESSES
C
      CALL FORCE(CON,PROP,FORC,REAC,X,Y,AL,NODES)
```

```
C
C    OUTPUT
C
      CALL OUTPT(AL,FORC,REAC)
C
      CALL EXIT
      END
```

Program 51   Input program (INPUT)

The input cards are basically the same as before, except that the basic parameters card also include the Poisson's coefficient, and that the element property
read is the element thickness. The FORTRAN code is the following:

```
      SUBROUTINE INPUT(X,Y,CON,PROP,AL,IB,REAC)
C
C
C          PROGRAM 51 - INPUT PROGRAM
C
      COMMON NRMX,NCMX,NDFEL,NN,NE,NLN,NBN,NDF,NNE,N,MS,IN,IO,E,G
      DIMENSION X(1),Y(1),CON(1),PROP(1),AL(1),IB(1),REAC(1),W(3),IC
C
C    W =   AN AUXILIARY VECTOR TO TEMPORARELY STORE A SET OF
C          NODAL LOADS AND PRESCRIBED UNKNOWN VALUES
C    IC =  AUXILIARY ARRAY TO STORE TEMPORARELY THE CONNECTIVITY
C          OF AN ELEMENT, AND THE BOUNDARY UNKNOWNS STATUS
C          INDICATORS
C
C    READ BASIC PARAMETERS
C
C    NN  = NUMBER OF NODES
C    NE  = NUMBER OF ELEMENTS
C    NLN = NUMBER OF LOADED NODES
C    NBN = NUMBER OF BOUNDARY NODES
C    E   = MODULUS OF ELASTICITY
C
      WRITE(IO,20)
   20 FORMAT(' ',130('*'))
      READ (IN,1) NN,NE,NLN,NBN,E,G
      WRITE(IO,21) NN,NE,NLN,NBN,E,G
   21 FORMAT(//' INTERNAL DATA'//' NUMBER OF NODES         :',I5/' NU
     *R OF ELEMENTS         :',I5/' NUMBER OF LOADED NODES   :',I5/' NUMB
     *OF SUPPORT NODES :',I5/' MODULUS OF ELASTICITY :',F15.0/' POISS
     *COEFFICIENT      :',F15.4//' NODAL COORDINATES'/7X,'NODE',6X,'X',9
     *'Y')
    1 FORMAT(4I10,2F10.2)
C
C    READ NODAL COORDINATES IN ARRAY X AND Y
C
      READ(IN,2) (I,X(I),Y(I),J=1,NN)
      WRITE(IO,2) (I,X(I),Y(I),I=1,NN)
    2 FORMAT(I10,2F10.2)
C
```

```
C   READ ELEMENT CONNECTIVITY IN ARRAY CON
C   AND ELEMENT THICKNESS IN ARRAY PROP
C
      WRITE(IO,22)
   22 FORMAT(/' ELEMENT CONNECTIVITY AND PROPERTIES'/4X,'ELEMENT',14X,
     *'NODES',16X,'THICKNESS')
      DO 3 J=1,NE
      READ(IN,4) I,IC(1),IC(2),IC(3),PROP(I)
      WRITE(IO,34) I,IC(1),IC(2),IC(3),PROP(I)
      N1=NNE*(I-1)
      CON(N1+1)=IC(1)
      CON(N1+2)=IC(2)
    3 CON(N1+3)=IC(3)
    4 FORMAT(4I10,F10.3)
   34 FORMAT(4I10,F15.5)
C
C   COMPUTE N, ACTUAL NUMBER OF UNKNOWNS, AND CLEAR THE LOAD
C   VECTOR
C
      N=NN*NDF
      DO 5 I=1,N
    5 AL(I)=0.
C
C   READ THE NODAL LOADS AND STORE THEM IN ARRAY AL
C
      WRITE(IO,23)
   23 FORMAT(/' NODAL LOADS'/7X,'NODE',5X,'PX',8X,'PY')
      DO 6 I=1,NLN
      READ (IN,2) J,(W(K),K=1,NDF)
      WRITE(IO,2) J,(W(K),K=1,NDF)
      DO 6 K=1,NDF
      L=NDF*(J-1)+K
    6 AL(L)=W(K)
C
C   READ BOUNDARY NODES DATA. STORE UNKNOWNS STATUS
C   INDICATORS IN ARRAY IB, AND PRESCRIBED UNKNOWNS
C   VALUES IN ARRAY REAC
C
      WRITE(IO,24)
   24 FORMAT(/' BOUNDARY CONDITION DATA'/23X,'STATUS',14X,'PRESCRIBED VA
     *LUES'/15X,'(0:PRESCRIBED, 1:FREE)'/7X,'NODE',8X,'U',9X,'V',16X,'U'
     *,9X,'V')
      DO 7 I=1,NBN
      READ(IN,8) J,(IC(K),K=1,NDF),(W(K),K=1,NDF)
      WRITE(IO,9) J,(IC(K),K=1,NDF),(W(K),K=1,NDF)
      L1=(NDF+1)*(I-1)+1
      L2=NDF*(J-1)
      IB(L1)=J
      DO 7 K=1,NDF
      N1=L1+K
      N2=L2+K
      IB(N1)=IC(K)
    7 REAC(N2)=W(K)
    8 FORMAT(3I10,2F10.4)
    9 FORMAT(3I10,10X,2F10.4)
C
      RETURN
      END
```

Program 52   Element stiffness matrix (STIFF)

The subprogram STIFF computes the coefficients of the element stiffness
matrix, as given in Table 6.3. Its FORTRAN code is the following:

```
SUBROUTINE STIFF (NEL,X,Y,PROP,CON,ELST,AL)
COMMON NRMX,NCMX,NDFEL,NN,NE,NLN,NBN,NDF,NNE,N,MS,IN,IO,E,G
DIMENSION X(1),Y(1),CON(1),PROP(1),ELST(NDFEL,NDFEL),AL(1),
*ROT (6,6),V(6),
*B(3),D(3)
  L = NNE * (NEL − 1)
  N1 = CON (L+1)
  N2 = CON (L+2)
  N3 = CON (L+3)
  B(1) = Y(N2) − Y(N3)
  B(2) = Y(N3) − Y(N1)
  B(3) = Y(N1) − Y(N2)
  D(1) = X(N3) − X(N2)
  D(2) = X(N1) − X(N3)
  D(3) = X(N2) − X(N1)
  A = (B(1) * D(2) − B(2) * D(1))/2
  DO 1 I = 1,3
  B(I) = B(I)/(2.*A)
1 D(I) = D(I)/(2.*A)
  ALA = E*PROP(NEL)*A/(1 − G**2.)
  AX = (1.−G)/2
  DO 2 I = 1,6
  DO 2 J = 1,6
2 ELST (I,J) = 0
  DO 3 I = 1,3
  IX = (I − 1) * 2 + 1
  IZ = (I − 1) * 2 + 2
  DO 3 J = 1,3
  JX = (J − 1) * 2 + 1
  JZ = (J − 1) * 2 + 2
  ELST (IX,JX) = (B(I) * B(J) + AX * D(I) * D(J)) * ALA
  ELST (IX,JZ) =(G*B(I) * D(J) + AX * D(I) * B(J)) * ALA
  ELST (IZ,JZ) = (D(I) * D(J) + AX * B(I) * B(J)) * ALA
3 ELST (IZ,JX) = (G*D(I) * B(J) + AX * B(I) * D(J)) * ALA
  RETURN
  END
```

The integers N1, N2, and N3, and the arrays **B** and **D** have the same meaning as in Program 47. The variable A contains twice the element area, ANUP contains the value $(1 - v)/2$, and C contains the constant D_t of Table 6.3. The coefficients of the element stiffness matrix are computed and stored in array ELST.

Program 53 Evaluation of element stresses (FORCE)

The subprogram FORCE evaluates the element stresses according to the expression

$$\sigma = D \, B \, U^e$$

and the average nodal stresses. Its FORTRAN code is the following:

```
      SUBROUTINE FORCE(CON,PROP,FORC,REAC,X,Y,AL,NODES)
C
C            PROGRAM 53
C  COMPUTATION OF ELEMENT STRESSES
C
      COMMON NRMX,NCMX,NDFEL,NN,NE,NLN,NBN,NDF,NNE,N,MS,IN,IO,E,ANU
      DIMENSION CON(1),PROP(1),FORC(1),REAC(1),X(1),Y(1),AL(1),B(3),
     *D(3),NODES(1)
C
C  NEL  = NUMBER OF CURRENT ELEMENT
C  N1,N2,N3 = NUMBERS OF FIRST, SECOND, AND THIRD ELEMENT NODE
C  D1,D2,D3 = LENGTH OF FIRST, SECOND, AND THIRD ELEMENT SIDES
C
      DO 100 NEL=1,NE
      L=NNE*(NEL-1)+1
      N1=CON(L)
      N2=CON(L+1)
      N3=CON(L+2)
C
C  DEFINE TRIANGULAR COORDINATES (SEE EQUATION 6.23)
C  ARRAYS B AND D CONTAIN SECOND AND THIRD COLUMN OF MATRIX C
C  A    = AREA OF ELEMENT TIMES 2
C
      B(1)=Y(N2)-Y(N3)
      B(2)=Y(N3)-Y(N1)
      B(3)=Y(N1)-Y(N2)
      D(1)=X(N3)-X(N2)
      D(2)=X(N1)-X(N3)
      D(3)=X(N2)-X(N1)
      A=(B(1)*D(2)-B(2)*D(1))
      DO 1 I=1,3
      B(I)=B(I)/A
    1 D(I)=D(I)/A
C
C  COMPUTE STRESSES FOR ELEMENT NEL
C
      ANUP=(1-ANU)/2
      K1=NDF*(N1-1)
      K2=NDF*(N2-1)
```

```
      K3=NDF*(N3-1)
      C=E/(1-ANU*ANU)
      L=3*(NEL-1)
      FORC(L+1)=C+( B(1)*AL(K1+1)+B(2)*AL(K2+1)+B(3)*AL(K3+1)+
     *ANU*(D(1)*AL(K1+2)+D(2)*AL(K2+2)+D(3)*AL(K3+2)))
      FORC(L+2)=C*(ANU*(B(1)*AL(K1+1)+B(2)*AL(K2+1)+B(3)*AL(K3+1))
     *+D(1)*AL(K1+2)+D(2)*AL(K2+2)+D(3)*AL(K3+2))
      FORC(L+3)=C*(1-ANU)*(D(1)*AL(K1+1)+D(2)*AL(K2+1)+D(3)*
     *AL(K3+1)+B(1)*AL(K1+2)+B(2)*AL(K2+2)+B(3)*AL(K3+2))/2.
C
C     COMPUTE THE STRESSES NODAL AVERAGES, IN ARRAY REAC
C
      K1=3*(N1-1)
      K2=3*(N2-1)
      K3=3*(N3-1)
      DO 50 I=1,3
      L1=K1+I
      L2=K2+I
      L3=K3+I
      L4=L+I
      REAC(L1)=REAC(L1)+FORC(L4)
      REAC(L2)=REAC(L2)+FORC(L4)
   50 REAC(L3)=REAC(L3)+FORC(L4)
C
C     ARRAY NODES CONTAINS THE NUMBER OF ELEMENTS CONNECTED TO EACH NOD
C
      NODES(N1)=NODES(N1)+1
      NODES(N2)=NODES(N2)+1
  100 NODES(N3)=NODES(N3)+1
      DO 150 I=1,NN
      K1=3*I-2
      K2=K1+2
      DO 150 J=K1,K2
  150 REAC (J)=REAC(J)/NODES(I)
      RETURN
      END
```

The variables N1, N2, N3, A, and ANUP, and the arrays **B** and **D**, are the same as in Program 52. The integers K1, K2, and K3 are used to retrieve the displacements for the three element nodes. Next the element stresses are computed, according to the formulae

$$\sigma_x = \frac{E}{1 - v^2} \, [u_1 \, c_{21} + u_2 \, c_{22} + u_3 \, c_{23} + v \, v_1 \, c_{31}$$

$$+ v \, v_2 \, c_{32} + v \, v_3 \, c_{33}]$$

$$\sigma_y = \frac{E}{1 - v^2} \, [v \, u_1 \, c_{21} + v \, u_2 \, c_{22} + v \, u_3 \, c_{23} + v_1 \, c_{31}$$

$$+ v_2 \, c_{32} + v_3 \, c_{33}]$$

$$\tau = \frac{E}{2(1+v)} \left[u_1 \, c_{31} + u_2 \, c_{32} + u_3 \, c_{33} + v_1 \, c_{21} \right.$$

$$\left. + v_2 \, c_{22} + v_3 \, c_{23} \right]$$

and are stored in array FORC.

The element stresses are also added to the array REAC, by components, for each of the three element nodes. Every time this is done the values in array NODES corresponding to each of the element nodes are increased by one unit. When the loop on the elements is completed the array NODES contain the number of elements incident on each node, while REAC contains the sum of the stresses for the elements incident on each node. From them the average nodal stresses are computed and stored in array REAC.

As we have mentioned before the element stresses are constant over the element. Their values are normally assigned to the element baricenter. The nodal average stresses generally give a better estimate for the real stresses, and are also easier to work with because they correspond to the nodal points, whose coordinates are known exactly before hand.

Program 54   Output program (OUTPT)

The output program OUTPT, which prints the nodal displacement, element stresses, and nodal average stresses, is the following:

```
      SUBROUTINE OUTPT(AL,FORC,REAC)
C
C            PROGRAM 54 - OUTPUT PROGRAM
C
      COMMON NRMX,NCMX,NDFEL,NN,NE,NLN,NBN,NDF,NNE,N,MS,IN,IO,E,G
      DIMENSION AL(1),REAC(1),FORC(1)
C
C  WRITE NODAL DISPLACEMENTS
C
      WRITE(IO,1)
    1 FORMAT(//1X,130('*')//' RESULTS'//' NODAL DISPLACEMENTS'/7X,'NODE'
     *,11X,'U',14X,'V')
      DO 10 I=1,NN
      K1=NDF*(I-1)+1
      K2=K1+NDF-1
   10 WRITE(IO,2) I,(AL(J),J=K1,K2)
    2 FORMAT(I10,6F15.4)
C
C  WRITE THE ELEMENT STRESSES
C
      WRITE(IO,3)
    3 FORMAT(///' ELEMENT STRESSES'/5X,'ELEMENT',8X,'S11',12X,'S22',12X,
     *'S12')
      DO 20 I=1,NE
```

```
       K1=3*(I-1)+1
       K2=K1+2
   20 WRITE(IO,4) I,(FORC (J),J=K1,K2)
    4 FORMAT(I10,3F15.4)
C
C   WRITE THE STRESSES NODAL AVERAGES
C
       WRITE(IO,6)
    6 FORMAT(//' NODAL AVERAGE STRESSES'/7X,'NODE',9X,'S11',12X,'S22
      *12X,'S12')
       DO 30 I=1,NN
       K1=3*(I-1)+1
       K2=K1+2
   30 WRITE(IO,4) I,(REAC(J),J=K1,K2)
       WRITE(IO,5)
    5 FORMAT(//1X,130('*'))
       RETURN
       END
```

Example 6.3

We consider the case of the cantilever beam of Figure 6.14(a). Its dimensions are 3 by 9, and its thickness is 1.

Young's modulus and the Poisson's coefficient are taken as $E = 1$ and $\nu = 0$. A concentrated load of unit intensity is applied to the free end. This load will be considered applied by halves at each of the corners at the free end.

Figure 6.14(b) gives the finite element mesh selected, including the node and element numbering. Since this problem is anti-symmetric only one half of it needs to be considered. For that, however, the proper boundary conditions, as shown, should be imposed on the nodes lying on the symmetry line.

The input and the output for this problem are given below.

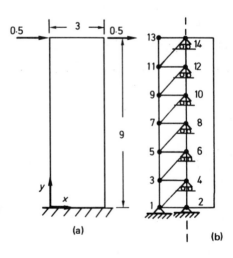

Figure 6.14

```
***** INPUT FOR PLANE STRESS ANALYSIS PROGRAM
     14          12           1          8      1.
      1
      2   1.5
      3                1.5
      4   1.5         1.5
      5                3.
      6   1.5         3.
      7                4.5
      8   1.5         4.5
      9    .          6.
     10   1.5         6.
     11                7.5
     12   1.5         7.5
     13                9.
     14   1.5         9.
      1           1           2          4      1.
      2           1           4          3      1.
      3           3           4          6      1.
      4           3           6          5      1.
      5           5           6          8      1.
      6           5           8          7      1.
      7           7           8         10      9      1.
      8           7          10          9      1.
      9           9          10         12      1.
     10           9          12         11      1.
     11          11          12         14      1.
     12          11          14         13      1.
     13   0.5
      1
      2
      4           1
      6           1
      8           1
     10           1
     12           1
     14           1
```

The output produced by the program is the following:

```
****************************************************************

INTERNAL DATA

NUMBER OF NODES

NUMBER OF NODES        :   14
NUMBER OF ELEMENTS     :   12
NUMBER OF LOADED NODES :    1
NUMBER OF SUPPORT NODES:    8
MODULUS OF ELASTICITY :              1.
POISSON COEFFICIENT    :          0.0000

NODAL COORDINATES
     NODE        X              Y
      1        0.00           0.00
      2        1.50           0.00
```

3	0.00	1.50
4	1.50	1.50
5	0.00	3.00
6	1.50	3.00
7	0.00	4.50
8	1.50	4.50
9	0.00	6.00
10	1.50	6.00
11	0.00	7.50
12	1.50	7.50
13	0.00	9.00
14	1.50	9.00

ELEMENT CONNECTIVITY AND PROPERTIES

ELEMENT		NODES		THICKNESS
1	1	2	4	1.00000
2	1	4	3	1.00000
3	3	4	6	1.00000
4	3	6	5	1.00000
5	5	6	8	1.00000
6	5	8	7	1.00000
7	7	8	10	1.00000
8	7	10	9	1.00000
9	9	10	12	1.00000
10	9	12	11	1.00000
11	11	12	14	1.00000
12	11	14	13	1.00000

NODAL LOADS

NODE	PX	PY
13	0.50	0.00

BOUNDARY CONDITION DATA

	STATUS (0: PRESCRIBED, 1: FREE)		PRESCRIBED VALUES	
NODE	U	V	U	V
1	0	0	0.0000	0.0000
2	0	0	0.0000	0.0000
4	1	0	0.0000	0.0000
6	1	0	0.0000	0.0000
8	1	0	0.0000	0.0000
10	1	0	0.0000	0.0000
12	1	0	0.0000	0.0000
14	1	0	0.0000	0.0000

****

RESULTS

NODAL DISPLACEMENTS

NODE	U	V
1	0.0000	0.0000
2	0.0000	0.0000
3	3.2550	4.4183
4	3.1634	0.0000
5	10.4779	8.0199
6	10.3786	0.0000
7	20.8983	10.8200
8	20.7981	0.0000
9	33.7200	12.8206

10	33.6170	0.0000
11	48.1601	14.0271
12	48.0246	0.0000
13	63.6195	14.5045
14	63.0969	0.0000

ELEMENT STRESSES

ELEMENT	S11	S22	S12
1	0.0000	0.0000	1.0545
2	−0.0611	2.9455	−0.3878
3	−0.0611	0.0000	0.9323
4	−0.0662	2.4010	−0.2656
5	−0.0662	0.0000	0.7999
6	−0.0668	1.8668	−0.1332
7	−0.0668	0.0000	0.6663
8	−0.0686	1.3337	0.0004
9	−0.0686	0.0000	0.5290
10	−0.0903	0.8043	0.1377
11	−0.0903	0.0000	0.3484
12	−0.3484	0.3183	0.3183

NODAL AVERAGE STRESSES

NODE	S11	S22	S12
1	−0.0305	1.4728	0.3333
2	0.0000	0.0000	1.0545
3	−0.0628	1.7822	0.0930
4	−0.0407	0.9818	0.5330
5	−0.0664	1.4226	0.1337
6	−0.0645	0.8003	0.4889
7	−0.0674	1.0668	0.1778
8	−0.0666	0.6223	0.4443
9	−0.0759	0.7127	0.2223
10	−0.0680	0.4446	0.3986
11	−0.1763	0.3742	0.2681
12	−0.0831	0.2681	0.3384
13	−0.3484	0.3183	0.3183
14	−0.2193	0.1591	0.3333

•••

When the horizontal displacement at the tip of the beam is computed according to theory of elasticity, the result is $u = 117$. The computed results show that the horizontal displacements obtained for nodes 13 and 14 are not very good approximations. This is due to the fact that the simple triangular element is not sufficiently adequate, when using coarse meshes to model problems with a high gradient for the displacements. However, as the finite element mesh is refined, using more nodes and elements, the approximate results will converge toward the analytical results. We conclude, then, that a more refined mesh is required, in order to obtain reasonable results for this problem.

6.6 HIGHER ORDER ELEMENTS

In Section 6.4 the formulation for a simple triangular element, based on linear expansions for the displacements u and v, was presented. As it was

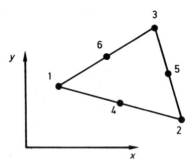

Figure 6.15   Six nodes triangle

explained, the fact that strains and stresses are constant makes that element relatively inefficient. To implement better elements one can approximate the displacements, working with polynomials of higher order.

One higher order triangular element is the six nodes triangle shown in Figure 6.15. In this case second order polynomials are used, expressing the displacements by

$$u = \phi_1 u_1 + \phi_2 u_2 + \phi_3 u_3 + \phi_4 u_4 + \phi_5 u_5 + \phi_6 u_6$$

$$v = \phi_1 v_1 + \phi_2 v_2 + \phi_3 v_3 + \phi_4 v_4 + \phi_5 v_5 + \phi_6 v_6 \qquad (6.87)$$

where

$$\phi_1 = \xi_1(2\xi_1 - 1)$$

$$\phi_2 = \xi_2(2\xi_2 - 1)$$

$$\phi_3 = \xi_3(2\xi_3 - 1)$$

$$\phi_4 = 4\xi_1 \xi_2$$

$$\phi_5 = 4\xi_2 \xi_3$$

$$\phi_6 = 4\xi_3 \xi_1 \qquad (6.88)$$

These shape functions are shown in Figure 6.16.

For this element, since the displacements are approximated by second order polynomials, the strains and stresses will vary linearly over the element. Fewer elements will be needed to achieve a given precision, than in the case of the linear triangular element. However, the computation of the element stiffness matrices will require more computer time. In addition, due to the presence

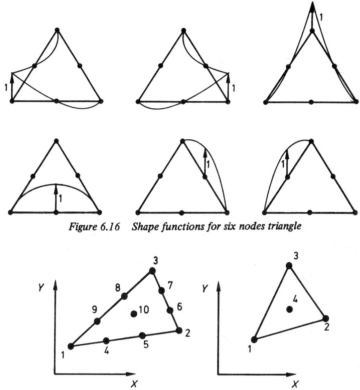

Figure 6.16 Shape functions for six nodes triangle

Figure 6.17 Cubic order triangular element

of the mid-side nodes this element will lead to systems of equations having a larger bandwidth which, again, require more time to be solved.

Obviously, one can in general obtain elements of better quality by using even higher order polynomials for the displacement expansions. However, the labour involved in the element implementation increases considerably. In some cases, the additional element degrees of freedom required to work with a polynomial expansion with more terms are obtained considering additional nodal points, as is the case of the cubic order triangle of Figure 6.17(a), having ten nodal points, including an internal one at its baricentre, totalling 20 degrees of freedom. Alternatively, a cubic order triangle can be implemented with just four nodes, as indicated in Figure 6.17(b). In this case, for nodes 1, 2 and 3 the nodal displacement vector used is

$$\mathbf{u} = \left\{ u \quad v \quad \frac{\partial u}{\partial x} \quad \frac{\partial v}{\partial x} \quad \frac{\partial u}{\partial y} \quad \frac{\partial v}{\partial y} \right\}$$

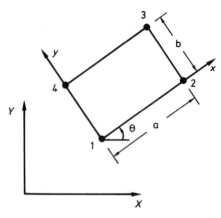

Figure 6.18

while for the internal node 4, the displacement vector remains

$$u = \{u \quad v\}$$

Triangular elements of order higher than the third could eventually be developed. For plane stress problems, however, the need to use such elements very seldom arises.

Thus far most of the elements studied were of triangular shape but finite elements of any practical geometric shape could be implemented. For instance a very popular element is the simple rectangle shown in Figure 6.18. The formulation of this element is normally developed, because of convenience, for the local coordinate axes x and y. The relation of these with the global coordinate axes X and Y is given by the angle θ. Considering the local reference system, and using an undetermined parameters expansion, the displacements are approximated by

$$u = \alpha_1 + \alpha_2 x + \alpha_3 y + \alpha_4 xy$$

$$v = \alpha_5 + \alpha_6 x + \alpha_7 y + \alpha_8 xy \qquad (6.89)$$

This gives a linear variation for the displacements on the element boundary, but a slightly non linear one inside the element, due to the presence of the terms xy. Also, due to these terms this element gives much better results than the simple triangle. The problem is, however, that it can only be used for integration domains of rectangular shape, unless the finite element mesh could be constructed mixing triangular and rectangular elements.

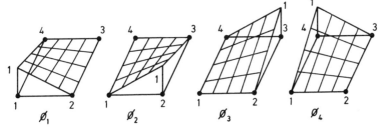

Figure 6.19   Shape functions

To implement this element it is more efficient to start with an expansion of the type

$$u = \phi_1 u_1 + \phi_2 u_2 + \phi_3 u_3 + \phi_4 u_4$$

$$v = \phi_1 v_1 + \phi_2 v_2 + \phi_3 v_3 + \phi_4 v_4 \qquad (6.90)$$

where

$$\phi_1 = \left(1 - \frac{x}{a}\right)\left(1 - \frac{y}{b}\right)$$

$$\phi_2 = \frac{x}{a}\left(1 - \frac{y}{b}\right)$$

$$\phi_3 = \frac{x}{a}\frac{y}{b}$$

$$\phi_4 = \left(1 - \frac{x}{a}\right)\frac{y}{b} \qquad (6.91)$$

These shape functions are shown schematically in Figure 6.19. Taking into account Equation (6.91) it is easy to compute

$$\epsilon = \begin{Bmatrix} \epsilon_x \\ \epsilon_y \\ \gamma \end{Bmatrix} = \frac{1}{ab} \begin{vmatrix} -(b-y) & 0 & (b-y) & 0 & y & 0 & -y & 0 \\ 0 & -(a-x) & 0 & -x & 0 & x & 0 & (a-x) \\ -(a-x) & -(b-y) & -x & (b-y) & x & y & (a-x) & -y \end{vmatrix} \begin{Bmatrix} u_1 \\ v_1 \\ u_2 \\ v_2 \\ u_3 \\ v_3 \\ u_4 \\ v_4 \end{Bmatrix}$$

$$(6.92)$$

Table 6.4   STIFFNESS MATRIX FOR SIMPLE RECTANGULAR ELEMENT

$$K = \frac{Et}{12(1-\nu^2)}$$

$$
\begin{bmatrix}
4\beta+2(1-\nu)/\beta & \tfrac{3}{2}(1+\nu) & 2\beta-2(1-\nu)/\beta & \tfrac{3}{2}(1-3\nu) & -2\beta-(1-\nu)/\beta & -3(1+\nu)/2 & -4\beta+(1-\nu)/\beta & -3(1-3\nu)/2 \\[4pt]
 & 4/\beta+2(1-\nu)\beta & -\tfrac{3}{2}(1-3\nu) & -4/\beta+(1-\nu)\beta & -3(1+\nu)/2 & -2/\beta-(1-\nu)\beta & 3(1-3\nu)/2 & 2/\beta-2(1-\nu)\beta \\[4pt]
 & & 4\beta+2(1-\nu)/\beta & -\tfrac{3}{2}(1+\nu) & -4\beta+(1-\nu)/\beta & 3(1-3\nu)/2 & -2\beta-(1-\nu)/\beta & 3(1+\nu)/2 \\[4pt]
 & & & 4/\beta+2(1-\nu)\beta & -3(1-3\nu)/2 & 2/\beta-2(1-\nu)\beta & 3(1+\nu)/2 & -2/\beta-(1-\nu)\beta \\[4pt]
 & & \text{Symm.} & & 4\beta+2(1-\nu)/\beta & 3(1+\nu)/2 & 2\beta-2(1-\nu)/\beta & 3(1-3\nu)/2 \\[4pt]
 & & & & & 4/\beta+2(1-\nu)\beta & -3(1-3\nu)/2 & -4/\beta+(1-\nu)\beta \\[4pt]
 & & & & & & 4\beta+2(1-\nu)/\beta & -3(1+\nu)/2 \\[4pt]
 & & & & & & & 4/\beta+2(1-\nu)\beta
\end{bmatrix}
$$

$$\beta = b/a$$

or

$$\epsilon = B\, U^e \tag{6.93}$$

The element stiffness matrix can then be computed using the expression

$$K^e = t \iint B^T\, D\, B\, dx\, dy \tag{6.94}$$

and is given in Table 6.4.

It should be noticed, however, that this stiffness matrix is referred to the local reference frame. Before assembling the total stiffness matrix, it must be transformed to be referred to the global reference system. This is done using the formula

$$K^{e,g} = R^T K^e\, R \tag{6.95}$$

where

$$R = \begin{bmatrix} r & 0 & 0 & 0 \\ 0 & r & 0 & 0 \\ 0 & 0 & r & 0 \\ 0 & 0 & 0 & r \end{bmatrix} \tag{6.96}$$

and

$$r = \begin{bmatrix} \cos\theta & \sin\theta \\ -\sin\theta & \cos\theta \end{bmatrix} \tag{6.97}$$

Once the problem system of equations is solved, the nodal displacements are known with reference to the global frame. For each element they are rotated with respect to the local frame using the expression

$$U^{e,L} = R\, U^{e,G} \tag{6.98}$$

and the strains and stresses are computed using the expressions

$$\epsilon^L = B\, U^{e,L} \tag{6.99}$$

$$\sigma^L = D\, \epsilon^L \tag{6.100}$$

These values are referred to the local axis. When they need to be referred

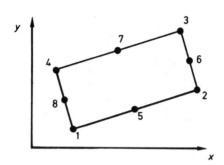

Figure 6.20   Eight nodes rectangular element

to the global axis we have to multiply $\boldsymbol{\epsilon}$ by the following rotation matrix

$$
\mathbf{R}_\epsilon = \begin{bmatrix} \cos^2\theta & \sin^2\theta & -\dfrac{\sin(2\theta)}{2} \\[2ex] \sin^2\theta & \cos^2\theta & \dfrac{\sin(2\theta)}{2} \\[2ex] \sin(2\theta) & -\sin(2\theta) & \cos(2\theta) \end{bmatrix} \tag{6.101}
$$

The stresses are multiplied by

$$
\mathbf{R}_\sigma = \begin{bmatrix} \cos^2 & \sin^2\theta & -\sin(2\theta) \\[2ex] \sin^2\theta & \cos^2\theta & \sin(2\theta) \\[2ex] \dfrac{\sin(2\theta)}{2} & -\dfrac{\sin(2\theta)}{2} & \cos(2\theta) \end{bmatrix} \tag{6.102}
$$

As in the case of triangular elements, rectangular elements of higher order can also be implemented. One common element is the 8 nodes rectangular element of Figure 6.20, based on displacement expansions of the type

$$
u = \alpha_1 + \alpha_2 x + \alpha_3 y + \alpha_4 x^2 + \alpha_5 y^2 + \alpha_6 xy + \alpha_7 x^2 y + \alpha_8 xy^2
$$

$$
v = \alpha_9 + \alpha_{10} x + \alpha_{11} y + \alpha_{12} x^2 + \alpha_{13} y^2 + \alpha_{14} xy + \alpha_{15} x^2 y + \alpha_{16} xy^2
$$

Other higher order rectangular elements can be found in the technical literature.

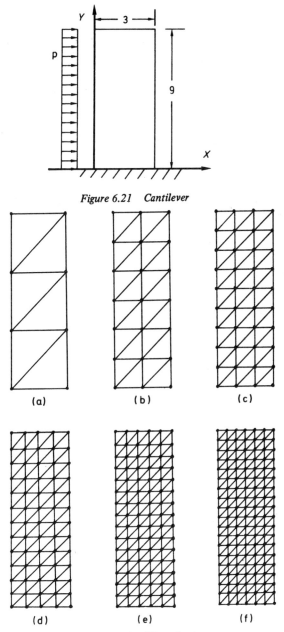

Figure 6.21   Cantilever

Figure 6.22   Discretization

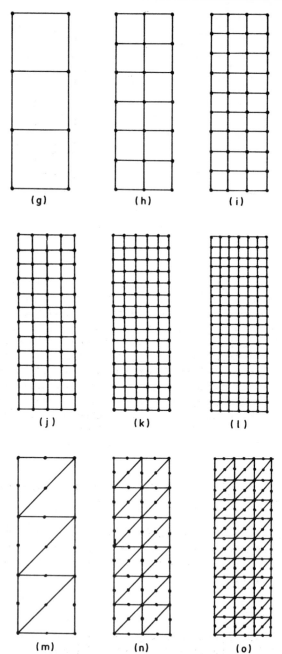

Figure 6.22 (contd.)

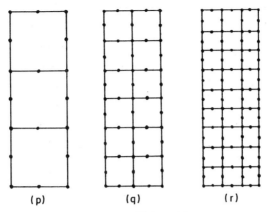

(p)   (q)   (r)

Figure 6.22 (contd.)

Example 6.4

As an illustrative example of the application of some of the plane stress elements discussed above, let us take the case of the beam shown in Figure 6.21. This cantilever beam of dimensions 3×9, and thickness $t = 1$, is loaded by a uniformly distributed load of intensity $p = 1$. It is considered that Young's modulus is $E = 1$, and that Poisson's coefficient is $\nu = 0$. This problem is solved using the finite element meshes, shown in Figure 6.22, of first and second order triangular and rectangular elements. Table 6.5 gives a summary of the characteristics of each mesh, the results obtained, and the computer time required for solution by the LORANE system[15]. The results are compared with the analytical solution presented by Timoshenko and Goodier. The finite element results are plotted in Figure 6.23, against the number of nodes, and in Figure 6.24, against the computer time required.

6.7 HIGHER ORDER FINITE ELEMENT PROGRAM FOR THE SOLUTION OF THE LAPLACE EQUATION

In what follows a computer program for the solution of Laplace's equation is presented. The code uses the second order triangular elements described in Section 6.6. The program uses the same assembler and solver subroutines as previously presented.

6.7.1 Data structure

The meaning of the variables is unchanged by comparison with the code presented in 6.3. The meaning of the arrays is as follows.

Table 6.5 CANTILEVER BEAM RESULTS

Element	Mesh	Nodes	Elements	Analysis time (s)	Total time (s)	u Displacement for x = 1.5, y = 9
Triangle 3 nodes	a	8	6	2	4	186.84
	b	21	24	4	10	223.65
	c	40	54	11	22	292.29
	d	65	96	22	43	329.73
	e	96	150	40	79	351.02
	f	133	216	67	119	363.94
Rectangle 4 nodes	g	8	3	2	4	270.76
	h	21	12	5	10	351.94
	i	40	27	12	23	375.48
	j	65	48	26	44	384.76
	k	96	75	51	83	386.28
	l	133	108	80	129	389.67
Triangle 6 nodes	m	21	6	4	9	387.14
	n	65	24	28	43	396.14
	o	133	54	92	131	396.31
Rectangle 8 nodes	p	18	3	7	11	391.28
	q	53	12	41	54	396.22
	r	106	27	116	146	396.48
Exact solution (Timoshenko and Goodier)						396.8

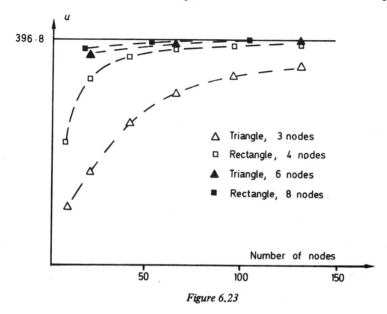

Figure 6.23

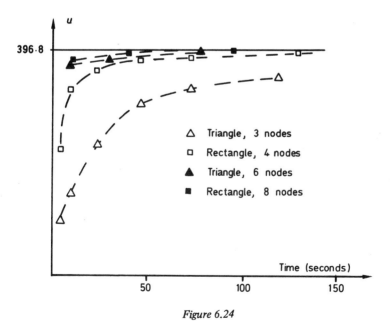

Figure 6.24

KON(300)   Connectivity vector, changed to an integer variable instead of CON as it is more appropriate to standard FORTRAN. It is also extended to 300 to take into consideration the larger number of nodes per element.

BI: Unchanged except its dimension has been extended to 60.

X: Unchanged

Y: Unchanged

PROP: Unchanged

AL: Unchanged

RENO(300) This vector replaces the previous REAC(200) and stresses the prescribed boundary unknowns, i.e. one value per boundary node.

V(30)   The same vector as previously but extended to 30 instead of 20.

Program 55   Main program

The variable ALA of Program 55 has been taken out of the common as this program solves the Poisson's equation rather than the extended Laplace's equation. The rest of the common remains unchanged except that NRMX = 100 and NCMX = 10, in this case, are defined inside the main program only. The same with the variable NDFEL = NDF*NNE (i.e. number of degrees of freedom per node multiplied by number of nodes per element).

The main program also tests for errors before proceeding to the analysis, in particular

(1) If there are too many nodes in the data
(2) If there are too many elements
(3) If the data has too many boundary nodes

```
C***********************************************************************
C
C                        PROGRAM 55
C                     MAIN PROGRAM FOR
C          SOLUTION OF POISSON EQUATION
C
C          COMMON NN,NE,NLN,NBN,NDF,NNE,N,MS,IN,IO,E,G,C
C          DIMENSION X(100),Y(100),KON(300),PROP(300),IB(60),
C         *TK(100,20),AL(100),RENO(300),ELST(6,6),V(20)
C
C INITIALIZATION OF PROGRAM PARAMETERS
C
C   MNN   = MAXIMUM NUMBER OF NODES ALLOWED
C   MNE   = MAXIMUM NUMBER OF ELEMENT ALLOWED
C   MNB   = MAXIMUM NUMBER OF BOUNDARY NODES ALLOWED
C   NRMX  = ROW DIMENSION FOR THE TOTAL MATRIX OF THE PROBLEM
C   NCMX  = COLUMN DIMENSION FOR THE TOTAL MATRIX
C           OR MAXIMUN BAND-WIDTH ALLOWED
C   NDF   = NUMBER OF DEGREES OF FREEDOM PER
C           NODE
C   NNE   NUMBER OF NODES PER ELEMENT
```

```
C   NDFEL = TOTAL NUMBER OF DEGREES OF FREEDOM
C           FOR ONE ELEMENT
C
        MNN=100
        MNE=100
        MNB=60
        NRMX=100
        NCMX=20
        NDF=1
        NNE=6
        NDFEL=NDF*NNE
C
C   ASSIGN DATA SET NUMBERS TO IN, FOR INPUT,
C   AND IO, FOR OUTPUT
C
        IN=5
        IO=6
C
C   APPLY THE ANALYSIS STEPS
C
C   INPUT
C
        CALL INPUT(X,Y,KON,PROP,AL,IB,RENO)
C
C   CHECK FOR LIMITS
C
        IF(MNN-NN)1,2,2
    1   WRITE(IO,101)
  101   FORMAT(/' **** TOO MANY NODES **** '/)
        GO TO 999
    2   IF(MNE-NE)3,4,4
    3   WRITE(IO,103)
  103   FORMAT(/' **** TOO MANY ELEMENTS ****'/)
        GO TO 999
    4   IF(MNB-NBN)5,6,6
    5   WRITE(IO,105)
  105   FORMAT(/' **** TOO MANY BOUNDARY NODES ****'/)
        GO TO 999
C
C   ASSEMBLING OF THE TOTAL MATRIX FOR THE PROBLEM
C
    6   CALL ASSEM(X,Y,KON,PROP,TK,ELST,AL,NRMX,NCMX,NDFEL)
C
C   CHECK FOR ERROR CONDITIONS
C
        IF(MS)7,7,8
    7   WRITE(IO,107)
  107   FORMAT(/' **** ERRORS DETECTED PREVENT ANALYSIS ****'/)
        GO TO 999
C
C   INTRODUCTION OF BOUNDARY CONDITIONS
C
    8   CALL BOUND(TK,AL,RENO,IB,NRMX,NCMX)
C
C   SOLUTION OF THE SYSTEM OF EQUATIONS
C
        CALL SLBSI(TK,AL,V,N,MS,NRMX,NCMX)
C
C   CHECK FOR ERROR CONDITIONS
C
        IF(MS)7,9,9
C
C   COMPUTATION OF SECONDARY RESULTS
C
    9   CALL FORCE(KON,PROP,RENO,X,Y,AL)
```

```
C
C    OUTPUT
C
      CALL OUTPT(AL,RENO,KON)
C
  999 CALL EXIT
      END
```

Program 56   Input program (INPUT)

The input codes are now as follows:

    (1) *Basic parameters card*   It includes the number of nodes NN, number of elements NE, the number of boundary nodes NBN, the constants h_x and h_y (stored in E and G) and the value of C. The format is 3I10, 3F10.2.

    (2) *Nodal coordinates cards*   Unchanged.

    (3) *Element connectivity and properties cards*   There will be a card for each element containing the element number, the numbers of the six element nodes and the values of the normal flux vector $\bar{q}$ for each of the three element sides, with format 7I5, 3F10.4.

    (4) *Boundary data cards*   Considering one unknown per node. There will be one card for each boundary node when the variable is prescribed containing the node number and the value of the prescribed unknown. The format is I10, F10.4.

The FORTRAN code for the subprogram is given below.

```
      SUBROUTINE INPUT(X,Y,KON,PROP,AL,IB,RENO)
C
C               PROGRAM 56
C          PROGRAM FOR DATA INPUT
C
      COMMON NN,NE,NLN,NBN,NDF,NNE,N,MS,IN,IO,E,G,C
      DIMENSION X(1),Y(1),KON(1),PROP(1),AL(1),IB(1),RENO(1),W(3),IC(6)
C
C    W = AN AUXILIARY VECTOR TO TEMPORARELY STORE A SET OF
C        FLUXES ON ELEMENT SIDES
C    IC = AUXILIARY ARRAY TO STORE TEMPORARELY THE CONNECTIVITY
C         OF AN ELEMENT.
C
C    READ BASIC PARAMETERS
C
C    NN  = NUMBER OF NODES
C    NE  = NUMBER OF ELEMENTS
C    NBN = NUMBER OF BOUNDARY NODES
      WRITE(IO,20)
   20 FORMAT(' ',79('*'))
      READ(IN,1)  NN,NE,NBN,E,G,C
      WRITE(IO,21) NN,NE,NBN,E,G,C
   21 FORMAT(///' INTERNAL DATA'///' NUMBER OF NODES         :',I5/
     *' NUMBER OF ELEMENTS         :',I5/
     *' NUMBER OF BOUNDARY NODES:',I5/
     *' CONSTANT HX             :',F10.2/
     *' CONSTANT HY             :',F10.2/
     *' CONSTANT C              :',F10.2/
     *' NODAL COORDINATES'/7X,'NODE',6X,'X',9X,'Y')
    1 FORMAT(3I10,3F10.2)
```

```
C
C   READ NODAL COORDINATES IN ARRAY X AND Y
C
      READ(IN,2) (I,X(I),Y(I),J=1,NN)
      WRITE(IO,2) (I,X(I),Y(I),I=1,NN)
    2 FORMAT(I10,2F10.2)
C
C   READ ELEMENT CONNECTIVITY IN ARRAY KON
C   AND THE BOUNDARY FLUXES ON EACH OF THE ELEMENT SIDES
C
      WRITE(IO,22)
   22 FORMAT(/' ELEMENT CONNECTIVITY AND FLUXES'/4X,'ELEMENT',16X,
     *'NODES',9X,'QN1', 7X,'QN2', 7X,'QN3')
      DO 3 J=1,NE
      READ(IN,4) I,(IC(II),II=1,6),W(1),W(2),W(3)
      WRITE(IO,34) I,(IC(II),II=1,6),W(1),W(2),W(3)
      N1=3*(I-1)
      PROP(N1+1)=W(1)
      PROP(N1+2)=W(2)
      PROP(N1+3)=W(3)
      N1=NNE*(I-1)
      DO 1004 II=1,6
 1004 KON(N1+II)=IC(II)
    3 CONTINUE
    4 FORMAT(7I5,3F10.4)
   34 FORMAT(7I5,3F10.3)
C
C   COMPUTE N, ACTUAL NUMBER OF UNKNOWNS
C   AND CLEAR THE RIGHT HAND SIDE VECTOR
C
      N=NN*NDF
      DO 5 I=1,N
    5 AL(I)=0.
C
C   COMPUTE HALF BAND WIDTH
C
      CALL BAND(NE,NDF,NNE,MS,IO,KON)
C
C   READ BOUNDARY NODE DATA AND STORE THE
C   PRESCRIBED UNKNOWN VALUE IN ARRAY RENO
C
      WRITE(IO,24)
   24 FORMAT(/' BOUNDARY CONDITION DATA'/8X,'NODE',5X,
     *'PRESCRIBED VALUES')
      DO 7 I=1,NBN
      READ(IN,8) J,RENO(J)
      WRITE(IO,9) J,RENO(J)
      IB(2*I-1)=J
    7 IB(2*I)=0
    8 FORMAT(I10,F10.4)
    9 FORMAT(I10,10X,F10.4)
      RETURN
      END
```

Program 57 Computation of element matrices (STIFF)

This subroutine computes the element matrices $\mathbf{K}^e$ and $\mathbf{P}_c$ and $\mathbf{P}_q$ based on the integrals

$$\mathbf{K}^e = \int\int h_x \left(\frac{\partial \boldsymbol{\phi}}{\partial x}\right)\left(\frac{\partial \boldsymbol{\phi}}{\partial x}\right)^T + h_y \left(\frac{\partial \boldsymbol{\phi}}{\partial y}\right)\left(\frac{\partial \boldsymbol{\phi}}{\partial y}\right)^T \quad \mathrm{d}x \, \mathrm{d}y \qquad (6.103)$$

$$\mathbf{P}_c = \int \int c\,\phi\,\mathrm{d}x\,\mathrm{d}y \tag{6.104}$$

$$\mathbf{P}_q = \int \bar{q}\,\phi\,\mathrm{d}S \tag{6.105}$$

where ϕ^T is given by Equation (6.88), i.e.

$$\phi^T = \{\xi_1(2\xi_1 - 1),\, \xi_2(2\xi_2 - 1),\, \xi_3(2\xi_3 - 1),\, 4\xi_1\xi_2,\, 4\xi_2\xi_3,\, 4\xi_3\xi_1\} \tag{6.106}$$

The derivatives are given by the chain rule (Equation (6.39)), i.e.

$$\frac{\partial \phi}{\partial x} = \frac{\partial \phi}{\partial \xi_1}\frac{\partial \xi_1}{\partial x} + \frac{\partial \phi}{\partial \xi_2}\frac{\partial \xi_2}{\partial x} + \frac{\partial \phi}{\partial \xi_3}\frac{\partial \xi_3}{\partial x} = c_{21}\frac{\partial \phi}{\partial \xi_1} + c_{22}\frac{\partial \phi}{\partial \xi_2} + c_{23}\frac{\partial \phi}{\partial \xi_3} \tag{6.107}$$

$$\frac{\partial \phi}{\partial y} = c_{31}\frac{\partial \phi}{\partial \xi_1} + c_{32}\frac{\partial \phi}{\partial \xi_2} + c_{33}\frac{\partial \phi}{\partial \xi_3}$$

The final matrices are obtained by substituting (6.106) and (6.107) into Equations (6.103) to (6.105) and integrating using rules (6.40) and (6.41).

The main difference between this subroutine and 47 is that it works with any C2 and C3 instead of B and D as previously.

Notice that the program computes the matrix expression $\mathbf{K}^e$ storing the coefficients in array ELST(6,6) and then the vectors $\mathbf{P}_c$ and $\mathbf{P}_q$ which are added into the array AL.

```
      SUBROUTINE STIFF(NEL,X,Y,PROP,KON,ELST,AL,NDFEL)
C
C                 PROGRAM 57
C  COMPUTATION OF THE ELEMENT MATRIX EQUATION
C
      COMMON NN,NE,NLN,NBN,NDF,NNE,N,MS,IN,IO,E,G,C
      DIMENSION X(1),Y(1),KON(1),PROP(1),ELST(NDFEL,NDFEL),AL(1),
     *C2(4),C3(4)
C
C  NEL   = NUMBER OF CURRENT ELEMENT
C  N1,N2,N3 = NUMBERS OF FIRST, SECOND, AND THIRD ELEMENT NODE
C  D1,D2,D3 = LENGTH OF FIRST, SECOND, AND THIRD ELEMENT SIDES
C
      L=NNE*(NEL-1)+1
      N1=KON(L)
      N2=KON(L+1)
      N3=KON(L+2)
      N4=KON(L+3)
      N5=KON(L+4)
      N6=KON(L+5)
      D1=SQRT((X(N2)-X(N1))**2+(Y(N2)-Y(N1))**2)
      D2=SQRT((X(N3)-X(N2))**2+(Y(N3)-Y(N2))**2)
      D3=SQRT((X(N1)-X(N3))**2+(Y(N1)-Y(N3))**2)
C
C  COMPUTE SECOND ROW (C2), AND THIRD ROW (C3), OF MATRIX C
```

```
C
C   A   = AREA OF ELEMENT
C
      C2(1)=Y(N2)-Y(N3)
      C2(2)=Y(N3)-Y(N1)
      C2(3)=Y(N1)-Y(N2)
      C3(1)=X(N3)-X(N2)
      C3(2)=X(N1)-X(N3)
      C3(3)=X(N2)-X(N1)
      A=(C2(1)*C3(2)-C2(2)*C3(1))/2.
      DO 5 I=1,3
      C2(I)=C2(I)/2./A
    5 C3(I)=C3(I)/2./A
      C2(4)=C2(1)
      C3(4)=C3(1)
C
C   CHECK FOR ERROR CONDITIONS
C
      IF(A)1,1,2
    1 WRITE(IO,101) NEL
  101 FORMAT(/' **** ZERO OR NEGATIVE AREA FOR ELEMENT :',I5,' ****'/)
      MS=0
      GO TO 999
C
C   COMPUTE ELEMENT MATRIX
C
    2 C2122=C2(1)*C2(2)*E+C3(1)*C3(2)*G
      C2123=C2(1)*C2(3)*E+C3(1)*C3(3)*G
      C2223=C2(2)*C2(3)*E+C3(2)*C3(3)*G
      DO 10 I=1,3
   10 ELST(I,I)=8.*A*(C2(I)*C2(I)*E+C3(I)*C3(I)*G)
      ELST(I+3,I+3)=8./3.*A*((C2(I)*C2(I)+C2(I)*C2(I+1)+C2(I+1)*C2(I+1))
     **E+(C3(I)*C3(I)+C3(I)*C3(I+1)+C3(I+1)*C3(I+1))*G)
      ELST(1,2)=20./3.*A*C2122
      ELST(1,3)=20./3.*A*C2123
      ELST(1,4)=ELST(1,2)/5.
      ELST(1,5)=0.
      ELST(1,6)=ELST(1,3)/5.
      ELST(2,3)=20./3.*A*C2223
      ELST(2,4)=ELST(1,4)
      ELST(2,5)=ELST(2,3)/5.
      ELST(2,6)=0.
      ELST(3,4)=0.
      ELST(3,5)=ELST(2,5)
      ELST(3,6)=ELST(1,6)
      ELST(4,5)=4./3.*A*((C2(1)*(C2(2)+2.*C2(3))+C2(2)*(C2(2)+C2(3)))*E+
     *            (C3(1)*(C3(2)+2.*C3(3))+C3(2)*(C3(2)+C3(3)))*G)
      ELST(4,6)=4./3.*A*((C2(1)*(C2(1)+C2(3))+C2(2)*(C2(1)+2.*C2(3)))*E+
     *            (C3(1)*(C3(1)+C3(3))+C3(2)*(C3(1)+2.*C3(3)))*G)
      ELST(5,6)=4./3.*A*((C2(2)*(2.*C2(1)+C2(3))+C2(3)*(C2(1)+C2(3)))*E+
     *            (C3(2)*(2.*C3(1)+C3(3))+C3(3)*(C3(1)+C3(3)))*G)
      DO 11 I=2,6
      IM1=I-1
      DO 11 J=1,IM1
   11 ELST(I,J)=ELST(J,I)
C
C   COMPUTE ELEMENT VECTOR
C
      K=3*(NEL-1)
      CC=-A*C/3.
      AL(N1)=AL(N1)+(PROP(K+1)*D1+PROP(K+3)*D3)/6.
      AL(N2)=AL(N2)+(PROP(K+1)*D1+PROP(K+2)*D2)/6.
      AL(N3)=AL(N3)+(PROP(K+2)*D2+PROP(K+3)*D3)/6.
      AL(N4)=AL(N4)+PROP(K+1)*D1*2./3.+CC
      AL(N5)=AL(N5)+PROP(K+2)*D2*2./3.+CC
      AL(N6)=AL(N6)+PROP(K+3)*D3*2./3.+CC
  999 RETURN
      END
```

Program 58

This program computes the derivatives of the nodal variables in x and y direction. It is similar to subroutine 48 but working now with 6 instead of 3 nodes.

$$\frac{\partial u}{\partial x} = \sum_{i=1}^{3} \frac{\partial u}{\partial \xi_i} \frac{\partial \xi_i}{\partial x} = \sum_{i=1}^{3} \left\{ \frac{\partial u}{\partial \xi_i} c_{2i} \right\} = \sum_{i=1}^{3} \left\{ \frac{\partial \boldsymbol{\phi}^T}{\partial \xi_i} \mathbf{u}_n c_{2i} \right\} \tag{6.108}$$

$$\frac{\partial u}{\partial y} = \sum_{i=1}^{3} \left\{ \frac{\partial \boldsymbol{\phi}^T}{\partial \xi_i} \mathbf{u}^n c_{3i} \right\} \tag{6.109}$$

```
      SUBROUTINE FORCE(KON,PROP,RENO,X,Y,AL)
C
C                 PROGRAM 58
C  COMPUTATION OF SECONDARY RESULTS
C
      COMMON NN,NE,NLN,NBN,NDF,NNE,N,MS,IN,IO,E
      DIMENSION KON(1),PROP(1),RENO(1),X(1),Y(1),AL(1),C2(3),C3(3)
C
C  NEL  = NUMBER OF CURRENT ELEMENT
C  N1,N2,N3 = NUMBERS OF FIRST, SECOND, AND THIRD ELEMENT NODE
C
      E=0.
      DO 100 NEL=1,NE
      L=NNE*(NEL-1)+1
      N1=KON(L)
      N2=KON(L+1)
      N3=KON(L+2)
      N4=KON(L+3)
      N5=KON(L+4)
      N6=KON(L+5)
C
C  COMPUTE SECOND ROW (C2), AND THIRD ROW (C3),OF MATRIX C
C  A    = AREA OF ELEMENT TIMES 2
C
      C2(1)=Y(N2)-Y(N3)
      C2(2)=Y(N3)-Y(N1)
      C2(3)=Y(N1)-Y(N2)
      C3(1)=X(N3)-X(N2)
      C3(2)=X(N1)-X(N3)
      C3(3)=X(N2)-X(N1)
      A=(C2(1)*C3(2)-C2(2)*C3(1))
      DO 5 I=1,3
      C2(I)=C2(I)/A
    5 C3(I)=C3(I)/A
C
C  COMPUTE DERIVATIVES OF PROBLEM VARIABLE
C  FOR EACH ELEMENT
C
      L=6*(NEL-1)
      RENO(L+1)=4.*(C2(2)*(AL(N4)-AL(N2))+C2(3)*(AL(N6)-AL(N3)))
      RENO(L+2)=4.*(C3(2)*(AL(N4)-AL(N2))+C3(3)*(AL(N6)-AL(N3)))
      RENO(L+3)=4.*(C2(1)*(AL(N4)-AL(N1))+C2(3)*(AL(N5)-AL(N3)))
      RENO(L+4)=4.*(C3(1)*(AL(N4)-AL(N1))+C3(3)*(AL(N5)-AL(N3)))
      RENO(L+5)=4.*(C2(1)*(AL(N6)-AL(N1))+C2(2)*(AL(N5)-AL(N2)))
      RENO(L+6)=4.*(C3(1)*(AL(N6)-AL(N1))+C3(2)*(AL(N5)-AL(N2)))
  100 E=E+A*(AL(N4)+AL(N5)+AL(N6))/6.
      RETURN
      END
```

Program 59   Output program (OUTPT)

The program OUTPT prints the values of the nodal variables, of its derivatives and of the integral.

$$\int u \, dt \qquad (6.110)$$

according to the following FORTRAN code.

```
      SUBROUTINE OUTPT(AL,RENO,KON)
C
C             PROGRAM 59
C          OUTPUT OF RESULTS
C
      COMMON NN,NE,NLN,NBN,NDF,NNE,N,MS,IN,IO,E
      DIMENSION AL(1),RENO(1),KON(1)
C
C   WRITE VALUES OF PROBLEM VARIABLE AT NODAL POINTS
C
      WRITE(IO,1)
    1 FORMAT(//1X,130('*')//' RESULTS'//' NODAL VARIABLES    '//7X,'NODE'
     *,6X,'VARIABLE')
      WRITE(IO,2) (I,AL(I),I=1,NN)
    2 FORMAT(I10,F15.4)
C
C   WRITE DERIVATIVES OF PROBLEM VARIABLE
C
      WRITE(IO,3)
    3 FORMAT(//' DERIVATIVES OF THE PROBLEM VARIABLE OVER EACH ELEMENT
     *4X,'ELEMENT',6X,'NODE',9X,'X',14X,'Y',14X,'N')
      DO 4 I=1,NE
      WRITE(IO,33)
   33 FORMAT(/)
      K=NNE*(I-1)
      DO 4 J=1,3
      KK=K+(J-1)*2+1
      DN=SQRT(RENO(KK)**2+RENO(KK+1)**2)
    4 WRITE(IO,5) I,KON(K+J),RENO(KK),RENO(KK+1),DN
    5 FORMAT(2I10,3F15.5)
      WRITE(IO,7) E
    7 FORMAT(//' VALUE OF THE INTEGRAL :',F15.4)
      WRITE(IO,8)
    8 FORMAT(//1X,130('*'))
      RETURN
      END
```

Example 6.5

This example is similar to Example 6.2 in this chapter, i.e. study of the torsion of a prismatic beam of elliptical cross section. The problem is now solved using 24 six noded elements and 65 nodes. The constants $h_x = h_y = 1$ and $c = -2$ are as previously. The potential at all boundary nodes is taken to be zero (see Figure 6.25).

The input for the program is given below followed by the corresponding output. Notice that the value of the area integral (6.110) is now 2.3767, which gives the following value for the torsional rigidity, τ

$$\tau = 2 \int u \, dA = 4.7534$$

This value is higher than the one previously obtained, $\tau = 4.56$, using linear elements. The analytical solution for this problem gives $\tau = 5.026$, which means that our numerical solution produces 5% of error, rather than 10% as previously. In order to improve these results one needs to use curved elements as most of the error is now due to the fact that the shape of the section is not properly represented.

Input

		24		32	1.		1.		−2.
65									
1	−2.								
2									
3									
4	−1.8		0.4359						
5									
6	−1.8		−0.4359						
7									
8									
9	−1.5								
10									
11									
12									
13									
14	−1.2		0.8						
15									
16	−1.2		−0.8						
17									
18									
19	−0.9								
20									
21									
22	−0.6		0.954						
23									
24									
25									
26	−0.6		−0.954						
27									
28									
29									
30									
31	0.		1.						
32									
33									
34									
35	0.		−1.						
36									
37									
38									
39									
40									
41	0.6		0.954						
42									
43	0.9								
44									
45	0.6		−0.954						
46									
47									
48									
49									
50									
51	1.2		0.8						

52		
53	1.5	
54		
55	1.2	−0.8
56		
57		
58		
59		
60	1.8	0.4359
61		
62	1.8	−0.4359
63		
64		
65	2.	

1	1	9	4	5	8	2
2	9	1	6	5	3	10
3	9	14	4	12	7	8
4	6	16	9	11	13	10
5	9	19	14	15	18	12
6	19	9	16	15	13	20
7	22	14	19	17	18	23
8	16	26	19	21	25	20
9	19	33	22	24	28	23
10	33	19	26	24	25	29
11	22	33	31	28	32	27
12	33	26	35	29	30	34
13	33	41	31	37	36	32
14	45	33	35	39	34	40
15	33	43	41	38	42	37
16	43	33	45	38	39	44
17	51	41	43	46	42	47
18	43	45	55	44	50	49
19	43	53	51	48	52	47
20	53	43	55	48	49	54
21	51	53	60	52	57	56
22	53	55	62	54	59	58
23	53	65	60	61	63	57
24	65	53	62	61	58	64

1
2
3
4
6
7
11
14
16
17
21
22
26
27
30
31
35
36
40
41
45
46
50
51
55
56
59
60

```
62
63
64
65
```

Output

INTERNAL DATA

NUMBER OF NODES	:	65
NUMBER OF ELEMENTS	:	24
NUMBER OF BOUNDARY NODES	:	32
CONSTANT HX	:	1.00
CONSTANT HY	:	1.00
CONSTANT C	:	−2.00

NODAL COORDINATES

NODE	X	Y
1	−2.00	0.00
2	0.00	0.00
3	0.00	0.00
4	−1.80	0.44
5	0.00	0.00
6	−1.80	−0.44
7	0.00	0.00
8	0.00	0.00
9	−1.50	0.00
10	0.00	0.00
11	0.00	0.00
12	0.00	0.00
13	0.00	0.00
14	−1.20	0.80
15	0.00	0.00
16	−1.20	−0.80
17	0.00	0.00
18	0.00	0.00
19	−0.90	0.00
20	0.00	0.00
21	0.00	0.00
22	−0.60	0.95
23	0.00	0.00
24	0.00	0.00
25	0.00	0.00
26	−0.60	−0.95
27	0.00	0.00
28	0.00	0.00
29	0.00	0.00
30	0.00	0.00
31	0.00	1.00
32	0.00	0.00
33	0.00	0.00
34	0.00	0.00
35	0.00	−1.00
36	0.00	0.00
37	0.00	0.00
38	0.00	0.00
39	0.00	0.00
40	0.00	0.00
41	0.60	0.95
42	0.00	0.00
43	0.90	0.00
44	0.00	0.00
45	0.60	−0.95
46	0.00	0.00
47	0.00	0.00
48	0.00	0.00

49	0.00	0.00
50	0.00	0.00
51	1.20	0.80
52	0.00	0.00
53	1.50	0.00
54	0.00	0.00
55	1.20	−0.80
56	0.00	0.00
57	0.00	0.00
58	0.00	0.00
59	0.00	0.00
60	1.80	0.44
61	0.00	0.00
62	1.80	−0.44
63	0.00	0.00
64	0.00	0.00
65	2.00	0.00

ELEMENT CONNECTIVITY AND FLUXES

ELEMENT				NODES			QN1	QN2	QN3
1	1	9	4	5	8	2	0.000	0.000	0.000
2	9	1	6	5	3	10	0.000	0.000	0.000
3	9	14	4	12	7	8	0.000	0.000	0.000
4	6	16	9	11	13	10	0.000	0.000	0.000
5	9	19	14	15	18	12	0.000	0.000	0.000
6	19	9	16	15	13	20	0.000	0.000	0.000
7	22	14	19	17	18	23	0.000	0.000	0.000
8	16	26	19	21	25	20	0.000	0.000	0.000
9	19	33	22	24	28	23	0.000	0.000	0.000
10	33	19	26	24	25	29	0.000	0.000	0.000
11	22	33	31	28	32	27	0.000	0.000	0.000
12	33	26	35	29	30	34	0.000	0.000	0.000
13	33	41	31	37	36	32	0.000	0.000	0.000
14	45	33	35	39	34	40	0.000	0.000	0.000
15	33	43	41	38	42	37	0.000	0.000	0.000
16	43	33	45	38	39	44	0.000	0.000	0.000
17	51	41	43	46	42	47	0.000	0.000	0.000
18	43	45	55	44	50	49	0.000	0.000	0.000
19	43	53	51	48	52	47	0.000	0.000	0.000
20	53	43	55	48	49	54	0.000	0.000	0.000
21	51	53	60	52	57	56	0.000	0.000	0.000
22	53	55	62	54	59	58	0.000	0.000	0.000
23	53	65	60	61	63	57	0.000	0.000	0.000
24	65	53	62	61	58	64	0.000	0.000	0.000

—— HALF-BANDWIDTH IS EQUAL TO 15——

BOUNDARY CONDITION DATA

NODE	PRESCRIBED VALUES
1	0.0000
2	0.0000
3	0.0000
4	0.0000
6	0.0000
7	0.0000
11	0.0000
14	0.0000
16	0.0000
17	0.0000
21	0.0000
22	0.0000
26	0.0000
27	0.0000
30	0.0000
31	0.0000

35	0.0000
36	0.0000
40	0.0000
41	0.0000
45	0.0000
46	0.0000
50	0.0000
51	0.0000
55	0.0000
56	0.0000
59	0.0000
60	0.0000
62	0.0000
63	0.0000
64	0.0000
65	0.0000

****

RESULTS

NODAL VARIABLES

1	0.0000
2	0.0000
3	0.0000
4	0.0000
5	0.1618
6	0.0000
7	0.0000
8	0.1930
9	0.3178
10	0.1930
11	0.0000
12	0.2782
13	0.2782
14	0.0000
15	0.4850
16	0.0000
17	0.0000
18	0.4306
19	0.6136
20	0.4306
21	0.0000
22	0.0000
23	0.4872
24	0.7404
25	0.4872
26	0.0000
27	0.0000
28	0.5839
29	0.5839
30	0.0000
31	0.0000
32	0.5855
33	0.7812
34	0.5855
35	0.0000
36	0.0000
37	0.5839
38	0.7404
39	0.5839
40	0.0000
41	0.0000
42	0.4872
43	0.6136
44	0.4872

45	0.0000
46	0.0000
47	0.4306
48	0.4850
49	0.4306
50	0.0000
51	0.0000
52	0.2782
53	0.3178
54	0.2782
55	0.0000
56	0.0000
57	0.1930
58	0.1930
59	0.0000
60	0.0000
61	0.1618
62	0.0000
63	0.0000
64	0.0000
65	0.0000

DERIVATIVES OF THE PROBLEM VARIABLE OVER EACH ELEMENT

ELEMENT	NODE	X	Y	N
1	1	0.65880	−1.61459	1.74382
1	9	−1.29440	0.88020	1.56532
1	4	0.90840	−0.41679	0.99945
2	9	0.61240	−0.88020	1.07228
2	1	0.65880	0.30227	0.72483
2	6	0.90840	0.41679	0.99945
3	9	0.57880	1.52504	1.63119
3	14	0.78070	−1.28651	1.50486
3	4	0.44603	−0.73501	0.85976
4	6	0.44603	0.73501	0.85976
4	16	1.90157	0.51509	1.97010
4	9	0.57880	0.01780	0.57908
5	9	0.62167	−0.58850	0.85604
5	19	−2.70367	1.13912	2.93384
5	14	0.52300	−1.18988	1.29974
6	19	0.36433	−0.54325	0.65411
6	9	2.21067	−0.56200	2.28098
6	16	0.52300	1.18988	1.29974
7	22	0.39077	−1.52246	1.57181
7	14	−3.01286	−0.21482	3.02051
7	19	0.37886	−0.00593	0.37890
8	16	0.32451	1.26431	1.30529
8	26	3.18940	0.47298	3.22428
8	19	0.37886	0.00593	0.37890

9	19	0.37733	−0.35681	0.51932
9	33	−2.60889	0.80741	2.73097
9	22	0.24356	−1.47617	1.49613
10	33	−0.00489	−0.16422	0.16429
10	19	2.42267	−1.28092	2.74045
10	26	0.24356	1.47617	1.49613
11	22	0.11917	−1.55440	1.55896
11	33	−0.16889	2.34200	2.34808
11	31	0.11966	−1.56080	1.56538
12	33	0.01079	−2.34200	2.34202
12	26	0.11917	1.55440	1.55896
12	35	0.11966	1.56080	1.56538
13	33	−0.01079	2.34200	2.34202
13	41	−0.11917	−1.55440	1.55896
13	31	−0.11966	−1.56080	1.56538
14	45	−0.11917	1.55440	1.55896
14	33	0.16889	−2.34200	2.34808
14	35	−0.11966	1.56080	1.56538
15	33	0.00489	0.16422	0.16429
15	43	−2.42267	1.28092	2.74045
15	41	−0.24356	−1.47617	1.49613
16	43	−0.37733	0.35681	0.51932
16	33	2.60889	−0.80741	2.73097
16	45	−0.24356	1.47617	1.49613
17	51	−0.32451	−1.26431	1.30529
17	41	−3.18940	−0.47298	3.22428
17	43	−0.37886	−0.00593	0.37890
18	43	−0.37886	−2.09304	2.12706
18	45	−0.39077	1.52246	1.57181
18	55	−0.32451	1.26431	1.30529
19	43	−0.36433	0.54325	0.65411
19	53	−2.21067	0.56200	2.28098
19	51	−0.52300	−1.18988	1.29974
20	53	−0.62167	0.58850	0.85604
20	43	2.70367	−1.13912	2.93384
20	55	−0.52300	1.18988	1.29974
21	51	−0.78070	−0.51509	0.93531
21	53	0.35745	1.52504	1.56637
21	60	−0.44603	−0.73501	0.85976
22	53	−0.57880	−1.52504	1.63119

22	55	−0.78070	1.28651	1.50486
22	62	−0.44603	0.73501	0.85976
23	53	−0.61240	0.88020	1.07228
23	65	−0.65880	−0.30227	0.72483
23	60	−0.90840	−0.41679	0.99945
24	65	−0.65880	1.61459	1.74382
24	53	1.29440	−0.88020	1.56532
24	62	−0.90840	0.41679	0.99945

VALUE OF THE INTEGRAL :   2.3767

****

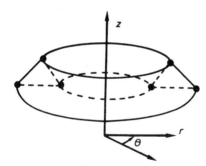

Figure 6.25

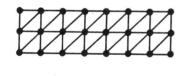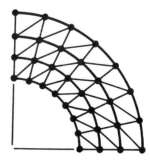

Figure 6.26

Exercises

(1) Consider the axisymmetric element shown in Figure 6.25. If u does not vary with respect to θ Laplace's equation takes the form

$$\frac{\partial^2 u}{\partial r^2} + \frac{1}{r}\frac{\partial u}{\partial r} + \frac{\partial^2 u}{\partial z^2} = 0$$

Propose a linear u function for this element and develop the corresponding matrices.

(2) Compare the classical Rayleigh-Ritz and Galerkin methods with the finite element technique. Discuss their differences.

(3) Discuss the differences between 3-node triangular and 4-node rectangular elements. Why does the rectangular element give better convergence?

(4) Comment on how you would number the nodes and elements for the plane grids in Figure 6.26. Also, determine the bandwidth for your choice, assuming two unknowns per node.

(5) Define and discuss the different steps of a two-dimensional elastic-structures computer program. Assume that stiffness matrix (6×6) is already known and develop a macro flow chart.

(6) If the displacement expansions contain all the rigid body modes, what does this imply about the singular nature of the element stiffness matrix?

References

1. Turner, M. J., Clough, R. W., Martin, H. C. and Topp, L. J., 'Stiffness and deflection analysis of Complex Structures', *J. Aero. Sci.* **23**, 805–823 (1956)
2. Argyris, J. H., 'Energy Theorems and Structural Analysis', *Aircraft Eng.* **26** (1954), 27 (1955)
3. Melosh, R. J., 'A Stiffness Matrix for the Analysis of Thin Plates in Bending', *J. Aero. Sci.*, **28**, No.34 (1961)
4. Adini, A., 'Analysis of Shell Structures by the Finite Element Method', Thesis, Univ. of Berkeley, Calif. (1961)
5. Clough, R. W. and Tocher, J. L., 'Finite Element Stiffness Matrices for Analysis of Plate Bending', *Proc. Conf. Matrix Methods in Structural Mechanics*, Dayton, Ohio (1965)
6. Argyris, J. H., 'Continua and Discontinua', *Proc. Conf. Matrix Methods in Structural Mechanics*, Dayton, Ohio (1965)
7. Grafton, P. E. and Strome, D. R., 'Analysis of Axisymmetric Shells by the Direct Stiffness Method', *A.I.A.A. Journal*, **1**, No.10 (1963)
8. Percy, J. W., Pian, T. H., Klein, S. and Navaratna, D. R., 'Application of the Matrix Displacement Method of Linear Elastic Analysis of Shells of Revolution', *A.I.A.A. Journal*, **3**, No.11 (1965)

9.   Connor, J. J. and Brebbia, C. A., 'Stiffness Matrix for Shallow Rectangular Shell Element', *J. Eng. Mech. Div.*, ASCE, 93, No.5 (1967)

10.   Melosh, R. J., 'Development of the Stiffness Method to Define Bounds on Elastic Behaviour of Structures', PhD. Dissertation, Univ. of Washington (1962)

11.   Pian, T. H. and Tong, P. 'Basis of Finite Element Methods for Solid Continua', *Int. J. Numerical Methods in Engineering*, 1, No.1 (1969)

12.   Oden, J. T., Finite Elements of Nonlinear Continua, McGraw-Hill (1972)

13.   Arantes de Oliveira, E. R., 'Completeness and Convergence in the Finite Element Method', Proc. 2nd *Conf. Matrix Methods in Structural Mechanics*, Dayton, Ohio (1968)

14.   Strang, G. and Fix, G. J., *An Analysis of the Finite Element Method*, Prentice-Hall (1973)

15.   Ferrante, A. J., 'On General Systems for Finite Element Analysis: The POL Approach', *Int. Symp. Discrete Methods in Engng, CISE Milan*, Etas-Libris (1974)

Bibliography

Brebbia, C. A. and Connor, J. J., *Fundamentals of Finite Element Techniques for Structural Engineers*, Butterworths, London (1973)

Connor, J. J. and Brebbia, C. A., *Finite Element Techniques for Fluid Flow*, Newnes–Butterworths, London (1976)

Desai, C. S. and Abel, J. F. *Introduction to the Finite Element Method*, Van Nostrand, Reinhold (1972)

Norrie, D. H. and Vries, D. De, *The Finite Element Method, Fundamentals and Applications*, Academic Press (1973)

Chapter 7

Fluid mechanics

7.1 INTRODUCTION

In order to understand the difference between the behaviour of a fluid and
that of a solid, let us consider a container filled with a fluid, such as the one
shown in Figure 7.1. Assume that the face A of a body suspended in the fluid
is the only surface of the body in contact with the fluid. If we apply a horizontal force P to the body this will start to move with a certain velocity v
which is transmitted to the fluid and diminishes in the vertical direction until
it reaches zero at the bottom of the container. The relationship between the
stresses on A and the velocities were given by Newton as

$$\tau = \mu \frac{\partial v}{\partial y} \qquad (7.1)$$

where $P/A = \tau$ is the shear stress on the face of the body, $\partial v/\partial y$ indicates the
variation of the velocity v with respect to depth and μ is a characteristic of
the fluid, called the viscosity coefficient. We can in addition exert pressure
forces on the fluid, which would react by changing its volume in accordance

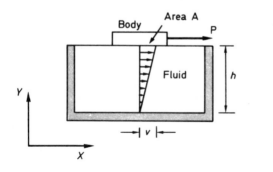

Figure 7.1

338

with a certain compressibility coefficient. However in most engineering problems the fluids can be considered for all practical purposes incompressible. We will also assume that they are inviscid, because for our applications their viscosity is very small and can be neglected. With these hypotheses the fluids will only withstand pressure forces and will react without changing volume.

It is clear from Equation (7.1) and the above discussion that a fluid can undergo significant deformation. Hence the problem will become non-linear as far as geometry is concerned.

We will refer the particle to a special system different from the system used for solids (for which the deformations were very small), called an Eulerian reference system which is well suited for fluids. Consider a particle shown in Figure 7.2 and take as independent variables the actual coordinates x, y and the time t. Any quantity such as pressure, velocity, etc. are functions of position and time. In general

$$f = f(x, y, t) \tag{7.2}$$

where f denotes the variable associated with the point (x, y) at time t. During the Δt time increment, the material point moves from x to $x + \Delta x$ and $y + \Delta y$ and the variable f changes to $f + \Delta f$.

Assuming f to be a continuous function we can write

$$\Delta f = \frac{\partial f}{\partial x} \Delta x + \frac{\partial f}{\partial y} \Delta y + \frac{\partial f}{\partial t} \Delta t \tag{7.3}$$

In the limit as $\Delta t \to 0$, we have

$$\frac{Df}{Dt} = \lim_{\Delta t \to 0} \frac{\Delta f}{\Delta t} = \lim \left(\frac{\partial f}{\partial x} \frac{\Delta x}{\Delta t} + \frac{\partial f}{\partial y} \frac{\Delta y}{\Delta t} + \frac{\partial f}{\partial t} \right)$$

$$= \frac{\partial f}{\partial x} \lim_{\Delta t \to 0} \left(\frac{\Delta x}{\Delta t} \right) + \frac{\partial f}{\partial y} \lim_{\Delta t \to 0} \left(\frac{\Delta y}{\Delta t} \right) + \frac{\partial f}{\partial t} \tag{7.4}$$

The velocity vector components are,

$$u = \lim_{\Delta t \to 0} \left(\frac{\Delta x}{\Delta t} \right), \quad v = \lim_{\Delta t \to 0} \left(\frac{\Delta y}{\Delta t} \right) \tag{7.5}$$

Hence Equation (7.4) can be written

$$\frac{Df}{Dt} = u \frac{\partial f}{\partial x} + v \frac{\partial f}{\partial y} + \frac{\partial f}{\partial t} \tag{7.6}$$

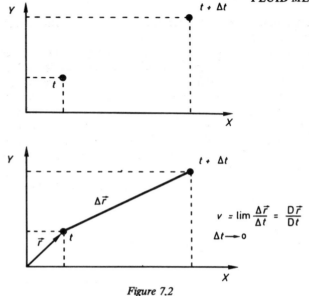

Figure 7.2

Since we are now actually following the material point or particle, the term 'material' derivative is used for $D(\ )/Dt$. The first two terms in (7.6) are the convective terms and the last the 'local' time derivative.

7.2   GOVERNING EQUATIONS

The two dimensional state of stress for a fluid can be defined as for a solid, in terms of σ_x, σ_y, τ stress components at any point and time. Hence the equilibrium equations are the same as those previously seen, i.e.

$$\frac{\partial \sigma_x}{\partial x} + \frac{\partial \tau}{\partial y} + \rho b_x = 0$$

$$\frac{\partial \tau}{\partial x} + \frac{\partial \sigma_y}{\partial y} + \rho b_y = 0 \tag{7.7}$$

where b_x, b_y are body forces per unit volume and mass.

The stress boundary conditions are,

$$p_x = \sigma_x \, \alpha_{nx} + \tau \, \alpha_{ny} = \overline{p}_x$$

$$p_y = \tau \, \alpha_{nx} + \sigma_y \, \alpha_{ny} = \overline{p}_y \tag{7.8}$$

where α_n is the direction cosine of the outward normal n with respect to the x, y axis (see Figure 7.3).

Figure 7.3 shows the classification of the boundary regions. The zone S_v represents the surface area on which the velocities are prescribed. If S_v is a physical boundary such as a wall, the velocity components are $v_n = 0$ (no-slip) and $v_s \neq 0$ (the fluid can slip in the tangential direction, which occurs in the inviscid case). Symmetry and inflow velocity constraints are also included in S_v. S_p is defined as the surface zone on which prescribed boundary forces are acting, for instance normal and tangential forces acting on the boundary of the fluid (if the fluid is inviscid, tangential forces cannot be imposed and we can only apply normal pressures).

Equations (7.7) are time independent. In order to include the dynamic forces we can use D'Alembert's principle, i.e.

$$\rho b_x \rightarrow \rho b_x - \frac{D}{Dt}(\rho u), \quad \rho b_y \rightarrow \rho b_y - \frac{D}{Dt}(\rho v) \tag{7.9}$$

If the fluid is *incompressible* its density ρ will remain constant and Equations (7.9) can be written as,

$$\rho b_x - \rho \frac{Du}{Dt}, \quad \rho b_y - \rho \frac{Dv}{Dt} \tag{7.10}$$

This gives the following equilibrium equations

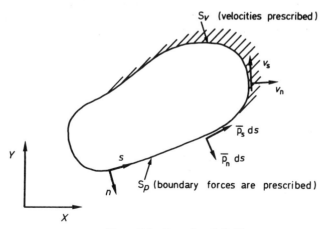

Figure 7.3   Boundary definition

$$\frac{\partial \sigma_x}{\partial x} + \frac{\partial \tau}{\partial y} + \rho b_x = \rho \frac{Du}{Dt}$$

$$\frac{\partial \tau}{\partial x} + \frac{\partial \sigma_y}{\partial y} + \rho b_y = \rho \frac{Dv}{Dt} \qquad (7.11)$$

7.3   PERFECT FLUIDS

Fluids which are inviscid and incompressible are called perfect fluids and obey mathematical relationships which greatly simplify the problem. Let us first deduce the incompressibility relation by considering an element of fluid at time t, as shown in Figure 7.4. In order to have continuity it is necessary that the amount of fluid entering the differential volume be equal to the amount leaving it. This flow-in, flow-out relationship implies that

$$u\, dy + v\, dx = \left(u + \frac{\partial u}{\partial x} dx \right) dy + \left(v + \frac{\partial v}{\partial y} dy \right) dx \qquad (7.12)$$

That is

$$\frac{\partial u}{\partial x} + \frac{\partial v}{\partial y} = 0 \qquad (7.13)$$

Equation (7.13) is called the continuity or incompressibility condition.

Let us now consider what happens when the fluid is inviscid, i.e. the vis-

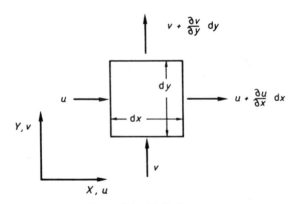

Figure 7.4   Fluid elements

cosity $\mu = 0$. We cannot apply shear forces to such a fluid, which physically means that the fluid particle cannot equilibrate any shear forces. Hence we cannot produce rotation in an inviscid fluid particle. This condition is called the irrotationality condition and can be written mathematically as (see Figure 7.5)

$$\frac{\partial v}{\partial x} = \frac{\partial u}{\partial y} \qquad\qquad (7.14)$$

or

$$\omega = \frac{1}{2}\left(\frac{\partial v}{\partial x} - \frac{\partial u}{\partial y}\right) = 0 \qquad\qquad (7.15)$$

where ω is the angular velocity about the z axis.

A fluid that is incompressible and inviscid, that is, satisfies Equations (7.13) and (7.14), is called a 'perfect fluid'.

The state of stress for a perfect fluid is simply

$$\sigma_x = -p, \qquad \sigma_y = -p$$

$$\tau = 0 \qquad\qquad (7.16)$$

i.e. the state of stress is defined by a single variable p (pressure). Note that by convention, p is taken positive when compressive.

The equilibrium Equations (7.11) are now

$$-\frac{\partial p}{\partial x} + \rho b_x = \rho\,\frac{Du}{Dt}$$

$$-\frac{\partial p}{\partial y} + \rho b_y = \rho\,\frac{Dv}{Dt} \qquad\qquad (7.17)$$

with boundary conditions

$$\overline{p}_n = -p \text{ or } \overline{v}_n = v_n \qquad\qquad (7.18)$$

$\overline{p}_s$ or tangential velocity, $\overline{v}_s$ are not needed, as there is no possibility of applying tangential forces to an inviscid fluid.

Steady flow occurs when the derivatives with respect to time can be neglected. Hence

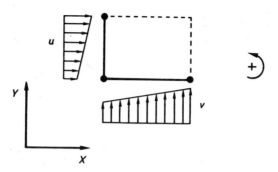

Figure 7.5

$$\frac{Du}{Dt} \Rightarrow u \frac{\partial u}{\partial x} + v \frac{\partial v}{\partial y} \qquad (7.19)$$

For steady flow the equilibrium Equations (7.17) can be written as

$$-\frac{1}{\rho} \frac{\partial p}{\partial x} + b_x - \frac{\partial}{\partial x} \left\{ \frac{1}{2}(u^2 + v^2) \right\} = 0$$

$$-\frac{1}{\rho} \frac{\partial p}{\partial y} + b_y - \frac{\partial}{\partial y} \left\{ \frac{1}{2}(u^2 + v^2) \right\} = 0 \qquad (7.20)$$

Equation (7.20) has been deduced taken into account the irrotationality condition.

7.3.1 Bernouilli's principle

When a body force potential exists we can write

$$b_x = \frac{\partial \Omega}{\partial x}, \qquad b_y = \frac{\partial \Omega}{\partial y} \qquad (7.21)$$

If this potential is due to gravity forces acting in the opposite direction to the y axis we have

$$\Omega = -y\,g \qquad (7.22)$$

where y is the elevation of the point.

Equations (7.20) can now be written as a single formula

$$d \left[\frac{p}{\gamma} + y + \frac{1}{2g} (u^2 + v^2) \right] = 0 \tag{7.23}$$

where $\gamma = \rho g$. This equation implies that

$$\frac{p}{\gamma} + y + \frac{1}{2g} V^2 = \text{constant} \tag{7.24}$$

with $V^2 = u^2 + v^2$. Equation (7.24) is valid for any point in the fluid and is called Bernouilli's formula for two dimensional, inviscid, incompressible and steady flow. We generally call p/γ: pressure head, y: elevation head, $V^2/2g$: velocity head.

Let us consider how Equation (7.24) can be interpreted for the pipe and channel shown in Figure 7.6. Consider in each case two sections with elevations y_a and y_b. The pressure at each end is usually found by a manometer reading (pipe) or simply by knowing the position of the free surface (channel). The velocity head is $V_a^2/2g$ and $V_b^2/2g$ for the two sections. Bernouilli's equation implies that the energy for sections A and B is the same. In practice, this constant energy principle is not valid as there is always a loss of energy due to friction, turbulence or other causes. This loss is represented in Figure 7.6 by the difference between the dashed horizontal line and the dashed inclined line representing the sum of energies which is called the 'hydraulic grade line'. This loss is called the 'head loss'.

7.3.2 Potential flows

Bernouilli's equation substitutes the two momentum equations relating velocities to pressure. In addition the fluid has to satisfy continuity and be inviscid. If the flow of an inviscid fluid is at any time irrotational it will remain so and we will always have

$$\omega = 0 \tag{7.25}$$

which allows us to define a potential ζ which identically satisfies (7.25), i.e.

$$u = \frac{\partial \zeta}{\partial x}, \quad v = \frac{\partial \zeta}{\partial y} \tag{7.26}$$

Having a potential greatly simplifies the problem as now a single function ζ replaces the two velocities. The incompressibility or continuity condition (7.13) becomes

$$\nabla^2 \zeta = 0 \tag{7.27}$$

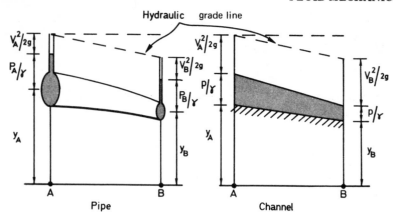

Figure 7.6   Flow in a pipe and a channel

7.3.3 Streamlines formulation

For non-turbulent perfect fluids of the type considered in this chapter the
flow particles follow a series of lines, called streamlines, Figure 7.7. (A line
drawn in the fluid so that its tangent at each point is in the direction of
the fluid velocity at that point is called a streamline.) Streamline flow is also
called laminar flow. For steady flow streamlines do not change with time.

The fluid loss between two streamlines at a differential distance is

$$d\psi = u\,dy - v\,dx \qquad (7.28)$$

where $d\psi$ is an exact differential which can also be written

$$d\psi = \frac{\partial \psi}{\partial y}\,dy + \frac{\partial \psi}{\partial x}\,dx \qquad (7.29)$$

Thus

$$u = \frac{\partial \psi}{\partial y}, \quad v = -\frac{\partial \psi}{\partial x} \qquad (7.30)$$

ψ is called the stream function and is such that the incompressibility condition is identically satisfied. For a streamline $d\psi \equiv 0$, hence we have $\psi =$
constant. Consider now a series of streamlines as shown in Figure 7.7 and a
curve AB which intercepts them. (In what follows we assume unit depth.)
The flux through AB is given by

$$Q = \int_A^B v_n \, dS = \int_A^B (u \, \alpha_{nx} + v \, \alpha_{ny}) dS \qquad (7.31)$$

where α_{nx}, α_{ny} are the direction cosines of the normal to **AB** with respect to the x and y axis. Thus

$$\alpha_{nx} = \frac{dy}{dS}, \qquad \alpha_{ny} = -\frac{dx}{dS} \qquad (7.32)$$

Hence Equation (7.31) becomes

$$Q = \int_A^B (u \, dy + v \, dx) \qquad (7.33)$$

Substituting the velocities using (7.30) we obtain

$$Q = \int_A^B \left(\frac{\partial \psi}{\partial y} \, dy + \frac{\partial \psi}{\partial x} \, dx \right) = \int_A^B d\psi \qquad (7.34)$$

$$Q = \psi(B) - \psi(A)$$

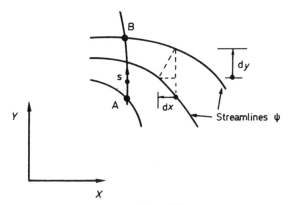

Figure 7.7

The total flux is then equal to the difference between the stream functions A and B.

Note that by definition, streamlines and equipotentials are orthogonal. To show this, consider an increment of the ψ and ζ functions

$$d\psi = \frac{\partial \psi}{\partial x}\, dx + \frac{\partial \psi}{\partial y}\, dy$$

$$d\zeta = \frac{\partial \zeta}{\partial x}\, dx + \frac{\partial \zeta}{\partial y}\, dy \tag{7.35}$$

We can now define two vectors tangent to the streamlines and equipotentials respectively,

$$e_\psi = \left(\frac{\partial \psi}{\partial x}, \frac{\partial \psi}{\partial y}\right), \quad e_\zeta = \left(\frac{\partial \zeta}{\partial x}, \frac{\partial \zeta}{\partial y}\right) \tag{7.36}$$

By using Equations (7.26) and (7.30) these vectors can be shown to be orthogonal, i.e.

$$e_\psi \cdot e_\zeta = 0 \tag{7.37}$$

Hence streamlines and equipotentials are orthogonal.

When working with streamlines we identically satisfy the compressibility condition but the condition of irrotationality gives a Laplacian

$$\omega = \nabla^2 \psi = 0 \tag{7.38}$$

with boundary conditions $\psi = \overline{\psi}$ on S_1 and $\overline{v}_s = -(\partial \psi / \partial n)$ on S_2.

7.3.4 Galerkin formulation

Let us first study the formulation corresponding to the potential function ζ. For this case the continuity equation becomes

$$\nabla^2 \zeta = 0 \tag{7.39}$$

with boundary conditions $\zeta = \overline{\zeta}$ on S_1 and $\overline{v}_n = (\partial \zeta / \partial n)$ on S_2. We want to find the function ζ and its derivatives with respect to x and y (that is the velocities). The pressure can afterwards be found through Bernouilli's equation.

We can write the following variational statement,

$$\iint (\nabla^2 \zeta) \delta \zeta \, dA = \int\limits_{S_2} \left(\frac{\partial \zeta}{\partial n} - \bar{v}_n \right) \delta \zeta \, dS \qquad (7.40)$$

where the ζ function is assumed to satisfy the boundary conditions on S_1.

Integrating Equation (7.40) by parts we have

$$-\iint \vec{\nabla} \zeta \cdot \vec{\nabla} \delta \zeta \, dA + \int\limits_{S_2} \bar{v}_n \, \delta \zeta \, dS = 0 \qquad (7.41)$$

where

$$\nabla \zeta = \left(\frac{\partial \zeta}{\partial x}, \frac{\partial \zeta}{\partial y} \right).$$

The functional corresponding to (7.41) can be written

$$F(\zeta) = \frac{1}{2} \iint \nabla \zeta \cdot \nabla \zeta \, dA - \int\limits_{S_2} \bar{v}_n \zeta \, dS \qquad (7.42)$$

Note that the first term in Equation (7.42) is the kinetic energy of the system divided by the mass density. The second term can be related to the work done by the impulsive pressure which starts the system from rest.

Another Galerkin formulation can be based on the stream function ψ. Now the incompressibility condition is identically satisfied, but the irrotationality equation gives

$$\nabla^2 \psi = 0 \qquad (7.43)$$

with $\psi = \bar{\psi}$ on S_1 and $\bar{v}_s = -(\partial \psi / \partial n)$ on S_2. We can propose the following weighted residual expression

$$\iint (\nabla^2 \psi) \delta \psi \, dA = \int\limits_{S_2} \left(\frac{\partial \psi}{\partial n} + \bar{v}_s \right) \delta \psi \, dS \qquad (7.44)$$

Integrating by parts, we obtain

$$-\iint \vec{\nabla} \psi \cdot \vec{\nabla} \delta \psi \, dA = \int\limits_{S_2} \bar{v}_s \, \delta \psi \, dS \qquad (7.45)$$

which gives the functional

$$F(\psi) = \frac{1}{2} \iint \vec{\nabla}\psi \cdot \vec{\nabla}\psi \, dA + \int\limits_{S_2} \bar{v}_s \psi \, dS \qquad (7.46)$$

Example 7.1

Let us consider the case of *confined* flow around a cylinder (Figure 7.8a) and we assume that the flow is laminar and inviscid[1].

The smallest region that can be analysed by taking into account symmetry considerations is the region abcde, as shown in Figure 7.8a.

Due to symmetry we can choose $\psi = 0$ as a reference streamline ab and bc. We take $\psi = 2$ on ed and assume a linear variation of ψ between a and e. This gives the normal velocity on ae equal to 1.

Finally, the vertical component of velocity must vanish on cd, i.e.

$$v = \frac{\partial \psi}{\partial x} = 0, \text{ on } S_2 \qquad (a)$$

(where S_2 represents the part of the boundary from c to d).

The problem was solved by Martin[1] using three node triangular elements such that

$$\psi = \phi^T \psi^n \qquad (b)$$

where ϕ is the linear interpolation function. Equation (7.45) gives

$$\iint \vec{\nabla}\psi \cdot \vec{\nabla}\delta\psi \, dA = \int\limits_c^d \bar{v}_s \, \delta\psi \, dS = 0 \qquad (c)$$

Substituting (b) into (c) we have

$$\delta\psi^{n,T} \iint \phi \, \phi^T \, dA \, \psi^n = \delta\psi^{n,T} \int\limits_c^d \phi\bar{v}_s \, dS \qquad (d)$$

or

$$\mathbf{K}\psi^n = \mathbf{P} \qquad (e)$$

For the whole continuum we have, after applying boundary conditions $\psi_i = \bar{\psi}_i$.

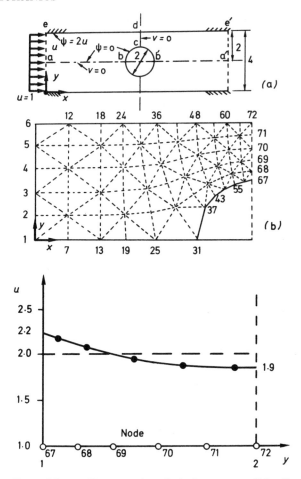

Figure 7.8   (a) Flow around a cylinder between parallel walls
(b) Velocity at vertical centreline (c-d)

$$K\psi = \mathscr{P} + \overline{\mathscr{P}} \tag{f}$$

Note that $\mathscr{P}$ is due to the known values of stream function S_1. This system of equations can be solved and the nodal values of the stream functions found.

Martin[1] has solved the example shown in Figure 7.8a, but using, instead of the abcde domain, the half-symmetry region abcb'a'e'e, on which the ψ values

Table 7.1

Node	X_2 coordinate	ψ
67	1.00	0
68	1.16	0.3419
69	1.33	0.6880
70	1.54	1.1072
71	1.76	1.5364
72	2.00	2.0000

are known at every point; hence the only boundaries are of type S_1. His results are shown in Table 7.1. The velocities at the vertical centre line can then be calculated. They are plotted in Figure 7.8b.

Martin also proposed applying exactly the $\partial\psi/\partial n$ condition on S_2 as a constraint equation relating the nodal unknowns. We usually prefer to apply them in a weighted residual fashion, as indicated in Equation (7.44). In this way the conditions are not generally identically satisfied but are more easily applied.

7.4 APPLICATIONS IN HYDRAULICS

Finite elements are becoming widely used for the solution of coastal engineering and other hydraulic problems. We will here consider only a few perfect fluid applications. The solution of more complex viscous cases is beyond the scope of this book.

Figure 7.9 defines the notation to be used. We define $\eta(x, y, t)$ as the elevation above sea level, and work with vertically integrated flow measures q_x and q_y which are functions of x, y, and t.

$$q_x = \int_{-h}^{\eta} u \, dz, \quad q_y = \int_{-h}^{\eta} v \, dz \tag{7.47}$$

As we are now working with 3 axes, we also have a third momentum equation in the z direction,

$$\rho \frac{Dw}{Dt} = -\frac{\partial p}{\partial z} + \rho b_z \tag{7.48}$$

In this case $b_z = -g$ (gravity force) and the terms in w can be neglected by comparison with the other terms. This gives

$$\frac{\partial p}{\partial z} + \rho g = 0 \tag{7.49}$$

After integration, Equation (7.49) gives the hydrostatic condition

$$p = g\rho(\eta - z) + p_a \tag{7.50}$$

where p_a is the atmospheric pressure which can be neglected if it is everywhere the same.

Equation (7.50) now replaces the third momentum equation. The other two equations after substituting (7.50) are

$$\frac{Du}{Dt} = -g\frac{\partial \eta}{\partial x}, \quad \frac{Dv}{Dt} = -g\frac{\partial \eta}{\partial y} \tag{7.51}$$

Continuity can generally be written as

$$\frac{\partial u}{\partial x} + \frac{\partial v}{\partial y} + \frac{\partial w}{\partial z} = 0 \tag{7.52}$$

We will now consider two different applications: (1) a steady state flow case. (2) a time dependent problem, which produces a wave equation.

First we will integrate Equations (7.52) with respect to z. To carry out this integration we define the 'kinematic' condition at the surface, which says that the vertical velocity at $z = \eta$ (Figure 7.9) is simply the time variation of the wave height, i.e.

$$w|_\eta = \frac{D\eta}{Dt}\bigg|_\eta = \frac{\partial \eta}{\partial t} + u\frac{\partial \eta}{\partial x} + v\frac{\partial \eta}{\partial y}\bigg|_\eta \tag{7.53}$$

We have a similar condition at the bottom, i.e.

$$w|_{-h} = \frac{Dh}{Dt}\bigg|_{-h} = \frac{\partial h}{\partial t} + u\frac{\partial h}{\partial x} + v\frac{\partial h}{\partial y}\bigg|_{-h} \tag{7.54}$$

where $\partial h/\partial t$ is zero for a stable bottom.

Let us now consider the integrated continuity equation,

$$\int_{-h}^{\eta} \left(\frac{\partial u}{\partial x} + \frac{\partial v}{\partial y} + \frac{\partial w}{\partial z}\right) dz = 0 \tag{7.55}$$

Integrating this expression using Leibnitz's theorem for integration between variable limits, plus Equations (7.53) and (7.54) we obtain

$$\frac{\partial q_x}{\partial x} + \frac{\partial q_y}{\partial y} + \frac{\partial \eta}{\partial t} = 0 \tag{7.56}$$

This equation can now be used to solve the steady state case.

7.4.1 Steady state case

For the steady state case we can consider the time independent version of Equation (7.56) i.e.

$$\frac{\partial q_x}{\partial x} + \frac{\partial q_y}{\partial y} = 0 \tag{7.57}$$

If we assume now that the wave height slopes and the bottom slopes are small we can write the integrated vorticity equations

$$2\omega = \frac{\partial q_y}{\partial x} - \frac{\partial q_x}{\partial y} = 0 \tag{7.58}$$

Defining a streamline ψ function such that

$$q_x = \frac{\partial \psi}{\partial y}, \quad q_y = -\frac{\partial \psi}{\partial x} \tag{7.59}$$

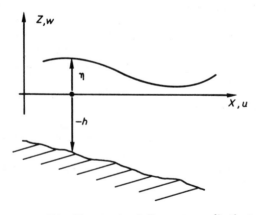

Figure 7.9   Notation for shallow water applications

Equation (7.57) is identically satisfied and the irrotationality condition gives

$$\frac{\partial}{\partial x}\left(\frac{\partial \psi}{\partial x}\right) + \frac{\partial}{\partial y}\left(\frac{\partial \psi}{\partial y}\right) = 0 \qquad (7.60)$$

with boundary conditions

$$\psi = \bar{\psi} \text{ on } S_1 \qquad (7.61)$$

The Galerkin type formula for the system (7.60), (7.61) is similar to the one seen previously for stream functions in confined flow problems. This gives, after integration by parts

$$\iint \left(\frac{\partial \psi}{\partial x}\frac{\partial \delta \psi}{\partial x} + \frac{\partial \psi}{\partial y}\frac{\partial \delta \psi}{\partial y}\right) dA = 0$$

The normal flux into the system is given by defining the values of ψ at the inlet and outlet of the system as will be seen in the next example. Note that Equation (7.60) is a particular case of Equation (6.15). Therefore we can use the extended Laplace's equation program of Chapter 6 to solve this type of problem by specifying $h_x = h_y = 1$ and $\lambda = c = 0$.

Example 7.2   Streamline flow in a river[2]
The following example was part of a study undertaken to analyse water circulation in the Guaiba River, Brazil. The region under study can be seen in Figure 7.10.

Figure 7.11(a) shows the finite element mesh used to study this river. The grid has 419 elements and 272 nodes and was chosen so as to have at least five nodes across the width of the river. The depth of each element was taken from the charts. The program produced the streamlines and nodal velocities.

The discharge into the Gauiba is due to a series of rivers called the Jacui complex and it was assumed to be 1000 m^3/s, which is a low value. This discharge was modelled by giving the values of the stream functions at the entrance and exit nodal points in such a way as to have a total discharge of 1000 m^3/s at each section.

The outlet of a new effluent discharge pipe will be situated at the northern entrance of the river. It is then of practical interest to determine the time that the water takes to leave the system. This can be done by integrating T along the streamlines, Figure 7.11(b), i.e.

$$T = \int \frac{ds}{v}$$

Figure 7.10   General view of the region

where s is the length measured along a given streamline. The time for most of the discharge was found to be in the order of 2 days.

7.4.2 Time dependent case

We will consider now a special time dependent problem. This is the problem c harbour resonance, which produces a wave equation.

We start by neglecting the convective terms in Equation (7.51). This gives

$$\frac{\partial u}{\partial t} = -g\frac{\partial \eta}{\partial x}, \quad \frac{\partial v}{\partial t} = -g\frac{\partial \eta}{\partial y} \tag{7.62}$$

Note that this assumption is valid in many harbours where the velocities of the water and their derivatives will be small.

The continuity equation is given by Equation (7.56), i.e.

$$\frac{\partial q_x}{\partial x} + \frac{\partial q_y}{\partial y} + \frac{\partial \eta}{\partial t} = 0 \tag{7.63}$$

We can now integrate (7.62) with respect to z considering that the bottom and wave slopes are small and can be neglected. This gives

$$\frac{\partial q_x}{\partial t} = -gh\frac{\partial \eta}{\partial x}, \quad \frac{\partial q_y}{\partial t} = -gh\frac{\partial \eta}{\partial y} \tag{7.64}$$

Note that we have also taken $h \simeq h + \eta$.

Differential Equations (7.63) and (7.64) can be transformed into a single differential equation of the second order by differentiating both of Equations (7.62) with respect to x and y respectively and substituting them into the continuity equation (7.63) differentiated with respect to t. This gives

$$\frac{\partial}{\partial x}\left(h\frac{\partial \eta}{\partial x}\right) + \frac{\partial}{\partial y}\left(h\frac{\partial \eta}{\partial y}\right) - \frac{1}{g}\frac{\partial^2 \eta}{\partial t^2} = 0 \tag{7.65}$$

with boundary conditions

$$\eta = \overline{\eta} \text{ on } S_1$$

$$h\frac{\partial \eta}{\partial n} = -\frac{\partial}{\partial t}\left(\frac{1}{g}\overline{q}_n\right) \text{ on } S_2$$

Note that the second type of boundary condition is the equilibrium equation for the normal direction.

Harbour resonance and harmonic response due to surges can be investigated by expressing η as

$$\eta(x, y, t) = H(x, y)e^{i\omega t}$$

where ω is the circular frequency. Then (7.65) becomes

$$\frac{\partial}{\partial x}\left(h\frac{\partial H}{\partial x}\right) + \frac{\partial}{\partial y}\left(h\frac{\partial H}{\partial y}\right) + \frac{\omega^2}{g}H = 0 \tag{7.66}$$

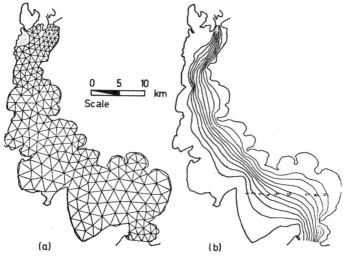

Figure 7.11   Gauiba river: (a) finite element mesh, (b) streamlines

with

$$H = \bar{H}, \text{ on } S_1$$

$$\frac{\partial H}{\partial n} = f, \text{ on } S_2 \tag{7.67}$$

We prescribe the following set of homogeneous boundary conditions for harmonic excitation,

$$H = 0, \text{ on } S_1$$

$$\frac{\partial H}{\partial n} = 0, \text{ on } S_2 \tag{7.68}$$

and determine the frequencies and modal shapes.

Finally if the horizontal displacements of the water particles need to be computed, we can obtain them from the velocities as follows. Given Equations (7.68) we can write

$$\frac{\partial^2 U}{\partial t^2} \simeq -g \frac{\partial H}{\partial x} e^{i\omega t}$$

$$\frac{\partial^2 V}{\partial t^2} \simeq -g \frac{\partial H}{\partial y} e^{i\omega t} \tag{7.69}$$

where U and V are displacements. Integrating over half a period we obtain,

$$|U| = \frac{g}{\omega^2} \frac{\partial H}{\partial x}$$

$$|V| = \frac{g}{\omega^2} \frac{\partial H}{\partial y} \tag{7.70}$$

These are the maximum horizontal displacements.

7.4.3 Finite element formulation of resonance problem

From Equation (7.66) and (7.67) we can write the following weighted residual expression

$$\iint \left\{ \frac{\partial}{\partial x} \left(h \frac{\partial H}{\partial x} \right) + \frac{\partial}{\partial y} \left(h \frac{\partial H}{\partial y} \right) + \frac{\omega^2}{g} H \right\} \, \delta H \, dx \, dy$$

$$= \int_{S_2} \left\{ h \frac{\partial H}{\partial n} - f \right\} \delta H \, dS \tag{7.71}$$

Integrating by parts we have

$$\iint \left\{ h \frac{\partial H}{\partial x} \frac{\partial \delta H}{\partial x} + h \frac{\partial H}{\partial y} \frac{\partial \delta H}{\partial y} - \frac{\omega^2}{g} H \delta H \right\} \, dx \, dy = \int_{S_2} f \delta H \, dS \tag{7.72}$$

Assume that the H variable can be approximated on each element by

$$H = \boldsymbol{\varphi}^T \, \mathbf{H}^n \tag{7.73}$$

where $\boldsymbol{\varphi}$ is an interpolation function, $\mathbf{H}^n$ are the nodal unknowns. We have for an element,

$$\delta \mathbf{H}^{n,T} \iint \left\{ h \left(\boldsymbol{\varphi}_x \boldsymbol{\varphi}_x^T + \boldsymbol{\varphi}_y \boldsymbol{\varphi}_y^T \right) - \frac{\omega^2}{g} \boldsymbol{\varphi} \boldsymbol{\varphi}^T \right\} \, dx \, dy \, \mathbf{H}^n = \delta \mathbf{H}^{n,T} \int \boldsymbol{\varphi} f \, dS \tag{7.74}$$

$$\boldsymbol{\varphi}_x = \frac{\partial \boldsymbol{\varphi}}{\partial x}, \quad \boldsymbol{\varphi}_y = \frac{\partial \boldsymbol{\varphi}}{\partial y}$$

We can write (7.74) as

$$\mathbf{K} \, \mathbf{H}^n - \omega^2 \, \mathbf{M} \, \mathbf{H}^n = \mathbf{P} \tag{7.75}$$

$$\mathbf{K} = \iint h(\boldsymbol{\varphi}_x \boldsymbol{\varphi}_x^T + \boldsymbol{\varphi}_y \boldsymbol{\varphi}_y^T) \, dx \, dy$$

$$\mathbf{M} = \frac{1}{g} \iint \boldsymbol{\varphi} \boldsymbol{\varphi}^T \, dx \, dy$$

$$\mathbf{P} = \int \boldsymbol{\varphi} f \, dS \tag{7.76}$$

We can then assemble Equations (7.76) for the whole continuum as usual.
Since Equation (7.66) is a particular case of Equation (6.15), we can again

use the extended Laplace's equation program of Chapter 6, to solve this type of problem. For this we need to specify $h_x = h_y = h$, $\lambda = \omega^2/g$, and $c = 0$.

Example 7.3. Harbour resonance

As an illustration, the case of the Duncan basin built during the Second World War in Table Bay Harbour, South Africa is analysed. This basin has been extensively studied[3] as the features of the Bay are such that some of the seiche frequencies are greatly amplified (Figure 7.12). This fact has been demonstrated by model experiment, harmonic analysis of seichograms and simple theoretical analysis, which can give reasonable results here, as the shape of the basin is approximately rectangular.

The finite element analysis was carried out by dividing the basin into 168 six node triangular elements, which give 377 nodal unknowns (second mesh). Periods T of 1, 2, 3, . . . min were chosen ($\omega = 2\pi/T$ = circular frequency) and an elevation $\bar{H} = 1$ assumed for the nodes at the entrance. The elevation outputs at berth E and O are shown in Figure 7.12(b).

Finally, the eigenvalues of the system were found employing the first mesh in order to use less computing time. These values were compared against those obtained by solving the equations for $T = 1, 2, 3, \ldots$ min, and good agreement was found (see Figure 7.12(b) which shows the finite element results).

Note that the first significant period of the basin is around $T = 11.45$ minutes (theoretical value) and that this period clearly shows in the experimental curve, although it is very much damped. This damping is expected because the period $T = 11.45$ min corresponds to the water flowing freely in and out of the basin which in practice does not happen.

7.5   FLOW THROUGH POROUS MEDIA

In what follows we will consider the type of flow that occurs in the saturated region of a porous soil. We assume that the pores in a medium such as sand are interconnected and that the fluid can flow through them. The ratio of these voids to the total volume is called porosity and denoted by

$$n = \frac{\text{Voids volume}}{\text{Total volume}} \tag{7.77}$$

The relationship governing groundwater flow is known as Darcy's law. Its form can be deduced from the general Navier-Stokes equations under the assumption that the flow is laminar and that inertial forces are negligible when compared to the viscous forces. Given the Reynolds number

$$\text{Re} = \frac{\text{Inertia forces}}{\text{Viscous forces}} = \frac{Vd\rho}{\mu} \tag{7.78}$$

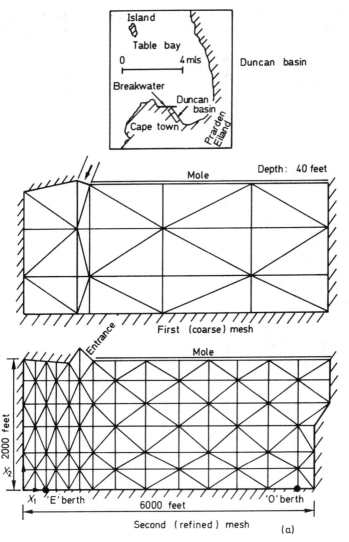

Figure 7.12   (a) Harmonic response analysis of Duncan Basin

where d is the average pore diameter, μ the fluid viscosity, ρ the density and V the velocity, the law applies for Re < 1.

Darcy's law can be written as

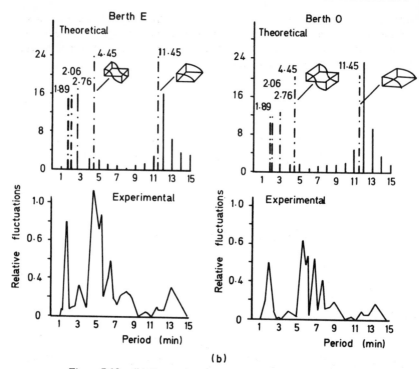

Figure 7.12   (b) Harmonic response analysis of Duncan Basin

$$\vec{v} = -K \vec{\nabla} \zeta \tag{7.79}$$

where K is the hydraulic conductivity and ζ is the head, with dimensions of length and equal to

$$\zeta = \frac{p}{\rho g} - \frac{\Omega}{g} \tag{7.80}$$

where p is pressure and Ω the potential of body forces. When only gravity is acting, we have

$$\Omega = -y \, g \tag{7.81}$$

For the two dimensional anisotropic case K in Equation (7.79) is a 2×2 array. If we assume that there is an orthogonal frame for which the conductivity array K has only 2 non-zero coefficients on the diagonal, the gradients of ζ and the corresponding velocities have the same directions, i.e. parallel to the axis of the frame. This system is called the *principal axis* of conductivity. One can write the velocity components as

$$u = -K_x \frac{\partial \zeta}{\partial x}, \quad v = -K_y \frac{\partial \zeta}{\partial y} \tag{7.82}$$

The balance of flow-in and flow-out of an element gives the same two-dimensional continuity equation as Equation (7.13), i.e.

$$\frac{\partial u}{\partial x} + \frac{\partial v}{\partial y} = 0 \tag{7.83}$$

As a function of the potential head this can be written

$$- \left\{ \frac{\partial}{\partial x} \left(K_x \frac{\partial \zeta}{\partial x} \right) + \frac{\partial}{\partial y} \left(K_y \frac{\partial \zeta}{\partial y} \right) \right\} = 0 \tag{7.84}$$

with boundary conditions of the type

$$\zeta = \bar{\zeta} \text{ on } S_1, v_n = - \left(K_x \frac{\partial \zeta}{\partial x} \alpha_{nx} + K_y \frac{\partial \zeta}{\partial y} \alpha_{ny} \right) = \bar{v}_n \text{ on } S_2 \tag{7.85}$$

Equations (7.84) and (7.85) can be expressed in a weighted residual way as follows:

$$\iint \left\{ \frac{\partial}{\partial x} \left(K_x \frac{\partial \zeta}{\partial x} \right) + \frac{\partial}{\partial y} \left(K_y \frac{\partial \zeta}{\partial y} \right) \right\} \delta \zeta \, dx \, dy$$

$$= \int_{S_2} \{ v_n - \bar{v}_n \} \delta \zeta \, dS \tag{7.86}$$

where ζ is assumed to be equal to $\bar{\zeta}$ on S_1 (hence $\delta \zeta \equiv 0$).
Integrating by parts we obtain

$$\iint \left(K_x \frac{\partial \zeta}{\partial x} \frac{\partial \delta \zeta}{\partial x} + K_y \frac{\partial \zeta}{\partial y} \frac{\partial \delta \zeta}{\partial y} \right) dx \, dy - \int_{S_2} \bar{v}_n \delta \zeta \, dS = 0 \tag{7.87}$$

This type of problem can also be solved using the extended Laplace equation program of Chapter 6.

7.5.1 Axisymmetric case

In many practical cases the analyst wants to solve axisymmetric seepage problems (e.g. in cases such as the flow towards a well). The statement corresponding to Equation (7.86) for the axisymmetric case (Figure 7.13) with axisymmetric loads is

$$\iiint \left\{ \frac{1}{r} \frac{\partial}{\partial r} \left(r K_{rr} \frac{\partial \zeta}{\partial r} \right) + \frac{\partial}{\partial r} \left(K_{zz} \frac{\partial \zeta}{\partial z} \right) \right\} \, \delta\zeta \, d\theta r \, dr \, dz$$

$$= \int_{S_2} \{ v_n - \overline{v}_n \} \, \delta\zeta \, d\theta \, dS \tag{7.88}$$

where

$$v_n = K_{rr} \frac{\partial \zeta}{\partial r} \alpha_{nr} + K_{zz} \frac{\partial \zeta}{\partial z} \alpha_{nz}$$

Integrating over θ $(0 < \theta < 2\pi)$, we obtain

$$2\pi \iint \left\{ \frac{\partial}{\partial r} \left(r K_{rr} \frac{\partial \zeta}{\partial r} \right) + r \frac{\partial}{\partial z} \left(K_{zz} \frac{\partial \zeta}{\partial z} \right) \right\} \, \delta\zeta \, dr \, dz$$

$$= 2\pi \int_{S_2} (v_n - \overline{v}_n) \, \delta\zeta \, dS \tag{7.89}$$

Finally we can integrate by parts Equation (7.89) and obtain the following expression

$$\iint r \left(K_{rr} \frac{\partial \zeta}{\partial r} \frac{\partial \delta\zeta}{\partial r} + K_{zz} \frac{\partial \zeta}{\partial z} \frac{\partial \delta\zeta}{\partial z} \right) \, dr \, dz - \int \overline{v}_n \, \delta\zeta \, dS = 0 \tag{7.90}$$

The essential difficulty with the finite element expression of Equation (7.90) is that it breaks down when $r = 0$. In order not to have this problem it is usual to take r on each element as a constant and equal to its value at the centroid of the element. Another possibility is to use a numerical integration

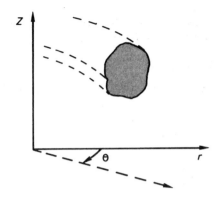

Figure 7.13   Axisymmetric case

scheme with integration points inside the elements only (i.e. none of them at $r = 0$).

Example 7.4. Flow towards a well

Let us consider the flow towards a well. The well has a radius of 1 m and completely penetrates the aquifer (Figure 7.14). The water level in the well is kept constant at 20 m above the impermeable base by pumping at a constant rate.

The Aquifer consists of homogeneous and isotropic soil ($K_1 = K_2$, $\alpha = 0$) for simplicity. The flow through the aquifer is confined at the bottom by an impermeable bed, while the top water surface remains free. The radius of influence of the well extended up to 370 m where the ground water level was taken at 80 m above the impervious boundary and the flow was assumed uniform and horizontal.

The problem was analysed by using cylindrical system of coordinates with the axisymmetric equations.

A top flow line was initially guessed and the flow region was divided into three node triangular elements, as shown in Figure 7.14. After each iteration the values of ζ at the free surface were compared with the elevation; if they were different the mesh was moved to satisfy the condition $\zeta = y$. Figure 7.15(a), represents the computer network of finite elements after five iterations. In the fifth iteration the difference between the computed potential head and the elevation of any potential along the free surface was less than 0.1% of the elevation. In fact the solution for the free surface was reasonably accurate after one or two iterations.

The area through which flow occurs is proportional to the radial distance from the well and the thickness of the flow region. Hence as the flow approaches the well this area becomes much smaller and the potential gradient is considerably increased (Figure 7.15(b)). Along the wall of the well the hydraulic gradient was estimated to be about three. This is three times

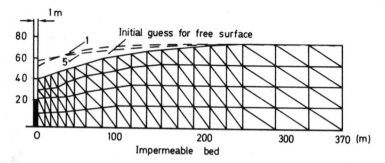

Figure 7.14   Initial element division with various top flow lines

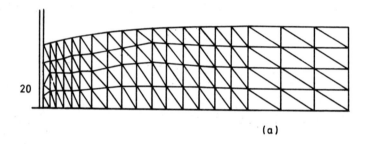

(a)

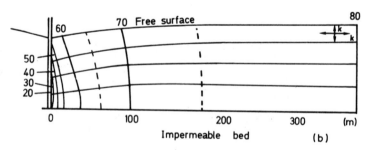

Figure 7.15   (a) Final element division for axisymmetric flow
(b) Seepage towards axisymmetric well

higher than the allowable value and therefore protection against 'piping'
would be needed in this case.

References

1. Martin, H. C., 'Finite element analysis of fluid flows', in Proc. 2nd Conf. Matrix Meth. Struct. Mech., AFFDL TR 68-150, Wright-Patterson Air Force Base, Ohio, USA (1969)

2. Bonilha, N., Brebbia, C. and Ferrante, A., 'Computational hydraulics studies in Rio Grande do Sul', in Proceedings 16th Congress of the International Association for Hydraulic Research, S. Paulo, Brasil, (1975)

3. Wilson, B. W., 'Research and model studies on range action in Table Bay Harbour, Cape Town'. Trans. South African Inst. Civil Eng., 1, No.6, 131–148; 1, No.7, 153–177 (1959)

4 Brebbia, C. A. and Spanous, K. A., 'Application of the finite element method to study irrotational flow', *Rev. Roumaine,* Sci. Tech. Series de Mecanique Appliquée, 19, No.3, (1973)

5. Brebbia, C. A. and Adey, R., 'Circulation studies by finite elements', in Proceedings Seminar on Finite Element Methods, Science Research Council, Abingdon, 1974. Published by Atlas Computer Laboratory, Didcot, Oxfordshire, U.K., (1975)

Index